MATHEMATICAL METHODS
IN
ENGINEERING

Edited by
GLYN A. O. DAVIES
Department of Aeronautics
Imperial College of Science and Technology

A Wiley–Interscience Publication

JOHN WILEY & SONS
Chichester · New York · Brisbane · Toronto · Singapore

Library of Congress Cataloging in Publication Data:

Main entry under title:

Mathematical methods in engineering.
   Guidebook 5 to Handbook of applicable mathematics.
   'A Wiley–Interscience publication.'
   Includes index.
   1. Engineering mathematics.   I. Davies, Glyn A. O.
II. Handbook of applicable mathematics.
TA330.M32   1984   620'.001'51   83-23250

ISBN 0 471 10331 4 (U.S.)

British Library Cataloguing in Publication Data

Mathematical methods in engineering.—(Handbook
   of applicable mathematics; guidebook 5)
   1. Engineering mathematics
   I. Davies, Glyn A. O.   II. Series
   510'.2462   TA330

ISBN 0 471 10331 4

Filmset by Eta Services (Typesetters) Ltd., Beccles, Suffolk.
Printed and bound in Great Britain.

D
620.0015'1
MAT

Contributing Authors

S. Barnett, Professor of Applied Mathematics, School of Mathematical Sciences, University of Bradford, West Yorkshire, U.K.

M. J. M. Bernal, Reader in Numerical Analysis, Department of Mathematics, Imperial College of Science and Technology, London, U.K.

P. Bettess, Senior Lecturer in Department of Civil Engineering, University College of Swansea, U.K.

R. Bettess, Senior Scientific Officer, Hydraulics Research Station, Wallingford, Oxfordshire, U.K.

G. A. O. Davies, Head of Department of Aeronautics, Imperial College of Science and Technology, London, U.K.

J. M. R. Graham, Reader in Department of Aeronautics, Imperial College of Science and Technology, London, U.K.

R. Hillier, Lecturer in Department of Aeronautics, Imperial College of Science and Technology, London, U.K.

D. P. Isherwood, Lecturer in Department of Mechanical Engineering, Imperial College of Science and Technology, London, U.K.

V. G. Jenson, Senior Lecturer in Department of Chemical Engineering, University of Birmingham, U.K.

N. B. Jones, Reader in Control Engineering, School of Engineering and Applied Sciences, University of Sussex, Brighton, U.K.

J. D. Lewins, University Lecturer in Nuclear Engineering, Engineering, Laboratories, University of Cambridge, U.K.

J. O. Medwell, Lecturer in Department of Mechanical Engineering, University College of Swansea, U.K.

J. Munro, Professor of Civil Engineering Systems and Head of Department of Civil Engineering, Imperial College of Science and Technology, London, U.K.

E. G. Prater, Senior Research Engineer, Institute of Foundation Engineering and Soil Mechanics, Federal Institute of Technology, Zurich, Switzerland.

C. Taylor, Reader in Department of Civil Engineering, University College of Swansea, U.K.

W. L. Wood, Honorary Fellow, formerly Senior Lecturer in Department of Mathematics, University of Reading, U.K.

v

Contents

Editorial Note

Mathematical skills and concepts are increasingly being used in a great variety of activities. Applications of mathematics range from intricate research projects to practical problems in commerce and industry.

Yet many people who are engaged in this type of work have no academic training in mathematics and, perhaps at a late stage of their careers, have neither the inclination nor the time to embark upon a systematic study of mathematics.

To meet the needs of these users of mathematics we have produced a series of texts or *guidebooks* with uniform title. One of these is *Mathematical Methods in Engineering* (with similar books on Medicine, Economics and so on). The purpose of this volume is to describe how mathematics is used as a tool in Engineering and to illustrate it with examples from this field. The guidebooks do not, as a rule, contain expositions of mathematics *per se*. A reader who wishes to learn about a particular mathematical topic, or consolidate his knowledge, is invited to consult the *core volumes* of the *Handbook of Applicable Mathematics*; they bear the titles

    I Algebra
    II Probability
    III Numerical Methods
    IV Analysis
    V Combinatorics and Geometry (Parts A and B)
    VI Statistics (Parts A and B)

The core volumes are specifically designed to elaborate on the mathematical concepts presented in the guidebooks. The aim is to provide information readily, to help towards an understanding of mathematical ideas and to enable the reader to master techniques needed in applications. Thus the guidebooks are furnished with references to the core volumes. It is essential to have an efficient reference system at our disposal. This system is explained fully in the Introduction to each core volume; we repeat here the following points. The core volumes are denoted by the Roman numerals mentioned above. Each mathematical item belongs to one of the six categories, namely:

(i) Definitions
(ii) Theorems, propositions, lemmas and corollaries
(iii) Equations and other displayed formulae
(iv) Examples
(v) Figures
(vi) Tables

A typical item is designated by a Roman numeral followed by three Arabic numerals a, b, c, where a refers to the chapter, b to the section and c to the individual item enumerated consecutively in each category. For example, 'IV Theorem 6.2.3' is the third theorem of section 2 in Chapter 6 of the core volume IV. We refer to equation 6.2.3 of Volume IV as IV (6.2.3), and section 6.2 of Volume IV as IV §6.2.

We trust that these guidebooks will contribute to a deeper appreciation of the mathematical methods available for elucidating and solving problems in all the various disciplines covered by the series.

Preface and Acknowledgements

As explained in the introductory chapter of this book, the following chapters are a selection of those branches of engineering which particularly rely on the application of mathematical techniques. There is no coverage of the more discursive qualitative aspects of engineering, nor is there a chapter on electrical engineering or information technology which are separately identifiable and comprehensive disciplines which deserve guidebooks to themselves.

The chapters in this book are but brief summaries of various fields in engineering, just sufficiently long to describe the nature of the subject and to put into context the modelling which is necessary before a mathematical problem emerges. The mathematical solutions are not dwelt upon, but reference is simply made to the Handbook as described in the Editorial Notes. In this fashion it is hoped that engineers will find illumination where they look for it and, further, may discover helpful techniques which they had not looked for. There is often a singular lack of cross-fertilization between the various branches of engineering, let alone between engineers and other scientists. An equally serious communications gap also exists between practising engineers and applied mathematicians, and we hope that the Guidebook and Handbooks will contribute to the closure of this gap.

One consequence of the barriers that inevitably arise when we drift into separate areas of specialization is that of notation. The common symbols in use have been kept uniform wherever possible, but there is a limit—if only that imposed by the twenty-six characters in the Roman alphabet. In cases where traditions have become firmly entrenched, I have not striven to alter them simply for uniformity. For example, $\sqrt{-1}$ is denoted by 'i' in the core volumes and in all of these chapters with the exception of control engineering which adopts the time-honoured electrical practice of reserving that symbol for *current*.

It is not possible for a single engineer to present a precis of fifteen separate branches of engineering, and my first thanks must go to the authors of each chapter for their contributions and their willingness to shorten the content to a bare minimum. In many cases the engineering treatment is cursory and the reader is assumed to already have the knowledge or is directed to engineering references at the conclusion of each chapter. For the brevity, and for often wielding an editorial knife, I take complete responsibility.

I would especially like to thank my friend Roland Lewis, of the University College of Swansea, who undertook the task of compiling this guidebook in the first place, but who had eventually to bow to other pressures. He persuaded the authors of Chapters 3, 5, 6, 7, 8 and 9 to undertake their tasks, and for this gentle persuasion I am extremely grateful.

To Carol van der Ploeg I am indebted for inserting all the core volume references and for bravely reading all chapters as a lay person unbiased by specialized engineering prejudices. I am also grateful to Walter Ledermann for a similar role, but more important for conveying much enthusiasm for the total concept of guidebooks and handbooks.

Finally, of course, I am grateful to John Wiley for their forbearance over a long gestation period and their trust in believing that sixteen separate authors could be united in a common aim.

G. A. O. Davies
August 1983

1

Introduction

1.1 Objectives

The six volumes of the *Handbook of Applicable Mathematics* are a comprehensive account of the many branches of applicable mathematics. The word *applicable*—rather than 'applied'—has been chosen to emphasize that these core volumes are directed at *practitioners* who use mathematics to explain and solve problems, rather than as an abstract discipline. The terms 'pure' and 'applied' mathematics have been used in the past but such labels can rapidly become 'old hat'. The computer, for example, has dated many forms of analysis thought to be applied, whilst some pure abstract algebra has now been found to be applicable.

Of all practitioners who use mathematics as a working tool, engineers must surely form the largest group, and it is appropriate that a guidebook be devoted to engineering. Unfortunately the field of engineering today is so vast that two problems of scale present themselves immediately.

First, the scope of applicable mathematics ranges from the elementary to the sophisticated, and it is hoped that the six core volumes of the Handbook encompass something like the range currently used by modern engineers. This hope is probably a pious one. Many engineers are having to pioneer new mathematical techniques to meet the demand of high technology—particularly in the fields of numerical analysis and systems analysis, for example. These new techniques may be justified, developed and proven using physical arguments, and it may be some time before they are given mathematical rigor and catalogued in a cohesive way with familiar mathematics. Such is the nature of progress under pressure. The reader of this guidebook will therefore find techniques discussed which are not as yet incorporated into the core volumes, but references to other textbooks will be given where necessary.

Second, the field of engineering is itself so large that it might be a daunting task to include it all in a single guidebook. However, this text contains only those branches and aspects of engineering which specifically use mathematics as a routine tool. This leaves a smaller—and rather curious looking—field to cover and also explains why some non-mathematical areas in engineering are completely absent. There is also the problem of classification, of course. Should a guidebook have *vocational* chapters like mechanical engineering, civil

1

engineering, aeronautical, marine, and so on? It was decided against this in view of the various topics that should be covered. For example, the subject of Fluid Mechanics alone occurs in all of the divisions just mentioned, and many more. It would be absurd to suggest that mechanical and civil engineers use fluid mechanics in quite different ways—far better to present the topic and its associated mathematics in a general way if possible. A glance at the chapter headings will reveal that most of this book has an interdisciplinary flavour. Fluid mechanics, boundary layers, compressible flow, free-surface flow, heat transfer, structural mechanics and systems engineering are all subjects which stride across the various provinces of engineering. Hydrology and porous media are of interest mostly—but not exclusively—to civil engineers, and control engineering is not exclusively the domain of electrical engineers. The chapters on polymer engineering and nuclear power are, however, fairly specific. Notable omissions are the fields of electrical engineering and electronics, computers and information science, as it was felt that these readily identifiable subject areas deserved their own guidebooks.

It has not been the intention to produce an engineering textbook, but instead to limit discussion to an explanation of the problems which give rise to a mathematical need. Therefore chapters in general contain the physics of the problem or whatever is necessary to explain the modelling and put it into its engineering context. Whenever modelling is based on lengthy physical or logical arguments, the reader is referred to engineering references at the conclusion of that particular chapter. Cross-referencing to other chapters is also used, of course. Having described the engineering formulation of the problem and put the modelling into some form of quantitative perspective, the mathematical techniques are usually briefly described and often listed in some order of merit, like expense, accuracy or convenience. However, no mathematical details are entered into, but references are made to the core volumes, as explained in the introductory editorial note.

1.2 Modelling

The aim of mathematical modelling is to enable the engineer to *predict* without having actually to construct or reproduce. Although there is no substitute for actually building an aircraft, nuclear reactor, dam, chemical plant, transport system, and so on, the modern engineer has to get it right first time. If the economics or safety are wrong then such is the complexity of projects like those mentioned that only minor modifications can usually be tolerated before the financial viability of a project becomes at risk. Gone are the days when one could build a quick prototype and change it rapidly and often until it performed adequately. The ability to predict does not of course necessarily mean the exclusive use of mathematical models. Simulation may take the form of physical models or analogues, but even here the use of mathematics is likely to be important in validating the use of models which must differ from the real thing—otherwise they would have no advantage. If a reliable mathematical model or simulation can be found and tested then there is a valuable bonus in addition to the immediate cost saving. Other alternatives can be examined. The simulation of complex systems as an aid to *optimum design* is a rapidly growing industry not only in engineering but also in the less deterministic fields of socioeconomic and fiscal models, in project planning, and so on.

The formulation of a mathematical model involves three interrelated phases:

  (i) Define the important variables in the problem, their expected influence and the relations between them.

 (ii) Solve the relationship between the known inputs and the desired consequences.

(iii) Test the solutions for reliability and range of application.

The first stage is the true modelling stage and mostly consists in recognizing the nature of the problem so that existing theory, programs, etc., can be found and applied. Sometimes the problem may be entirely novel, of course, but the great majority of engineering problems have been met before, and any innovative analysis is usually an extension or improvement on established methods. It is this first stage which occupies most of the chapters in this book. Another glance at the chapter headings reveals that much of the material comes under the heading of 'applied mechanics' in the general fields of fluids, gases or solids. In principle many of the laws of mechanics governing the behaviour of fluids and solids are precise and exact, such as the equations of motion, conservation of mass, laws of thermodynamics, etc. These laws themselves are not usually sufficient and equations of state are also necessary. These latter equations usually involve some empirical relationship based on experience which relates *changes of state*, stress to strain rate, pressure to volume change and almost anything to temperature. Such laws are usually approximate and often linearized since all mathematical problems are simpler if they are linear. Outside the field of applied mechanics, some engineering problems may have to be formulated in an entirely empirical way—although analogies with applied mechanics are sometimes possible and helpful. Such empirical views of real life enter into several of the chapters in this book.

The second phase involves solving the equations after formulating them. Here further approximations have frequently to be introduced, even to the 'exact' governing equations. In practice the exact equations are usually non-linear and may have to be solved as such to preserve the expected behaviour of the physics. Often, however, it is possible to obtain approximate linearized solutions, and the engineer should know whether the exact equations are theoretically tractable, albeit expensively, say by using a tailor-made computer program (or by calling in a consultant). The approximations may be justified in mathematical terms, by using physical reasoning, or by both.

The final phase, having formulated a possible solution procedure, is to test it. The practising engineer will as often as not forego this stage if the problem and solution are familiar territory. If the modelling—or perhaps the range of validity of the solution—is doubtful, then assessment may be difficult. Comparison between a novel solution procedure, or modelling technique, and an alternative is one way of testing, but frequently the only way is to compare with the real thing if the experimental data are available. Perfect agreement can never be secured, of course, and mathematical techniques are often used in their own right to correlate observed results with the theoretical predictions.

The dangers and pitfalls in constructing reliable robust mathematical models, some useful rules and some cautionary tales can be found in Volume IV, Chapter 14.

1.3 Analytical resources

The mathematical techniques and descriptions referred to throughout this guidebook usually fall into one of two categories—*analytical* and *numerical*. Many engineering problems lead to simple governing equations which possess a closed-form analytical solution. The ubiquitous boundary-value problem is an example in which a 'field' behaviour is described by a differential equation and there are accompanying boundary (surface) conditions to be satisfied. To take a simplistic view, this class of problems is often solvable in a closed form in one dimension, and possibly also in two dimensions, but rarely in three dimensions. If the difficulties arise in the governing differential equations themselves, then the number of dimensions is often not the crucial factor. If the difficulties arise in the boundary conditions then they are often entirely geometric in nature, and an impasse in satisfying boundary conditions is no less important than in satisfying the differential equation. In many important engineering problems in the past fifty years or so, it has been necessary to accept crude analytical solutions solely because the geometry of the problem presented formidable difficulties.

Fortunately the advent of the computer has released us from the bondage of geometrical stumbling blocks. A numerical description and solution is nowadays entirely adequate and a closed-form approximate solution is no longer necessarily a desirable aim. The use of numerical methods aimed at awkwardly shaped regions is likewise true of awkward differential equations; again an analytical solution may actually be inferior, particularly if it has demanded questionable approximations. Unfortunately in the profession of engineering there is sometimes still a residual resistance to computational analysis, as if the computer were a crude blunt instrument, powerful but lacking finesse. This attitude dies hard even though the pre-computer solutions involved special functions which are just as expensive to compute as a numerical solution.

However, analytical solutions of even limited scope are extremely valuable in forming guidelines for a numerical solution in complex problems. They may tell us the order of approximations we may be able to tolerate and the likely errors. If the analysis reveals the presence of necessary singularities or discontinuities we would be foolish to ignore these features in a numerical description. The presence also of boundary layers and their extent can be predicted, which is of enormous help. The same is true of shock waves in fluids, gases, solids and traffic flow.

The engineer should therefore see classical analysis and numerical analysis as different but complementary. Superiority of one over the other is not a mark of esteem; both have their own logic, language and beauty. Also, nothing is static—not even attitudes. Our view of numerical analysis and modelling must undergo as many mutations as did the change from pre- to post-computer analysis. The advent of cheap reliable computer hardware—once the province of the large organization—has revolutionized analysis in non-linear problems, and the accessibility to vector and array processors is also colouring the way engineers are tackling complex modelling problems. The use of expert knowledge-based systems has started to make inroads into some computational packages.

Mathematical Methods in Engineering
Edited by G. A. O. Davies
© 1984, John Wiley & Sons, Ltd.

2

Engineering Fluid Mechanics

2.1 The Navier–Stokes equations

The fundamental relations expressing Newton's second law of motion for a fluid are the Navier–Stokes equations, valid for a *Newtonian fluid*. A Newtonian fluid is one having a linear relation between *stress* and *strain*, and is a fair approximation for many fluids and gases under certain conditions. Thus, in tensor notation [see V, Chapter 7],

$$\sigma_{ij} + p\delta_{ij} = \mu \left(\frac{\partial u_i}{\partial x_j} + \frac{\partial u_j}{\partial x_i} - \frac{2}{3} \delta_{ij} \frac{\partial u_k}{\partial x_k} \right) \tag{2.1}$$

where p is the normal (*isotropic*) stress or pressure, u_i is the velocity component, σ_{ij} the stress tensor, μ the coefficient of viscosity (a constant) and δ_{ij} is the *Kronecker delta* [see I, (6.2.8)].

Equating the stress tensor to the rate of change of momentum flux in the fluid leads to the Navier–Stokes equations, in *vector notation* [see IV, §17.2]:

$$\rho \frac{\partial \mathbf{U}}{\partial t} + \rho(\mathbf{U} \cdot \nabla)\mathbf{U} = -\nabla p + \mathbf{F} + \mu[\nabla^2 \mathbf{U} + \tfrac{1}{3}\nabla(\nabla \cdot \mathbf{U})]. \tag{2.2}$$

The vector $\mathbf{F}$ is any applied body force (e.g. gravity, ρ is the fluid density and ∇ is the *vector derivative* [see IV, Example 17.3.3] or 'grad'

$$\nabla = \mathbf{i}\frac{\partial}{\partial x} + \mathbf{j}\frac{\partial}{\partial y} + \mathbf{k}\frac{\partial}{\partial z}.$$

The expanded forms in *scalar* components are given in Section 6.2. These equations are usually associated with the equations expressing conservation of mass flux in the fluid:

$$\frac{\partial \rho}{\partial t} + \nabla \cdot (\rho \mathbf{U}) = 0. \tag{2.3}$$

The resulting set of equations with the appropriate *boundary conditions* on $\mathbf{U}$, p or ρ completely describe the flow of a Newtonian fluid. Moreover, to obtain a solution it is also necessary to provide an *equation of state*

$$\rho = \rho(p) \tag{2.4}$$

for the fluid. If, as frequently happens, this equation also depends on the fluid temperature, an additional equation relating the thermal and kinetic energies is required.

In the special case of *incompressible flow* these equations reduce to

$$\rho = \text{constant}, \tag{2.5}$$

$$\mathbf{V} \cdot \mathbf{U} = 0 \tag{2.6}$$

and

$$\frac{\partial \mathbf{U}}{\partial t} + (\mathbf{U} \cdot \mathbf{V})\mathbf{U} = -\frac{1}{\rho} \mathbf{V}p + v\mathbf{V}^2\mathbf{U} \tag{2.7}$$

where $v = \mu/\rho$ is called the *kinematic viscosity*.

The rest of this chapter looks at a series of special cases in which simplifying assumptions can be made because of the geometric nature of the flow or because of certain limitations on velocity, density, etc. The general mathematical techniques which will be employed throughout will be *ordinary differentiation, integration, partial differentiation, vector notation* and *vector calculus* [see IV, Chapters 3, 4 and 17]. A knowledge of *tensor* (suffix) *notation* is helpful [see V, Chapter 7].

2.2 Incompressible flow

An important class of flows are those where the force field is *conservative* [see IV, Example 17.3.3] so that the velocity $\mathbf{U}$ can be expressed as the gradient of a potential function ϕ; thus

$$\mathbf{U} = \mathbf{V}\phi \tag{2.8}$$

(in some texts $\mathbf{U} = -\mathbf{V}\phi$). The most important condition for this in an 'incompressible' flow (i.e. low Mach number <0.3) is that the *Reynolds number* R_e of the flow is high.

The Reynolds number is defined as

$$R_e = \rho \frac{Ul}{\mu}, \tag{2.9}$$

where U and l are characteristic velocities and lengths of the flow (for example the free stream velocity and the length of the body immersed in it) and ρ and μ are the density and viscosity of the fluid. The Reynolds number of the flow is considered to be high if

$$R_e > \text{about } 10^4$$

since the significant parameter is $\sqrt{R_e}$.

Irrotational steady flows

The *curl* [see IV, Chapter 13] of the steady Navier–Stokes equations gives the convection equation for the vorticity $\omega(\equiv \mathbf{V} \times \mathbf{U})$ as

$$\mathbf{U}\cdot\mathbf{V}\cdot\omega - \omega\cdot\mathbf{V}\cdot\mathbf{U} = O(R_e^{-1}) \tag{2.10}$$

using the notation explained in Volume IV, section 2.3. Neglecting the right-hand side in flows where R_e is large gives the solution, for a flow with *zero oncoming vorticity*:

$$\omega = \mathbf{0} \tag{2.11}$$

everywhere, and the flow is said to be 'irrotational'. Since $\omega = \mathbf{V} \times \mathbf{U}$ is zero, it follows that $\mathbf{U}$ can be written in terms of a potential (equation 2.8) provided the region is simply connected and ϕ is single-valued.

The equation for continuity of mass flow in an incompressible flow is

$$\mathbf{V}\cdot\mathbf{U} = 0. \tag{2.12}$$

Combining equations (2.8) and (2.12) the potential ϕ satisfies

$$\nabla^2\phi = 0 \tag{2.13}$$

which is *Laplace's equation* [see IV, §8.2].It is also possible from equation (2.12) to define a vector function ψ such that $\mathbf{V} \times \psi = \mathbf{U}$. Applying this to equation (2.10) with the further gauge condition that $\mathbf{V}\cdot\psi = 0$, we deduce that ψ also satisfies Laplace's equation

$$\nabla^2\psi = \mathbf{0}. \tag{2.14}$$

In two-dimensional flows, ψ becomes a scalar known as the *stream function*, constant along streamlines. The difference in the value of ψ between streamlines is equal to the volume flow between them.

The high Reynolds number approximation to the Navier–Stokes equation is, however, *non-uniform* because the small parameter R_e^{-1} multiplies the highest derivatives in the equations. Because of this all high Reynolds number flows contain regions which are *rotational* ($\omega \neq \mathbf{0}$) where the effect of viscosity is important.

For streamlined flow, these regions are confined to very thin layers surrounding the body (boundary layers, see Chapter 3) and trailing downstream (wakes). An example is the unstalled aerofoil.

Bodies are known as *bluff bodies* at high Reynolds numbers if the boundary layers on the two surfaces of the body separate from the surface before they meet up at the trailing edge. In such cases a wide wake is produced and the body has a high drag coefficient. Most bodies are bluff in this sense unless they have been purposely streamlined to minimize the adverse pressure gradients which cause boundary layer separation. High Reynolds number flows may be solved by the method of *matched asymptotic expansion* (MAE) (Nayfeh, 1973).

Qualitatively, a first-order outer potential flow solution is obtained which, because the boundary layers are thin, gives the pressure distribution on the body surface to within an error of order $R^{-1/2}$. This pressure (or the surface slip velocity) distribution is then used to calculate the boundary layer development which gives the distribution of shear stress (skin friction) on the surface and also the boundary layer displacement thickness δ_1 (see Chapter 3); δ_1 is then used to perturb the body surface. Alternatively, the stream function ψ may be matched at the outer edge of the boundary layer or a 'transpiration' velocity proportional to $\partial\delta_1/\partial s$ applied at the surface s where s is the surface coordinate in the direction of the stream. A corrected second-order outer potential solution is then calculated. This process can be carried on and will converge on the complete Navier–Stokes solution provided the boundary layers and wake remain thin. The MAE method breaks down near points of separation; strictly a more exact solution of the Navier–Stokes equations must be found for these regions, but in practice this is not often done (see, for example, Stewartson, 1969). It is not, however, possible to carry out the process with any confidence when the body is bluff.

Usually, at high Reynolds numbers, the first-order approximations are sufficiently accurate, except in some aerofoil calculations where greater accuracy is required and it is usual to go to higher approximations.

2.3 Solutions of the complete Navier–Stokes equations

A few very special analytical solutions of these equations do exist, e.g. laminar pipe and channel flow, couette flow and certain wedge flows (some of these are listed in Rosenhead, 1963). For the most part they are too specialized to be generally useful.

As an alternative, attempts have been made to integrate the equations numerically using either *finite difference* [see III, §9.1] or *finite element* (see Chapter 15) *techniques*. The results are fairly successful, including time-dependent flows, up to moderate values of R of 10^2 to 10^3 (see Bratanow, Ecer and Kobiske, 1973). Beyond that range, they meet a basic problem of high Reynolds number flows: if the flow is laminar and the wake thin, the MAE method is much more efficient, and if the wake is wide (bluff body) so that there are large regions of *turbulence* in the flow, the mesh (or element) size necessary to differentiate adequately the smallest scales of turbulence leads to an excessively large number N of meshes to cover the flow field, and the problem also becomes time-dependent.

Very approximately $N \sim R_e^{2+}$, which rapidly exceeds the capacity of present-day computers. Many solutions avoid this by using empirical formulae or artificial viscosities to describe the smallest scales of the flow and hence take a large mesh-size calculation to a high Reynolds number.

2.4 Potential flow methods (for the 'outer' inviscid irrotational flow)

Conformal transformation (two-dimensional flows only)

The functions ϕ and ψ satisfy the *Cauchy–Riemann equations* [see IV, Chapter 9] and therefore the complex function $W(z) = \phi(z) + i\psi(z)$ known as the *complex potential* is a *holomorphic function* of the *complex variable z* (the coordinate of a general point) [see IV, §9.1]. It should be noted that W remains *invariant* under *conformal transformation* between the z plane and any other (ζ) plane [see IV, §9.12].

The *method of conformal transformation* is used to transform the body contour $F(z)$ into a suitable contour $f(\zeta)$ for which $W(\zeta)$ can be calculated. It is usual to make the contour in the ζ plane either (i) a circle of radius a (say centred on $\zeta = 0$) for which the potential is

$$W(\zeta) = U_\infty(e^{-i\alpha}\zeta + a^2/\zeta e^{i\alpha}) \qquad (2.15)$$

where U_∞ is the transformed free stream in the ζ plane and α its incidence or (ii) a straight line parallel to the free stream for which $W(\zeta) = U_\infty\zeta$ [see IV, §9.2].

Two types of transformation are used:

(i) those which tend to identity far from the origin (i.e. $z = \zeta$ + smaller terms, when $|z| \to \infty$);
(ii) those which do not have this property (for which a finite point or points ζ_n in the ζ plane may correspond to the 'point at infinity' in the z plane [see IV, §9.16]).

Transformation of the first type (such as the *Joukowski* transformation $z = \zeta + a^2/\zeta$ [see IV, §9.13]) are usually used to solve problems consisting of a uniform free stream (at infinity) flowing past a single finite closed body.

Transformations of the second type (such as $z = \zeta^{1/2}$ and $z = k \log \zeta$) are used where either the body extends to infinity in some direction (e.g. flow past a wedge or in a long channel) or there are an infinite number of bodies in some regular sequence (such as a cascade of aerofoils). In the latter case suitable transformations make use of the *multivalued property of the complex logarithm* [see IV, §9.5] to transform the sequence of bodies in the z plane onto a single body in the ζ plane.

Conformal transformation preserves the angle between intersecting curves so that bodies with discontinuities in surface slope (corners, aerofoil trailing edges, etc.) can only be transformed into continuously smooth bodies (such as circles) by means of a transformation which is singular (non-conformal) at the corresponding points in the z plane. The Joukowski transformation given above, for example, is singular at $z = \pm 2a$ ($\partial\zeta/\partial z = \infty$ there) and more generally the *Schwarz–Christoffel* transformation can be constructed to be singular at an arbitrary number of points in the plane (Ahlfors, 1966, p. 228; Ramsey, 1935, p. 135).

Given, therefore, a suitable transformation to remove the discontinuities in surface slope it is always possible to transform any closed (non-intersecting) contour into a circle. In practice this may have to be done numerically by a process of iteration in the case of an arbitrary body and the reader is referred to Catherall, Foster and Sells (1969) for a description of a particular application of the technique. A detailed analysis of conformal transformation is given in Volume IV, section 9.12 and a discussion of useful transformations in Woods (1961).

Once the potential $W(\zeta)$ has been found in the transformed plane the complex velocity $U(\zeta) - iV(\zeta) = dW/d\zeta$ can be calculated.

The complex velocity in the z plane is then given by

$$U(z) - iV(z) = \frac{dW}{dz} = \frac{dW}{d\zeta} \frac{d\zeta}{dz}.$$

The overall forces on a two-dimensional body when the flow past it is describable by a complex potential, as above, may be calculated by the application of the *Blasius theorem* (Batchelor, 1967) in which the resulting integrals are most conveniently evaluated by *contour integration* [see IV, §9.3].

Two- and three-dimensional boundary conditions—Potential flow past a closed body

Because the exterior flow domain is *multiply connected* the problem of potential flow round a closed body is not completely defined by the basic boundary conditions that

(i) $\dfrac{\partial \phi}{\partial n} = 0$ (or $\psi = $ constant)

on the body surface (where n is the outward normal) and

(ii) ϕ (or ψ) tends to its free stream value far from the body.

An additional boundary condition is required to specify the circulation (or eigen-) solution associated with each closed body.

The condition which is usually applied is the *Kutta–Joukowski* condition that the rear separation streamline leaves the body at a specified point (usually the trailing edge) or other sharp corner. This is often specified by making the pressure difference (or loading) zero across this edge.

The Kutta–Joukowski condition replaces the neglected effect of viscosity (boundary layers) in the Navier–Stokes equations and fixed the circulation $\Gamma(= \oint \mathbf{U} \cdot d\mathbf{s}$ round the body) and hence the lift forces on each body.

In most cases of flow round bluff bodies where no single separation occurs it is usual to define the flow uniquely by specifying *zero mean circulation*. Recently, calculation methods have been devised specifying a Kutta–Joukowski condition at each separation point.

Inviscid theory for flow past a bluff body

Three methods are currently available, making different assumptions for the separated region behind the bluff body.

  (i) Free streamline theory, which is usually analysed by means of a conformal transformation of the *hodograph plane* (velocity variables) (Woods, 1961) assumes constant pressure in the separated region and for this reason also gives a fairly accurate prediction of *cavitating flows.*
 (ii) The wake/source model (Parkinson and Jandali, 1970) is probably the simplest to apply.
(iii) The shed vortex sheet (Clements and Maull, 1975) is in principle the most accurate.

However, all these methods are so far only applicable in two dimensions and the third is still in a development stage.

Singularities in the flow field

It is sometimes necessary to consider flows which contain singularities such as sources, vortices and multipoles. These correspond to the basic singular solutions of Laplace's equation [see IV, §8.2] but can also be identified physically with 'points' at which fluid is appearing or disappearing (sources and sinks) or at which a vortex is present. The far field of these physical phenomena is fairly well represented by the potential flow singularity, except in the case of a physical source where it is necessary to avoid it appearing as a jet of fluid.

Such singularities in the fluid have their appropriate associated complex potential functions and are usually treated in the presence of solid surfaces by *the method of images* (Ramsey, 1935, p. 47).

Singularities can also be placed suitably on or within the surface contours of bodies to represent the effect of that body surface on the exterior flow field. This is entirely a mathematical representation (when sources are used) although it has an obvious physical analogue in the case of vortices and solves the partial differential equation (in this case Laplace's) by means of the appropriate *Green functions* [see IV, Definition 8.2.1], as described below.

The surface singularity technique (boundary integral equation (BIE) method)

The aim of this technique is to find the 'Green equivalent layer' $\mu(s)$, say, which when integrated over the surface of the body provides the solution for the potential. Thus

$$\phi(z) = R_e \left(U_\infty z\, e^{-i\alpha} + \frac{i\Gamma}{2\pi} \log z \right) + \frac{1}{2\pi} \int_s \mu \log r\, ds, \qquad (2.16)$$

where r is the distance of z from s, a point on the body surface. Here U_∞ and α

are the free stream and its direction as before and Γ is the circulation to be specified by a Kutta–Joukowski or alternative condition.

It is usual to work numerically in terms of velocity rather than potential and the surface integral is evaluated by covering the body surface with a large number of elements (or facets) over which some prescribed variation (often constant) of μ is assumed.

The great advantage of the method is that it reduces the number of dimensions in the problem by one and that it is quite generally applicable, including cases governed by other linear equations (e.g. the Helmholtz wave equation for acoustics or water waves, see Chapter 5).

The method using sources is described in detail in Smith and Hess (1966). Using vertices is equivalent to the *collocation method* used in wing theory (see Falkner, 1953, and Garner, Hewitt and Labriyere, 1968) where the body surface is covered with either a vortex lattice or a continuous distribution and the normal velocity is evaluated at the centre of each mesh or collocation point.

Unlike conformal transformation, the method is not limited to two dimensions; however, three-dimensional problems with circulation are more difficult since step-by-step 'tracking' of the vortex wake is required. It is also applicable to unsteady flows (Giesing, 1968), but the same problem arises here (for two-dimensional flows also), and for this reason only the easier two-dimensional type of problem is usually calculated. A special analytical application of this method when the thickness of the body is much less than its length is known as *thin aerofoil theory* (Woods, 1961).

Separation of variables

Laplace's equation separates into eleven coordinate systems. However, only three (*cartesian, cylindrical polar* and *spherical polar* [see IV, §8.2 and examples 8.2.4 and 8.2.5] have general engineering applications. If the surfaces on which the boundary conditions are to be applied lie approximately along simple surfaces in any of these systems *and the flow field is expected to have a similar symmetry*, the method of *separation of variables* should be applied [see IV, §8.2]. An example would be certain flows inside a rectangular duct (cartesian). In practice the occurrence of both circulation and boundary layer separation makes all but the most obvious cases very difficult to solve since the expected symmetry of the flow field is removed.

The pressure equation

The pressure (p) equation for all potential flow regions of an incompressible flow is Bernoulli's steady flow equation (Chapter 5, equation 5.39). The $\rho g y$ term which occurs in this equation is usually negligible where no free surface is involved. Thus

$$p + \tfrac{1}{2}\rho q^2 = \text{constant} \tag{2.17}$$

where q is the local flow speed.

The absence of boundary layer effects (and hence skin friction separation and wakes) in the outer first-order potential flow solution means that the integrated drag on the body is zero:

$$D = 0.$$

The only exception to this is when the flow is either unsteady (see below), has more than one separation point (bluff) or is three dimensional *and* has circulation (induced drag due to trailing vorticity).

The integrated lift force (perpendicular to the stream) on the body is related directly to the circulation $\boldsymbol{\Gamma}$ (specified by the Kutta–Joukowski condition or otherwise) through the *Joukowski theorem*

$$\mathbf{L} = \rho \mathbf{U} \times \boldsymbol{\Gamma} \tag{2.18}$$

where $\mathbf{U}$ is the externally imposed velocity field (usually the free stream). This theorem can be derived from the *Blasius theorem* for two-dimensional flow (Ramsey, 1935, p. 109).

2.5 Unsteady incompressible irrotational flow

In the absence of any circulation round the body or bodies, the instantaneous kinematic problem is the same as for steady incompressible irrotational flow, i.e. the entire flow field responds instantaneously to any changes in the relative motion between each body and the stream. However, the pressure (Bernoulli's) equation (2.17) contains an additional acceleration term. Thus

$$p + \rho \left(\frac{\partial \phi}{\partial t} + \tfrac{1}{2}q^2 \right) = F(t \text{ only}), \tag{2.19}$$

where q is the local flow speed. For most problems it is permissible to take $F(t)$ as an absolute constant.

Except in cases of small amplitude unsteadiness it is necessary in evaluating the pressure either to calculate the potential ϕ with respect to non-accelerating inertial axes or, alternatively, if it is more convenient, to take an accelerating ($\mathbf{a}$) frame of reference $\mathbf{x}$ (as, for example, in the case of an accelerating body). An appropriate acceleration term must be included in the pressure equation (cf. $\rho g y$ in free-surface flow). Thus

$$p + \rho \left(\frac{\partial \phi}{\partial t} + \tfrac{1}{2}q^2 + \mathbf{x} \cdot \mathbf{a} \right) = \text{constant}. \tag{2.20}$$

Virtual mass effect

The $\partial\phi/\partial t$ term gives rise, among other things, to an overall drag on the body, even in potential flow. This drag is proportional to the acceleration $\dot{U}_\infty$ of the body relative to the free stream and can be written as

$$D = C_m \dot{U}_\infty m \qquad (2.21)$$

where m is the mass of *fluid displaced* by the body and C_m is the virtual mass coefficient. See Woods (1961) and Robinson and Laurmann (1956) for an analysis of some unsteady flows. An expression for the *drag force in unsteady flow* is then given by *Morison's equation*

$$D = \tfrac{1}{2}\rho U_\infty |U_\infty| SC_D + \dot{U}_\infty m C_m \qquad (2.22)$$

where C_D is a drag coefficient based on the area (usually frontal) S, and C_D and C_m are given empirically.

2.6 Rotational flows

When the oncoming flow is rotational, the solution of the vorticity convection equation is no longer zero. In general a *numerical step-by-step integration* of the steady or unsteady form of equation (2.10) is required:

$$\frac{\partial \boldsymbol{\omega}}{\partial t} + \mathbf{U}\cdot\mathbf{V}\cdot\boldsymbol{\omega} = \boldsymbol{\omega}\cdot\mathbf{V}\cdot\mathbf{U} + O(R^{-1}). \qquad (2.23)$$

This is infrequently done (usually by *finite element* or *difference* methods).[see III, §9.1]). The resulting distribution of vorticity forms the right-hand side of a *Poisson-type equation* [see IV, §8.2] for the stream function. In a few special cases where the rotational part of the flow is small or the perturbation of the stream by the body surface is small, these equations can be reduced to *Laplace's equation* [see IV, §8.2] for the perturbation stream function ψ' (see, for example, Townsend, 1976, which contains a calculation of duct flow through a gauze of variable shape).

Since the velocity field $\mathbf{U}$ is now a general *solenoidal vector*, i.e. $\mathbf{V}\cdot U = 0$, it can be written in the form

$$\mathbf{U} = \mathbf{V} \times \boldsymbol{\psi} + \mathbf{V}\cdot\phi.$$

The second part is the gradient of a potential function ϕ wich is used to satisfy the boundary conditions $\mathbf{U}\cdot\mathbf{n} = 0$ on the surface of the body. Thus

$$\frac{\partial \phi}{\partial n} = \mathbf{U}\cdot\mathbf{n} - (\mathbf{V} \times \boldsymbol{\psi})\cdot\mathbf{n},$$

where $\boldsymbol{\psi}$ has been calculated from the vorticity field $\boldsymbol{\omega}$. In order to satisfy continuity of mass flow ($\mathbf{\nabla \cdot U} = 0$), ϕ satisfies Laplace's equation

$$\nabla^2 \phi = 0,$$

and the same method of solution described above can be used.

2.7 Wakes

In this context a wake refers to the region of rotational fluid (i.e. containing vorticity) *downstream* of a body, disturbed by the relative motion of the body to that of the undisturbed fluid.

Inviscid

In inviscid (infinite Reynolds number) flow theory (see previous section) a wake of vorticity can only be generated by:

 (i) three-dimensional flows with lift forces present, in which case trailing vorticity (e.g. aircraft vortices) with the vortex vector in the stream direction is shed;
(ii) unsteady flows with fluctuating lift forces in which vorticity perpendicular to the stream direction is also shed (see Robinson and Laurmann, 1956, for an account of (i) and (ii));
(iii) an infinite Reynolds number representation of flow behind a bluff body in which the flow separates from more than one edge of the body and hence sheds vorticity at these edges.

In each of these cases the wake can be approximately accounted for by tracking the vortices in the wake and then calculating the additional velocity field $\mathbf{u}_w$ induced on the body by the wake by means of the *Biot–Savart law* (Woods, 1961):

$$\mathbf{u}_w(\mathbf{r}_0) = \frac{1}{4\pi} \int \frac{\gamma(\mathbf{r}) \times (\mathbf{r} - \mathbf{r}_0)\, dS_w(\mathbf{r})}{|\mathbf{r} - \mathbf{r}_0|^3},$$

where γ is the vorticity per unit area of the wake and the integral is taken over the surface S_w of the wake. It is often necessary to simplify this surface by, for example, assuming that it lies in the mid-plane of the wake.

The velocity $\mathbf{u}_w$ is then added to the free stream velocity so that the inviscid flow calculation round the body may be performed (as described above). This may sometimes require an *iterative process* [see III, §1.2].

The interaction of the inviscid flow field induced by a wake with its parent body (or with other adjacent bodies) may result in self-excited oscillations (flutter, aeroelastic oscillation) when the body is free to oscillate elastically with respect to the stream.

Viscous (laminar)

In this case the vorticity in the wake arises from the boundary layers of the body where fluid is retarded relative to the stream by the action of viscosity. Laminar wakes are very unstable above Reynolds numbers (based on thickness) of about 10^2. The behaviour of wakes at lower Reynolds numbers can be calculated by the methods of *Stokes* and *Oseen* (Batchelor, 1967).

Viscous (turbulent and bluff body wakes)

At higher Reynolds numbers, particularly behind bluff bodies, there is a tendency for periodic fluctuations of the separating shear layers to occur, leading to a rolling up of layers of vorticity of predominantly one sign into a quasi-periodic array of discrete vortices of alternate sign in the wake. With further increase of Reynolds number the wake also becomes turbulent closer to its point of origin until a Reynolds number is reached at which the boundary layers on the body become turbulent upstream of the wake. Because of the difficulty of predicting the behaviour of turbulence mathematically very little theoretical analysis is possible.

Far enough downstream for the turbulence to have decayed the Oseen approximation (see above) may be used to obtain a solution. Closer to the body, for wakes of streamlined bodies without a significant periodic component, methods similar to those used for turbulent boundary layers may be used. These require the empirical formulae for the Reynolds stress or higher order correlations of turbulent velocity in order to obtain a closed set of equations (see below). The resulting set of equations can then be solved by *finite difference techniques* [see III, §9.1] or *finite element techniques* (Chapter 15).

Wakes containing a significant periodic array of vortices are more difficult. They may be crudely represented by an inviscid vortex wake (as indicated above) or in the mean by the *source wake model*.

The characteristic frequency f of such a wake is found to be proportional to U_∞/b, where b is the body (or initial wake) width. Thus

$$\frac{fb}{U_\infty} = S, \tag{2.24}$$

the *Strouhal number*, which for a given body is approximately independent of the Reynolds number except in certain critical ranges.

The drag of a bluff body may also be approximately related to the strength of the vortices shed into the wake (Goldstein, 1938). Similarly, the drag of a three-dimensional lifting body is related to its trailing vortices. This is known as *induced drag*. Suitable *control volume analysis* (Goldstein, 1938) shows that the drag D due to the viscous wake of a body is related to the streamwise velocity U

far downstream by

$$D = \int_A \rho(U_\infty - U)U \, dA,$$

where A is the cross-sectional area of the wake. The effect of a viscous wake on the flow outside it is equivalent to an outward displacement of streamlines by an amount δ_1, the wake displacement thickness, defined as for a boundary layer and discussed previously.

2.8 Turbulence

Turbulence occurs mainly in flows containing a mean velocity shear (jets, wakes, boundary layers, pipe and duct flows, and the atmosphere). All these flows are unstable at high Reynolds numbers and the mean velocity shear then energizes the turbulence against the dissipative effects of viscosity.

When turbulence is present in a shear flow, additional turbulent stress terms (the Reynolds stresses) σ_T become important in the mean flow momentum equations. The Navier–Stokes equations (2.2) for the mean velocity $\bar{U}_i$ are then, in tensor notation,

$$\rho \bar{U}_j \frac{\partial \bar{U}_i}{\partial x_j} = -\frac{\partial p}{\partial x_i} + \frac{\partial}{\partial x_j}\left(\sigma_{T_{ij}} + \mu \frac{\partial \bar{U}_i}{\partial x_j}\right), \tag{2.25}$$

where $\sigma_{T_{ij}} = -\overline{\rho u_i u_j}$, a one-point *turbulent velocity* correlation.

The only extensive theoretical analysis (which is still far from complete) of turbulence is for *isotropic homogeneous turbulence*, in which the Reynolds shear stresses are zero (Townsend, 1976). This type of turbulent flow is most closely realized in the flow downstream an evenly spaced grid of bars (mono- or biplanar). Ranges of the development of this type of turbulence can be predicted theoretically and the analysis is helpful in understanding the behaviour of turbulence in more complex situations.

In general, however, turbulent flow calculations rely on empirical formulae for predicting σ_T. The simplest of these is the *mixing length assumption* which relates σ_T to the mean velocity shear by

$$\sigma_{T_{ij}} = \rho l^2 \left|\frac{\partial \bar{U}_i}{\partial x_j}\right| \frac{\partial \bar{U}_i}{\partial x_j}, \qquad \text{for } i \neq j,$$

$$= \rho \nu_T \frac{\partial \bar{U}_i}{\partial x_j}, \tag{2.26}$$

where l is the mixing length and ν_T the coefficient of eddy viscosity; l is related empirically to local length scales of the flow.

More sophisticated methods calculate σ_T from an equation for transport of σ_T

(or turbulence intensity) obtained by multiplying the Navier–Stokes equations for the instantaneous velocity component $(\bar{U}_i + u_i(t))$ by $(\bar{U}_j + u_j)$ and taking a time average. Since this equation contains further and higher order correlations of turbulent velocities, some of these terms are similarly modelled empirically. A comprehensive account of this is given in Townsend (1976). The resulting equations for σ_T and $\bar{U}$ can then be solved by either finite difference or finite element methods, except in simple cases when an analytical result can be obtained.

Response of bodies to turbulence

If a body is immersed in a turbulent stream, fluctuating forces are induced on it by the turbulent velocities. (The static pressure fluctuations are usually insignificant in this context.) If the scale of the turbulence (eddy size) is much larger than the body a quasi-steady approach based on Morison's equation (Section 2.5) is satisfactory. Thus

$$D(\text{streamwise force}) = \tfrac{1}{2}\sigma(\bar{U}_\infty + u)^2 C_D S + m C_m \dot{u}, \qquad (2.27)$$

where C_D, C_m, m and S are as defined before and $\dot{u}$ is the streamwise turbulent velocity component. Similarly, the force perpendicular to the stream is

$$L = \tfrac{1}{2}\rho(\bar{U}_\infty + u)^2 C_L S + \tfrac{1}{2}\rho v \bar{U}_\infty \frac{\partial C_L}{\partial \alpha} S + \tfrac{1}{2}\rho v \bar{U}_\infty C_D S, \qquad (2.28)$$

where C_L is the lift coefficient (Ashley and Landahl, 1965) and $\partial C_L/\partial \alpha$ is its derivative with respect to changes in incidence of the body in the direction of the turbulent velocity component v.

Similar results can also be deduced for the pressure distribution. These equations can be used to relate the statistical behaviour (or the spectrum) of unsteady forces on a body to the statistical description (or spectrum) of the turbulence (see, for example, Davenport, 1961).

Some more exact analyses have been attempted when the turbulence scale is comparable with the body scale, but these methods are still under development.

Apart from the general mathematical techniques listed earlier, an understanding of basic statistics, particularly *covariance, correlation* and *spectrum functions* [see II, §9.6, §9.8 and VI, §18.5.3], and a knowledge of *Fourier transforms* [see IV, §13.2] is useful.

2.9 Compressible flows (Mach number >0.3) (see also Chapter 4)

The equations governing the unsteady motion of a non-(heat-)conducting gas, under conditions of sufficiently high Reynolds number for the flow to be treated as inviscid, are:

From (2.2):
$$\frac{\partial \rho}{\partial t} + \mathbf{V} \cdot (\rho \mathbf{U}) = 0,$$

From (2.3):
$$\frac{\partial \mathbf{U}}{\partial t} + \mathbf{U} \cdot \mathbf{V} \mathbf{U} = -\frac{1}{\rho} \mathbf{V} p \qquad (2.29)$$

and
$$\frac{\partial S}{\partial t} + \mathbf{U} \cdot \mathbf{V} S = 0,$$

where ρ is the density, p the pressure, $\mathbf{U}$ the velocity and S the entropy of the gas.

If the Mach number $M = |U|/a$, where a is the local speed of sound, is sufficiently small everywhere, the flow may be considered to be incompressible: i.e.

$$\rho = \text{constant everywhere},$$

and the flow may be analysed as discussed in Section 2.2. The usual condition for incompressibility is

$$M^2 \ll 1 \qquad \text{(typically } M < 0.3)$$

(*but see also the comment under unsteady compressible flows/aero-acoustics in Section 2.10*).

Flow types

The terms *subsonic* and *supersonic* are strictly used to describe respectively flows having $M < 1$ and $M > 1$ everywhere. Flows containing regions with $M < 1$ as well as regions with $M > 1$ (mixed flows) are said to be *transonic*. This term is also used to describe subsonic and supersonic flows with Mach numbers close to one.

Subsonic and supersonic are also used more loosely to describe flows in which the Mach number M_∞ of the oncoming undisturbed stream is less than one or greater than one. These include some transonic mixed flows.

The term *hypersonic* is used to describe flows for which

$$M_\infty \, \delta = O(1),$$

where δ is the maximum flow deflection angle. In practice this usually implies high Mach numbers. For steady flows, regions with $M < 1$ are described by *elliptic equations* and regions with $M > 1$ by *hyperbolic equations* [see III, §9.1, and IV, §8.5].

The steady compressible flow equations (2.29) with $\partial/\partial t = 0$ can be integrated for 'one-dimensional flow', i.e.

$$\mathbf{U} = (U(x), 0, 0),$$

$$p = p(x),$$

$$\rho = \rho(x),$$

$$S = \text{constant},$$

to give the simple *isentropic flow* equations:

$$\frac{p}{p_0} = \left(1 + \frac{\gamma - 1}{2} M^2\right)^{-\gamma/(\gamma - 1)},$$

$$\frac{\rho}{\rho_0} = \left(1 + \frac{\gamma - 1}{2} M^2\right)^{-1/(\gamma - 1)},$$

where the suffix 0 indicates reservoir (stagnation) conditions. The thermodynamic relations for a gas are required to obtain these equations. The local Mach number $M = U/a$ is related to the cross-sectional area A of the flow (for example a pipe flow) by

$$A = \frac{W}{\rho_0 a_0 M} \left(1 + \frac{\gamma - 1}{2} M^2\right)^{(\gamma + 1)/[2(\gamma - 1)]}$$

and

$$\frac{a}{a_0} = \left(1 + \frac{\gamma - 1}{2} M^2\right)^{-1/2},$$

where W is the mass flow rate through A and γ is the ratio of specific heats. These equations can be used to calculate compressible flows in pipes and ducts where friction and heat transfer are unimportant. In particular, they can be used to calculate the area and pressure ratios necessary to generate supersonic flow in a duct. The equations are widely available in tabulated form (Keenan and Kaye, 1948), but can also be solved, if necessary, by any method for *transcendental equations* [see III, §5.1].

The analysis of one-dimensional compressible flows with friction, heat transfer and shock waves is discussed in Chapter 4.

Steady plane subsonic flow

In this case the continuity equation can be written in the form:

$$\left[a^2 - \left(\frac{\partial \phi}{\partial x}\right)^2\right] \frac{\partial^2 \phi}{\partial x^2} + \left[a^2 - \left(\frac{\partial \phi}{\partial y}\right)^2\right] \frac{\partial^2 \phi}{\partial y^2} - 2 \frac{\partial \phi}{\partial x} \frac{\partial \phi}{\partial y} \frac{\partial^2 \phi}{\partial x \, \partial y} = 0, \quad (2.30)$$

where ϕ is the velocity potential defined by

$$\nabla \phi = \mathbf{U},$$

as for incompressible flow. Since this equation is non-linear various approximate methods are used to obtain solutions.

(a) *Subsonic small disturbance theory.* If the flow is only slightly disturbed by the presence of the body (e.g. a thin aerofoil with thickness to chord ratio $\tau \ll 1$), the potential can be expressed as

$$\phi = U_\infty x + \phi',$$

where

$$\frac{|\nabla \phi'|}{U_\infty} \ll 1$$

and the x axis has been aligned with the free stream U_∞.

Provided $\tau \ll (1 - M_\infty)^{3/2}$, then the non-linear terms arising in equation (2.30) are negligible and the equation can be reduced to

$$(1 - M_\infty^2)\frac{\partial^2 \phi'}{\partial x^2} + \frac{\partial^2 \phi'}{\partial y^2} = 0.$$

This equation can be transformed to *Laplace's equation* [see IV, §8.2] and hence the compressible flow made similar to an incompressible flow by any *linear* (*affine*) *transformation*

$$\phi' = \beta^n \phi_0(x, \beta y)$$

(see Woods, 1961, for a detailed analysis). Using this transformation all the methods of Section 2.4 become available.

(b) *Transonic, small disturbance theory* $(M_\infty \sim 1)$. If in the above case $\tau \ll 1$ but $\tau = O(1 - M_\infty)^{3/2}$, the simplified equation becomes

$$\left(1 - M_\infty^2 - \frac{\gamma + 1}{U_\infty}\frac{\partial \phi'}{\partial x}\right)\frac{\partial^2 \phi'}{\partial x^2} + \frac{\partial^2 \phi'}{\partial y^2} = 0. \tag{2.31}$$

This non-linear equation has been successfully solved for mixed flows (with both elliptic and hyperbolic regions) using a finite difference technique with upstream biased x differences in the hyperbolic regions (see Murman and Cole, 1971). Many other 'shock-capturing' finite difference methods also exist for this regime.

(c) *Methods based on expansions in terms of the Mach number.* These are not widely used but are discussed in Woods (1961).

(d) *Transformation of the basic equations* (2.29). These methods are suitable for higher subsonic Mach numbers.

(i) *The hodograph transformation* (Woods, 1961). By rewriting equations (2.29) in terms of velocity variables $q = (U^2 + V^2)^{1/2}$ and $\theta = \tan^{-1} V/U$, a pair of linear equations can be derived for the potential ϕ and the compressible stream function ψ ($\nabla \times \psi = \rho/\rho_\infty \mathbf{U}$). These are

$$\frac{\partial \phi}{\partial \omega} = -\sigma \frac{\partial \psi}{\partial \theta}$$

and

$$\frac{\partial \phi}{\partial \theta} = \sigma \frac{\partial \psi}{\partial \omega}$$

where

$$\omega = \int \frac{1}{q} \sqrt{1 - M^2}\, dq,$$

$$\sigma = \frac{\rho_\infty}{\rho} \sqrt{1 - M^2}$$

and ρ_∞ is the free stream density.

Under certain conditions (such as the tangent gas assumption, see Woods, 1961) σ is constant and these equations become the *Cauchy–Riemann equations* [see IV, (9.1.2)] so that ϕ and ψ satisfy *Laplace's equation* in the transformed plane [see IV, §9.13].

However, the boundary conditions become very complicated in this plane and it is usual to make further approximations. A few solutions of special cases of the exact hodograph equations ($\sigma \neq$ constant) have been derived (Woods, 1961).

(ii) *Transformation to the interior of a circle.* This method has been derived for flow round streamlined bodies (see Sells, 1967). Equations (2.29) are transformed by means of a *conformal transformation* [see IV, §9.12] which takes the flow field exterior to the body into the interior of a circle. The resulting non-linear equation for the stream function is solved by a finite difference method [see III, §9.1] applied to a circular mesh in the transformed plane.

(e) *Similarity rules.* If we consider two thin profiles (thickness to chord ratio τ) which are related to each other by an invertible linear transformation (known in aeronautics as an *affine* relation), then the transonic flows about these profiles are similar if they both have the same value of the similarity parameter:

$$K = \frac{1 - M_\infty^2}{[\tau(\gamma + 1)M_\infty^2]^{2/3}}.$$

In this case the pressures at similar points in the flow are related as a consequence of

$$p = p_\infty + \frac{\rho_\infty \tau^{2/3} U_\infty^2}{2(\gamma + 1)^{1/3}}\, F(K, x).$$

These results follow from equation (2.31) above.

Table 2.1 Inviscid subsonic flow equations

Type of flow	Parametric restrictions	Potential equation	Pressure equation	Boundary conditions	
1(a). Quasi-steady, very low Mach number	$k \ll 1, M_\infty^2 \ll 1$	$\nabla^2 \phi = 0$	$p = p_\infty + \frac{1}{2}\rho(U_\infty^2 - (\nabla\phi)^2)$	$\dfrac{\partial\phi}{\partial n} = 0$	$\nabla\phi \to U_\infty$
1(b). Quasi-steady, low Mach number, thin aerofoil	$k \ll 1, M_\infty^2 \ll 1, \tau \ll 1$	$\nabla^2 \phi' = 0$	$p = p_\infty - \rho U_\infty \phi'_x$	$\phi'_y = -U_\infty F_x$	$\phi' \to 0$
2(a). Unsteady, low Mach number	$k \sim 1, M_\infty^2 \ll 1$	$\nabla^2 \phi = 0$	$p = p_\infty + \frac{1}{2}\rho(U_\infty^2 - (\nabla\phi)^2) - \rho\phi_t$	$\dfrac{\partial\phi}{\partial n} = 0$	$\nabla\phi \to U_\infty + f(r - a_\infty t)$
2(b). Unsteady, low Mach number, thin aerofoil	$k \sim 1, M_\infty^2 \ll 1, \tau \ll 1$	$\nabla^2 \phi' = 0$	$p = p_\infty - \rho U_\infty \phi'_x - \rho\phi'_t$	$\phi'_y = -U_\infty F_x$	$\phi' \to 0$
3. High frequency, low Mach number (classical acoustics)	$k \gg 1, M_\infty^2 \ll 1$	$\nabla^2 \phi - \dfrac{1}{a_\infty^2}\phi_{tt} = 0$	$p = p_\infty - \rho\phi_t$	$\dfrac{\partial\phi}{\partial n} = 0$	$\phi' \to f(r - a_\infty t)$
4. Quasi-steady, subsonic, thin aerofoil	$k \ll 1, \tau^{2/3} \ll 1 - M_\infty$	$(1 - M_\infty^2)\phi'_{xx} + \phi'_{yy} = 0$	$p = p_\infty - \rho U_\infty \phi'_x$	$\phi'_y = -U_\infty F_x$	$\phi' \to 0$
5. Unsteady, subsonic, thin aerofoil	$k \sim 1, \tau^{2/3} \ll 1 - M_\infty$	$(1 - M_\infty^2)\phi'_{xx} + \phi'_{yy} - \dfrac{2M_\infty}{a_\infty}\phi'_{xt} - \dfrac{1}{a_\infty^2}\phi'_{tt} = 0$	$p = p_\infty - \rho U_\infty \phi'_x - \rho\phi'_t$	$\phi'_y = -U_\infty F_x$	$\phi' \to f(r - a_\infty t)$
6. High frequency, subsonic	$k \gg 1, M_\infty \ll 1/k, \tau^{2/3} \ll 1 - M_\infty$	$(1 - M_\infty^2)\phi'_{xx} + \phi'_{yy} - \dfrac{1}{a_\infty^2}\phi'_{tt} = 0$	$p = p_\infty - \rho\phi'_t$	$\phi'_y = -U_\infty F_x$	$\phi' \to f(r - a_\infty t)$
7. Quasi-steady, transonic, thin aerofoil	$k \ll \tau^{2/3}, \tau \ll 1, M_\infty \sim 1$	$\left(1 - M_\infty^2 - \dfrac{\gamma+1}{U_\infty}\phi'_x\right)\phi'_{xx} + \phi'_{yy} = 0$	$p = p_\infty - \rho U_\infty \phi'_x$	$\phi'_y = -U_\infty F_x$	$\phi' \to 0$
8. Unsteady, transonic, thin aerofoil	$k \sim \tau^{2/3}, \tau \ll 1, M_\infty \sim 1$	$\left(1 - M_\infty^2 - \dfrac{\gamma+1}{U_\infty}\phi'_x\right)\phi'_{xx} + \phi'_{yy} - \dfrac{2M_\infty}{a_\infty}\phi'_{xt} = 0$	$p = p_\infty - \rho U_\infty \phi'_x - \rho\phi'_t$	$\phi'_y = -U_\infty F_x$	$\phi' \to 0$
9. High frequency, transonic, thin aerofoil (linearized)	$k \gg \tau^{2/3}, \tau \ll 1, M_\infty \sim 1$ and $k \gg 1 - M_\infty$	$\phi'_{yy} - \dfrac{2M_\infty}{a_\infty}\phi'_{xt} - \dfrac{1}{a_\infty^2}\phi'_{tt} = 0$	$p = p_\infty - \rho U_\infty \phi'_x - \rho\phi'_t$	$\phi'_y = -U_\infty F_x$	$\phi' \to f(r - a_\infty t)$

2.10 Three-dimensional subsonic flows

The continuity equation (2.30) for the potential ϕ can be simply generalized to include a third coordinate. Extensions of methods (a), (b), (c) and (e) to three dimensions are then available for obtaining solutions. In addition, if the body is slender (i.e. its cross-stream dimensions are much smaller than its streamwise dimension) the slender-body approximation may be applied.

Under this approximation the disturbance potential ϕ' can be obtained as a solution of Laplace's equation in the cross flow (y, z plane), depending only parametrically on the streamwise coordinate x through the boundary conditions. Solutions in the cross-flow plane are usually obtained by *conformal transformation* [see IV, §9.12]. Details of the theory are given in Woods (1961).

The remarks about the Reynolds number listed in Section 2.2 on incompressible flow apply also to subsonic compressible flows.

Unsteady compressible flows

Table 2.1 lists the approximate equations for the potential ϕ (or disturbance potential ϕ') which can be derived from the unsteady plane subsonic continuity equation under the given parametric restrictions. Here ϕ is the potential, ϕ' is the disturbance potential, $M_\infty = U_\infty/a_\infty$ is the free stream Mach number, $\tau = b/c$ is the thickness to chord ratio and $k = \omega c/2U_\infty$ is the reduced frequency.

Some care may be necessary in dealing with time-dependent flows to represent the flow singularities (vortices) which may now also occur in the wake of the body. This subject is dealt with in Robinson and Laurmann (1956).

In those unsteady flows in which compressibility is important $|M_\infty k = O(1)|$ the potential equation is either the *wave equation*

$$\nabla^2\phi - \frac{1}{a_\infty^2}\frac{\partial^2\phi}{\partial t^2} = 0$$

or a *convected wave equation* which can be transformed into the ordinary equation [see IV, §8.4].

J.M.R.G.

References

Ahlfors, L. V. (1966). *Complex Analysis*, 2nd ed., McGraw-Hill.

Ashley, H., and Landahl, M. T. (1965). *Aerodynamics of Wings and Bodies*, Addison-Wesley.

Batchelor, G. K. (1967). *Introduction to Fluid Dynamics*, Cambridge University Press.

Bratanow, T., Ecer, A., and Kobiske, M. (1973). AIAA paper No. 73–91, Eleventh Aerospace Sciences Meeting.

Catherall, D., Foster, D. N., and Sells, C. C. L. (1969). Roy. Aircraft Estab. Tech. Rep. No. 69118.

Clements, R. R., and Maull, D. J. (1975). *Progress in Aeronautical Sciences*, Vol. 16, p. 129, Pergamon.

Davenport, A. G. (1961). *Proc. Inst. Civ. Eng.*, **19**, 449.

Falkner, V. M. (1953). Aeron. Res. Council Rep. and Memo. No. 2591.

Garner, H. C., Hewitt, B. L., and Labrujere, T. E. (1968). Aeron. Res. Council Rep. and Memo. No. 3597.

Giesing, J. P. (1968). *J. Aircraft*, **5**, 135.

Goldstein, S. (1938). *Modern Developments in Fluid Dynamics*, Vol. 2, Oxford University Press.

Keenan, J. H., and Kaye, J. (1948). *Gas Tables*, Wiley.

Murman, E. M., and Cole, J. D. (1971). *AIAA J.*, **9**, 114.

Nayfeh, A. H. (1973). *Perturbation Methods*, Wiley.

Parkinson, G. V., and Jandali, T. (1970). *J. Fluid. Mech.*, **40**, 577.

Ramsey, A. S. (1935). *A Treatise on Hydrodynamics*, Vol. II, Bell.

Robinson, A., and Laurmann, J. A. (1956). *Wing Theory*, Cambridge University Press.

Rosenhead, L. (1963). *Laminar Boundary Layers*, Clarendon.

Sells, C. C. L. (1967). Roy. Aircraft Estab. Tech. Rep. No. 67146.

Smith, A. M. O., and Hess, J. L. (1966). *Progress in Aeronautical Sciences*, Vol. 8, Pergamon.

Stewartson, K. (1969). *Mathematika*, **16**, 106.

Townsend, A. A. (1976). *Structure of Turbulent Shear Flow*, 2nd ed., Cambridge University Press.

Woods, L. C. (1961). *Theory of Subsonic Plane Flow*, Cambridge University Press.

Mathematical Methods in Engineering
Edited by G. A. O. Davies
© 1984, John Wiley & Sons, Ltd.

3

Boundary Layers

3.1 Introduction

Before Prandtl introduced the concept of boundary layers in 1904, the study of flows of fluids with small viscosity was faced with an alarming and seemingly fundamental problem. The theories which neglected viscosity were often at variance with experimental observation, even in those situations where the effect of viscosity could reasonably be supposed to be small. We now realize that viscous forces, though small, play a significant role in the determination of the flow. Prandtl pointed out why viscous forces were important and how the inviscid solution could be used to determine the solution to the viscous problem.

Let us consider external flow around some body placed in an otherwise uniform stream. Inviscid theory allows a *slip velocity*; i.e. the velocity of the fluid on the surface of the body is non-zero. For viscous flow, however, the velocity on the surface must be zero, regardless of how small the viscosity is. If we are considering a body in a steady stream of fluid of undisturbed velocity U then, for fluids of small viscosity, the velocity adjusts rapidly, from being zero on the surface to the value U in a short distance. Thus the derivative of the velocity in this direction is locally large and the viscous terms may become significant, however small the viscosity. In the rest of the flow the velocity derivatives are, in general, small and the effects of viscosity may be ignored. Hence, we may be led to dividing the flow into two regions: one close to the body where the effects of viscosity cannot be ignored and the other, away from the body, where the flow is essentially inviscid. Flow in the former region should not be regarded as some insignificant alteration to the inviscid flow but as an important feature of the flow which can have a vital role. The region of flow in the neighbourhood of the body is called the *boundary layer*.

Boundary layers are not only formed on the surface of solid bodies but may be found anywhere where velocity gradients are sufficiently large, e.g. where a jet emerges into fluid otherwise at rest.

We have assumed that the motion of the fluid in the boundary layer is such that it remains small in thickness along the surface of the body. It may happen, however, that the boundary layer thickens and the retarded fluid in the boundary layer breaks away from the surface, disturbing large areas of the flow.

In this case we say that the flow is *separated*. Boundary layer theory can be used to study those flows where *separation* does not occur and may also be used to determine where separation will occur.

One may describe the procedure to determine the flow around a body as follows. First, determine the inviscid flow and then determine the boundary layer structure in the vicinity of the body which, in the far field, tends to the inviscid flow. The inviscid solution may then be modified to take account of the thickness of the boundary layer and a new boundary layer solution obtained and the process continued. In most cases, however, the process is halted once the initial boundary layer has been determined. For higher approximations to boundary layer theory see Van Dyke (1969).

We will now study the boundary layer equations which are the limiting form of the equations of motion as the Reynolds number R_e tends to infinity, or the kinematic viscosity v tends to zero.

3.2 Two-dimensional boundary layers

We consider the two-dimensional motion of fluid adjacent to a flat wall. The x axis is chosen in the plane of the wall and the y axis perpendicular to it (see Figure 3.1). We first consider the inviscid solution; this is that the velocity is U everywhere, with a slip condition at the wall. We must now determine the viscous flow which must satisfy the no-slip condition far from the wall. From Chapter 2, Section 2.1, the Navier–Stokes equation in the x direction is

$$\frac{\partial u}{\partial t} + u\frac{\partial u}{\partial x} + v\frac{\partial u}{\partial y} = -\frac{1}{\rho}\frac{\partial p}{\partial x} + v\left(\frac{\partial^2 u}{\partial x^2} + \frac{\partial^2 u}{\partial y^2}\right).$$

To simplify this equation we assume that the inertia terms are of the same order of magnitude as the friction terms and that $v(\partial^2 u/\partial x^2)$ is negligible in comparison with $v(\partial^2 u/\partial y^2)$. In the limit as the Reynolds number tends to infinity the

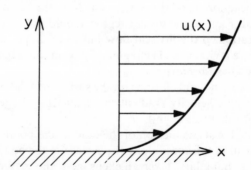

Figure 3.1: Velocity profile near the wall.

equations describing the motion become

$$\frac{\partial u}{\partial t} + u\frac{\partial u}{\partial x} + v\frac{\partial u}{\partial y} = -\frac{1}{\rho}\frac{\partial p}{\partial x} + v\frac{\partial^2 u}{\partial y^2}, \tag{3.1}$$

where p is the pressure derived from the inviscid solution. The continuity equation must also be satisfied:

$$\frac{\partial u}{\partial x} + \frac{\partial v}{\partial y} = 0. \tag{3.2}$$

We also have that the pressure across the boundary layer is constant. Since the velocity of the fluid far from the wall is $U(x, t)$ we have the boundary conditions

$$u = v = 0 \qquad \text{on } y = 0 \tag{3.3}$$

and

$$u \rightarrow U(x, t) \qquad \text{as } y \rightarrow \infty. \tag{3.4}$$

If the wall is permeable then the velocity at the wall may be non-zero and the boundary condition on the wall must be altered appropriately.

By comparing the boundary layer equation (3.1) with the inviscid Navier–Stokes equations, we see that boundary layers are an example of the mathematical *singular perturbation expansion* (Nayfeh, 1973) in which the inviscid solution is not valid in the neighbourhood of the wall as the no-slip condition cannot be applied.

For flow over a curved wall the equations take the same form, provided that the boundary layer thickness is small compared with the radius of curvature and that the radius of curvature does not change rapidly.

Invoking the same approximations as have already been used, we have that the shearing stress at the wall is

$$\tau_{\rm w} = \mu \left.\frac{\partial u}{\partial y}\right|_{y=0}. \tag{3.5}$$

The viscous drag is simply the integral of $\tau_{\rm w}$ over the surface.

If the boundary layer separates then the point of separation is identified with the point where the shearing stress vanishes, i.e. where

$$\left.\frac{\partial u}{\partial y}\right|_{y=0} = 0. \tag{3.6}$$

The boundary layer equations for steady flow are

$$u\frac{\partial u}{\partial x} + v\frac{\partial u}{\partial y} = -\frac{1}{\rho}\frac{\partial p}{\partial x} + v\frac{\partial^2 u}{\partial y^2} \tag{3.7}$$

and

$$\frac{\partial u}{\partial x} + \frac{\partial v}{\partial y} = 0. \tag{3.8}$$

It should be noted that these equations are of *parabolic type* [see IV, §8.5]. The appropriate boundary conditions are

$$u = v = 0 \qquad \text{on } y = 0 \tag{3.9}$$

and

$$u \to U(x) \qquad \text{as } y \to \infty \tag{3.10}$$

and a velocity profile $u(y)$ specified at some initial section, say at $x = 0$. The above boundary conditions assure the existence of a solution but there is no guarantee of uniqueness.

Blasius boundary layer on a semi-infinite flat plate

Consider a thin plate at zero incidence in a uniform stream with free stream velocity U. The leading edge of the plate is at the origin of the coordinates and the plate lies along the positive x axis. In the absence of a pressure gradient the equations of steady motion in the boundary layer are

$$u \frac{\partial u}{\partial x} + v \frac{\partial u}{\partial y} = v \frac{\partial^2 u}{\partial y^2} \tag{3.11}$$

and

$$\frac{\partial u}{\partial x} + \frac{\partial v}{\partial y} = 0, \tag{3.12}$$

with the boundary conditions

$$u = v = 0 \qquad \text{on } y = 0, \qquad x > 0, \tag{3.13}$$

$$u \to U \qquad \text{as } y \to \infty \tag{3.14}$$

and

$$u = U \qquad \text{at } x = 0. \tag{3.15}$$

We introduce the dimensionless coordinate

$$\eta = y \sqrt{\frac{U}{vx}} \tag{3.16}$$

(cf. the Reynolds number in Section 2.2) and write the stream function, $\psi(x, y)$ of Chapter 2, in terms of a dimensionless stream function f as

$$\psi(x, y) = \sqrt{vxU}\, f(\eta). \tag{3.17}$$

Substituting into the dynamic equation we have

$$ff'' + 2f''' = 0, \tag{3.18}$$

with the boundary conditions

$$f = 0, f' = 0 \qquad \text{on } \eta = 0 \tag{3.19}$$

and

$$f' \to 1 \qquad \text{as } \eta \to \infty. \tag{3.20}$$

Equations (3.18), (3.19) and (3.20) form an ordinary differential equation with *two-point boundary conditions* [see IV, §7.1]. For a given U they can be solved numerically (see, for example, Howarth, 1938). The shear stress is given by

$$\tau = \mu \left(\frac{U^3}{\nu x} \right)^{1/2} f''(0), \tag{3.21}$$

where $f''(0) = 0.332$. The skin friction on a plate of finite length l and unit width may then be calculated as

$$1.328 \sqrt{U^3 \mu \rho l},$$

or alternatively the drag coefficient is

$$C_D = \frac{1.328}{\sqrt{\dfrac{Ul}{\nu}}}$$

$$= \frac{1.328}{R_1^{1/2}} \tag{3.22}$$

where R_1 is the Reynolds number based on the length of the plate.

We have tacitly assumed, however, that the above boundary layer structure is valid everywhere. At the leading edge of the plate the present theory ceases to apply since the assumption

$$\left| \frac{\partial^2 u}{\partial x^2} \right| \ll \left| \frac{\partial^2 u}{\partial y^2} \right|$$

is not satisfied. For a solution valid in the region of the leading edge see Carrier and Lin (1958) and Van de Vooren and Dijkstra (1970). The boundary layer solution in the neighbourhood of the trailing edge must also be examined. At the trailing edge the restraining influence of the plate ends and the fluid at the bottom of the boundary layer is rapidly accelerated. This acceleration draws in fluid towards the inner layer of the boundary layer. This leads to a more complicated structure for the flow in the neighbourhood of the trailing edge. An analysis of this multistructure has been given by Stewartson (1974). This is not a minor refinement to the flow but may have a significant effect. The calculation of the asymptotic expansion for the drag coefficient C_D cannot be taken beyond the first term given above without considering this multistructure.

The connection between the equations for steady flow and the heat equation

Let us write the equations for steady flow in the form

$$u\frac{\partial u}{\partial x} + v\frac{\partial u}{\partial y} = U\frac{dU}{dx} + v\frac{\partial^2 u}{\partial y^2},$$

$$u = \frac{\partial \psi}{\partial y} \qquad \text{and} \qquad v = -\frac{\partial \psi}{\partial x}, \qquad (3.23)$$

where ψ is a stream function. If we now take x and ψ as independent variables, and use as a dependent variable $\chi = U^2 - u^2$, then equation (3.23) becomes

$$\frac{\partial \chi}{\partial x} = v(U^2 - \chi)\frac{\partial^2 \chi}{\partial \psi^2},$$

with

$$\chi = U^{2\cdot} \qquad \text{on } \psi = 0 \qquad\qquad (3.24)$$

and

$$\chi = 0 \qquad \text{on } \psi = \infty.$$

Equation (3.24) is a partial differential equation of *parabolic type in normal form*, similar to the *diffusion equation* except that it is non-linear [see IV, §8.6.5]. It has one *set of characteristics* [see III, §9.1], the lines $x = \text{constant}$. In order to determine the solution it is only necessary to give one boundary condition along any such line. If, for any value of x, χ is given as a function of ψ then $\partial\chi/\partial x$ is automatically known and hence $\chi(\psi)$ may be calculated for any neighbouring value of x. In terms of the initial variables, this implies that one velocity profile $u(y)$ at a certain value of x together with a knowledge of the mainstream velocity U determines the flow downstream. Numerical computations based on equation (3.24) are difficult because of the singularity on the surface of the body, where $\chi = U^2$.

Crocco (1939) gave a slightly different transformation. The equations are written in the form

$$\rho\left(u\frac{\partial u}{\partial x} + v\frac{\partial u}{\partial y}\right) = -\frac{dp}{dx} + \frac{\partial \tau}{\partial y}, \qquad (3.25)$$

where τ is the shear stress. We now take x and u to be independent variables, with τ as a dependent variable, and so we have

$$\tau = \mu\frac{\partial y}{\partial u} \qquad \text{and} \qquad y = \mu\int\frac{du}{\tau}. \qquad (3.26)$$

Hence the following equation is obtained:

$$\mu\rho u\frac{\partial}{\partial x}\left(\frac{1}{\tau}\right) + \frac{\partial^2 \tau}{\partial u^2} - \mu\frac{dp}{dx}\frac{\partial}{\partial u}\left(\frac{1}{\tau}\right) = 0. \qquad (3.27)$$

If the mainstream is uniform the equation simplifies to

$$\frac{\mu\rho}{\tau^2}\frac{\partial\tau}{\partial x} = \frac{1}{u}\frac{\partial^2\tau}{\partial u^2},$$

(3.28)

with the boundary conditions

$$\frac{\partial\tau}{\partial u} = 0 \qquad \text{on } u = 0$$

and

$$\tau = 0 \qquad \text{on } u = U.$$

Karman and Millikan (1934) used this approach as the basis for an approximation. In the outer regions of the boundary layer u does not change rapidly and so equation (3.24) is approximately linear and may be replaced by an equation of the form

$$\frac{\partial\chi}{\partial\phi} = v\frac{\partial^2\chi}{\partial\psi^2},$$

(3.29)

where

$$\phi = \int U(x)\,dx.$$

(3.30)

This can only give information, however, about the outer region and not about the flow in the proximity of the wall. This in itself leads to the problem of determining the appropriate boundary conditions at the wall since the equation is not a valid approximation.

Boundary layer thickness

In most cases, no definite limit can be given to the extent of the boundary layer since the velocity component u only attains the value of the mainstream flow U asymptotically. In practice, however, this is not particularly significant as the velocity in the boundary layer attains a value extremely close to U only a short distance from the wall.

To indicate the thickness of the boundary layer a number of different lengths may be defined:

(i) Displacement thickness:

$$\delta_1 = \int_0^\infty \left(1 - \frac{u}{U}\right) dy,$$

(3.31)

(ii) Momentum thickness:

$$\delta_2 = \int_0^\infty \frac{u}{U}\left(1 - \frac{u}{U}\right) dy \text{ and}$$

(3.32)

(iii) Energy thickness:

$$\delta_3 = \int_0^\infty \frac{u}{U}\left(1 - \frac{u^2}{U^2}\right) dy. \tag{3.33}$$

The limits of integration are 0 and ∞ but for numerical computations the integration may be terminated when the integrand becomes negligible [see IV, Chapter 4, and III, Chapter 7]. The displacement thickness indicates the distance by which the streamlines are displaced by the presence of the boundary layer. $U\delta_1$ is the reduction in volume flux across a normal to the surface due to the existence of the boundary layer. Similarly, $\rho U^2 \delta_2$ is the reduction in momentum flux and $\frac{1}{2}\rho U^3 \delta_3$ the reduction in kinetic energy flux.

If one is only interested in the gross features of the boundary layer one may integrate the momentum equation over the thickness of the boundary layer. The resulting equation,

$$\frac{\tau_w}{\rho} = \frac{\partial}{\partial x}(U^2\delta_2) + \delta_1 U \frac{\partial U}{\partial x} + \frac{\partial}{\partial t}(U\delta_1), \tag{3.34}$$

is called the momentum integral equation, which links the displacement and momentum thicknesses, where τ_w is the shearing stress at the wall.

Similarly, by multiplying the equation of motion by u and integrating over the boundary layer thickness we obtain the energy integral equation

$$\frac{\partial}{\partial x}(U^3\delta_3) = 2v \int_0^\infty \left(\frac{\partial u}{\partial y}\right)^2 dy. \tag{3.35}$$

Similarity solutions

It is possible to find *similarity solutions* of the boundary layer equations. In these solutions the dependent variables are a function of a single elementary function of the independent coordinates. The equations then reduce to *ordinary differential equations* [see VI, Chapter 7] which may be solved more easily than the original partial differential equations. Thus for these similarity solutions two velocity profiles $u(x, y)$ for different values of x differ only by a scaling factor of u and y.

Falkner and Skan (1930) studied the similarity solutions associated with flow external to a wedge. For such flows the mainstream velocity is given by

$$U(x) = \bar{U}\left(\frac{x}{L}\right)^m, \tag{3.36}$$

where L is some reference length and $\bar{U}$ and m are constants. The appropriate independent coordinate is then

$$\eta = y\sqrt{\frac{m+1}{2}\frac{U}{vx}} \tag{3.37}$$

and the boundary layer equation reduces to the *ordinary* differential equation:

$$f''' + ff' + \beta(1 - f'^2) = 0, \qquad (3.38)$$

where $\beta = 2m/(m + 1)$. This equation was studied by Falkner and Skan (1930), Hartree (1937) and Stewartson (1954). External flows given by equation (3.36) occur in the neighbourhood of the stagnation point of a wedge whose included angle is equal to $\pi\beta$. If $\beta = 0$ the problem reduces to that for a flat plate at zero incidence; $\beta = 1$ corresponds to the flow in the neighbourhood of the stagnation point on a plate held normal to the flow.

Two dimensional jet

We now treat a problem using boundary layer theory but not involving solid surfaces. We consider a steady jet emerging from a narrow slit. The x axis is taken to be the axis of the jet with the origin at the slit. The flux of momentum across any plane normal to the x axis is conserved, i.e.

$$M = 2\rho \int_0^\infty u^2 \, dy = \text{constant}. \qquad (3.39)$$

The appropriate boundary layer equation is

$$u \frac{\partial u}{\partial x} + v \frac{\partial u}{\partial y} = v \frac{\partial^2 u}{\partial y^2}, \qquad (3.40)$$

with the boundary condition

$$v = 0 \qquad \text{on } y = 0.$$

If

$$\eta = \frac{y}{x^{2/3}}$$

and we write the stream function $\psi(x, y)$ as

$$\psi = 6vx^{1/3}f(\eta) \qquad (3.41)$$

then the boundary layer equation reduces to

$$f''' + 2ff'' + 2f'^2 = 0, \qquad (3.42)$$

with the boundary conditions

$$f = f'' = 0 \qquad \text{on } \eta = 0$$

and

$$f' \to 0 \qquad \text{as } \eta \to \infty.$$

This has a solution of the form

$$f = \alpha \tanh (\alpha\eta), \qquad (3.43)$$

giving

$$u = 6v\alpha^2 x^{-1/3}(1 - \tanh^2 \alpha\eta). \tag{3.44}$$

The constant α is determined using the relationship (3.39). We have

$$M = 72\rho\alpha^3 v^2 \int_0^\infty (1 - \tanh^2 \xi)^2 \, d\xi$$

$$= 48\rho\alpha^3 v^2. \tag{3.45}$$

Separation

In all the cases considered so far we have assumed that the fluid in the boundary layer is never transported into the mainstream flow, i.e. the fluid in the boundary layer remains adjacent to the body. In many cases, however, this does not happen and the fluid suddenly breaks away from the body, setting up a region of *eddying flow* downstream. In such cases we say that *separation* has occurred and the point at which the boundary layer flow separates from the body is known as the separation point. Experimental and theoretical work indicate that the flow separates at or very near to the point on the body where the skin friction vanishes. Hence the separation point is normally defined as the point where

$$\left.\frac{\partial u}{\partial y}\right|_{y=0} = 0. \tag{3.46}$$

It may be inferred that in steady flow separation occurs only in decelerated flows, i.e. where $dp/dx > 0$.

Immediately downstream of separation the boundary layer becomes much thicker. This led Goldstein (1948) to suggest the existence of a singularity of the conventional boundary layer at this point. This singularity arises from the specification of the external pressure gradient. Recently, however, triple-deck theory has shown how the flow at separation may be studied and calculations carried on into the region of *reversed flow* downstream of the separation point. The theory shows that separation may be anticipated through upstream influence and that interaction between the pressure and displacement thickness leads to the appropriate thickening at separation. The flow acquires a three-region character. The two outer regions are essentially inviscid in character and the inner layer resembles a conventional boundary layer but instead of matching to the outer inviscid flow it matches to the flow in the middle layer. The flow in the middle layer remains relatively unaffected during separation. The upper deck, however, interacting with the lower deck, provides a pressure gradient which causes the lower deck to supply changes in the displacement thickness, inducing the pressure changes in the upper deck (Smith, 1977, and Stewartson, 1974). For a recent account of the impact of the triple-deck theory on all aspects of boundary layers, the reader is referred to Stewartson (1981).

3.3 Unsteady boundary layers

We will consider two different types of unsteady flow: impulsive motions and flows induced by small amplitude, periodic fluctuations. These two types of problem are represented by two classical solutions: one is Rayleigh's problem of the motion induced when an infinite plate is moved impulsively in its own plane with speed U and the other is the solution by Stokes of the flow induced by an infinite plate performing unidirectional harmonic oscillations in its own plane.

Impulsive motions

Let us consider the Rayleigh (1911) problem of an infinite plate moved impulsively in its own plane, with speed U. We take the coordinates x and y to be along and normal to the plate. The only non-zero component of velocity is parallel to the plate and is a function of y and t only. The appropriate equation is then

$$\frac{\partial u}{\partial t} = v \frac{\partial^2 u}{\partial y^2}, \tag{3.47}$$

with the boundary conditions

$$\begin{aligned} u = U &\quad \text{on } y = 0, \, t > 0, \\ u = 0 &\quad \text{for } y > 0, \, t = 0, \end{aligned} \tag{3.48}$$

and $u \to 0$ as $y \to \infty$ for all t. If we let $\eta = y/2(vt)^{1/2}$ then the solution is [see IV, §§8.3 and 13.4.1]

$$u = U(1 - \text{erf } \eta). \tag{3.49}$$

The Rayleigh problem may be generalized to that of an impulsively started cylinder moving parallel to its generators. This was considered by Batchelor (1954). If x and y are now cartesian coordinates in a plane normal to the axis of the cylinder the appropriate equation becomes

$$\frac{\partial u}{\partial t} = v \left(\frac{\partial^2 u}{\partial x^2} + \frac{\partial^2 u}{\partial y^2} \right), \tag{3.50}$$

with the boundary conditions

$$u = U \quad \text{on cylinder, } t > 0,$$

$$u = 0 \quad \text{exterior to cylinder, } t = 0, \tag{3.51}$$

and

$$u \to 0 \quad \text{as } (x^2 + y^2)^{1/2} \to \infty. \tag{3.52}$$

If the perimeter of the cylinder is denoted by $2\pi d$, the frictional force per unit length of the cylinder, F, is given by

$$\frac{F}{2\pi d} = \frac{\mu U}{d} \left[\frac{d}{(\pi vt)^{1/2}} + \tfrac{1}{2} + O\left(\frac{vt}{d^2} \right)^{1/2} \right]. \tag{3.53}$$

We again consider an impulsively started cylinder, but this time the motion is perpendicular to its generators. It can be expected that after some time the boundary layer will separate from the body. Thus interest has been centred on two problems: first, the motion for small time determining how and when the boundary layer first separates and second, the determination of the flow structure for large times.

Blasius (1908), Goldstein and Rosenhead (1936) and Collins and Dennis (1973) have studied the flow for small time by expanding the velocity components as a power series in t:

$$u = Uf_0(\eta) + tU \frac{dU}{dx} f_1(\eta) + O(t^2), \tag{3.54}$$

where

$$\eta = \frac{y}{2(\nu t)^{1/2}}. \tag{3.55}$$

For small t the flow appears to be that of an impulsively started plate and so $f_0(\eta) = \text{erf } \eta$.

Proudman and Johnson (1962) studied the structure of the flow at the rear stagnation point for large time. It had previously been known that no steady solution of the conventional boundary layer equations exists at a rear stagnation point. Proudman and Johnson were, however, able to give a valid steady structure for the flow in which the boundary layer equations were valid for all time. Idealizing the flow to be that away from a stagnation point of an infinite wall, they consider a thin layer adjacent to the wall where the flow reverses. By performing the transformation

$$x' = \frac{x}{L}, \qquad y' = y \left(\frac{U}{\nu L} \right)^{1/2}, \qquad t' = \frac{Ut}{L} \tag{3.56}$$

and

$$\psi = -2(\nu LU)^{1/2} x' f(y', t'), \qquad \text{where } L \text{ is a typical length,}$$

the equations reduce to

$$f_{y't'} = f_{y'y'y'} - ff_{y'y'} - 1 + f_{y'}^2, \tag{3.57}$$

with the boundary conditions

$$f = f_{y'} = 0 \qquad \text{on } y = 0, t > 0,$$

$$f_{y'} \to 1 \qquad \text{as } y \to \infty \tag{3.58}$$

and

$$f_{y'} = 1, \qquad t = 0, \qquad y' > 0.$$

For details of the solution see Proudman and Johnson (1962) and Robins and Howarth (1972). It is of interest to note that the flow in the neighbourhood of point A in Figure 3.2 resembles flow towards a forward stagnation point and the first term in the inner solution is in fact the forward stagnation point solution.

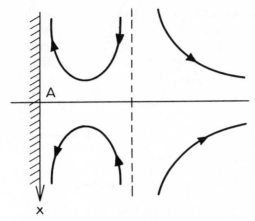

Figure 3.2: Flow near rear stagnation point.

Impulsive motion of a flat plate

We consider the flow generated when a semi-infinite flat plate is impulsively moved parallel to its own plane with speed U. The boundary layer equations become

$$\frac{\partial u}{\partial t} + u \frac{\partial u}{\partial x} + v \frac{\partial u}{\partial y} = v \frac{\partial^2 u}{\partial y^2} \qquad (3.59)$$

with the boundary conditions

$$
\begin{aligned}
u = v = 0 &\quad \text{on } y = 0, \\
u \to U &\quad \text{as } y \to \infty \text{ for } x > 0 \text{ and } t > 0, \\
u = U &\quad \text{for } x = 0, \ y > 0 \text{ and } t > 0
\end{aligned}
\qquad (3.60)
$$

and

$$u = U \qquad \text{for } x > 0, \ y > 0 \text{ and } t = 0.$$

Because of the parabolic nature of the boundary layer equations (3.59) the effects generated by the leading edge propagate down the plate with the maximum local velocity U. Hence for $x > Ut$ the flow, being unaffected by the leading edge, represents that for an impulsively started infinite plate, i.e. the Rayleigh solution. For $x < Ut$ the flow is akin to the Blasius flow for a semi-infinite plate with a transition between the two solutions in the region of $x = Ut$.

We now consider flows involving *periodic fluctuations*. The classical problem of this type is the one studied by Stokes of an infinite flat plate performing unidirectional *harmonic oscillations* in its own plane in a fluid otherwise at rest [see IV, §8.6.1]. Since the induced velocity in the fluid is only parallel to the plate the boundary layer equations reduce to

$$\frac{\partial u}{\partial t} = v \frac{\partial^2 u}{\partial y^2}, \tag{3.61}$$

subject to the boundary conditions

$$u = U \cos(\omega t) \qquad \text{on } y = 0$$

and

$$u \to 0 \qquad \text{as } y \to \infty. \tag{3.62}$$

The solution, as given by Stokes, is

$$u = U e^{-\eta} \cos(\omega t - \eta), \tag{3.63}$$

where

$$\eta = y \left(\frac{\omega}{2v}\right)^{1/2}. \tag{3.64}$$

3.4 Three-dimensional boundary layers

Let the surface of the body be denoted by S. The position of a point in space is described by its distance x_3 measured normal to S and the position of the foot of the normal on S, $\mathbf{r}$, given in terms of curvilinear coordinates x_1 and x_2 on S [see V, §13.6.2]. We have that

$$\mathbf{r} = \mathbf{r}(x_1, x_2)$$

and we denote the magnitude of the derivatives of $\mathbf{r}$ by h_1 and h_2 [see IV, §5.1]:

$$h_1 = \left|\frac{\partial \mathbf{r}}{\partial x_1}\right|, \qquad h_2 = \left|\frac{\partial \mathbf{r}}{\partial x_2}\right|.$$

The boundary layer equations are then given by

$$\frac{\partial u_1}{\partial t} + \frac{u_1}{h_1}\frac{\partial u_1}{\partial x_1} + \frac{u_2}{h_2}\frac{\partial u_1}{\partial x_2} + u_3\frac{\partial u_1}{\partial x_3} + \frac{u_1 u_2}{h_1 h_2}\frac{\partial h_1}{\partial x_2}$$
$$- \frac{u_2^2}{h_1 h_2}\frac{\partial h_2}{\partial x_1} = -\frac{1}{\rho}\frac{1}{h_1}\frac{\partial p}{\partial x_1} + v\frac{\partial^2 u_1}{\partial x_3^2}, \tag{3.65}$$

$$\frac{\partial u_2}{\partial t} + \frac{u_1}{h_1}\frac{\partial u_2}{\partial x_1} + \frac{u_2}{h_2}\frac{\partial u_2}{\partial x_2} + u_3\frac{\partial u_2}{\partial x_3} - \frac{u_1^2}{h_1 h_2}\frac{\partial h_1}{\partial x_2}$$
$$+ \frac{u_1 u_2}{h_1 h_2}\frac{\partial h_2}{\partial x_1} = -\frac{1}{\rho}\frac{1}{h_2}\frac{\partial p}{\partial x_2} + v\frac{\partial^2 u_2}{\partial x_3^2}, \tag{3.66}$$

and the equation of continuity is

$$\frac{1}{h_1 h_2}\left[\frac{\partial}{\partial x}(h_2 u_1) + \frac{\partial}{\partial x}(h_1 u_2)\right] + \frac{\partial u_3}{\partial x_3} = 0. \tag{3.67}$$

As in the two-dimensional case there are corresponding momentum integral equations:

$$\frac{\partial}{\partial t}\int_0^\infty (U_1 - u_1)\,dx_3 + \frac{1}{h_1}\frac{\partial}{\partial x_1}\int_0^\infty u_1(U_1 - u_1)\,dx_3 + \frac{1}{h_1}\frac{\partial U_1}{\partial x_1}\int_0^\infty (U_1 - u_1)\,dx_3$$

$$+ \frac{1}{h_2}\frac{\partial}{\partial x_2}\int_0^\infty u_2(U_1 - u_1)\,dx_3 + \frac{1}{h_2}\frac{\partial U_1}{\partial x_2}\int_0^\infty (U_2 - u_2)\,dx_3$$

$$+ \frac{1}{h_1 h_2}\frac{\partial h_1}{\partial x_2}\int_0^\infty [u_2(U_1 - u_1) + u_1(U_2 - u_2) + U_2(U_1 - u_1)]\,dx_3$$

$$+ \frac{1}{h_1 h_2}\frac{\partial h_2}{\partial x_1}\int_0^\infty [u_1(U_1 - u_1) - u_2(U_2 - u_2) - U_2(U_2 - u_2)]\,dx_3$$

$$= \frac{\tau_w}{\rho}, \tag{3.68}$$

where τ_w is the shear stress at the wall; and the corresponding equation with the subscripts 1 and 2 reversed, see Rosenhead (1963) for further details.

For a given surface the above equations must be solved numerically using appropriate *finite difference* [see III, Chapter 9] or *finite element* [see Chapter 15] techniques.

Rotationally symmetric flow past a blunt-nosed body of revolution

In this case the space coordinates that we shall use will be (r, θ, z), where θ is the angle measured from a fixed meridian plane, x is the distance measured from the nose of the body to the foot of the normal onto the surface and z is the normal distance of the point from the surface. The components of velocity at a point $(u, v$ and $w)$ are measured in the directions of increasing r, θ and z respectively [see IV, §17.3]. If $(r, \theta, 0)$ is a point P on the surface then r_0 is the distance of P from the axis of revolution. Since the motion is rotationally symmetric it is independent of θ. The boundary layer equations then become

$$\frac{\partial u}{\partial t} + u\frac{\partial u}{\partial x} + w\frac{\partial u}{\partial z} - \frac{v_1^2}{r_0}\frac{dr_0}{dx} = -\frac{1}{\rho}\frac{dp}{dx} + v\frac{\partial^2 u}{\partial z^2}, \tag{3.69}$$

$$\frac{\partial v}{\partial t} + u\frac{\partial v}{\partial x} + w\frac{\partial v}{\partial z} + \frac{uv}{r_0}\frac{dr_0}{dx} = v\frac{\partial^2 v}{\partial z^2} \tag{3.70}$$

and

$$\frac{\partial}{\partial x}(r_0 u) + \frac{\partial}{\partial z}(r_0 w) = 0. \tag{3.71}$$

These equations are only valid where the principal radii of curvature of the body [see V, §12.4.3] are large in comparison with the thickness of the boundary layer. The mainstream flow is determined by the equations

$$\frac{\partial U}{\partial t} + U\frac{\partial U}{\partial x} - \frac{V^2}{r_0}\frac{dr_0}{dx} = -\frac{1}{\rho}\frac{dp}{dx} \tag{3.72}$$

and

$$\frac{\partial V}{\partial t} + U\frac{\partial V}{\partial x} + \frac{UV}{r_0}\frac{dr_0}{dx} = 0. \tag{3.73}$$

In the neighbourhood of a forward stagnation point in steady axisymmetric flow the mainstream velocity is $(\bar{U}x, 0)$ where $\bar{U}$ is a constant depending on both the flow and the body. For a blunt-nosed body we have, to a first approximation, $r_0 = x$. Using the transformation

$$\eta = \left(\frac{\bar{U}}{v}\right)^{1/2} z, \qquad \psi = (\bar{U}v)^{1/2}\, x^2 f(\eta), \tag{3.74}$$

we have

$$u = \bar{U}xf'(\eta) \qquad \text{and} \qquad w = -2(\bar{U}v)^{1/2}f(\eta). \tag{3.75}$$

The boundary layer equations reduce to

$$f'^2 - 2ff'' = 1 + f''', \tag{3.76}$$

with the boundary conditions

$$f = f' = 0 \qquad \text{on } \eta = 0$$
$$\tag{3.77}$$
$$f'(\eta) \to 1 \qquad \text{as } \eta \to \infty.$$

and

Mangler (1945) showed that by performing the change of variables

$$\xi = \sqrt{2}\,\eta \qquad \text{and} \qquad g(\xi) = \sqrt{2}f(\eta) \tag{3.78}$$

the equation becomes

$$g''' + gg' = \tfrac{1}{2}(g'^2 - 1), \tag{3.79}$$

with the boundary conditions

$$g = g' = 0 \qquad \text{on } x = 0$$
$$\tag{3.80}$$
$$g'(\xi) \to 1 \qquad \text{as } \xi \to \infty.$$

and

This is a special case of the Falkner–Skan equations for flow over a wedge.

This is just one example of the relationship between axially symmetric boundary layers and two-dimensional boundary layers, which we now discuss in more detail.

Axially symmetric boundary layers on bodies of revolution transformed to two-dimensional boundary layers

We show that steady, axially symmetric boundary layers on bodies of revolution may be transformed to two-dimensional boundary layers (Mangler, 1945). The axial symmetry implies that $v = 0$ and hence the boundary layer equations become

$$u \frac{\partial u}{\partial x} + w \frac{\partial u}{\partial z} = U \frac{\partial U}{\partial x} + v \frac{\partial^2 u}{\partial z^2},$$

$$\frac{\partial}{\partial x} (r_0 u) + \frac{\partial}{\partial z} (r_0 w) = 0.$$

(3.81)

We now introduce the following transformations, where L is some constant length:

$$\bar{x} = \frac{1}{L^2} \int_0^x r_0^2 \, dx, \qquad \bar{z} = \frac{r_0 z}{L},$$

$$\bar{u} = u, \qquad \bar{w} = \frac{L}{r_0} \left(w + \frac{zu}{r_0} \frac{dr_0}{dx} \right)$$

(3.82)

and

$$\bar{U}(\bar{x}) = U(x).$$

(3.83)

This transformation reduces equations (3.81) to those for a two-dimensional boundary layer. The relation between the mainstream velocities is given by (3.83), which enables one to calculate the corresponding two-dimensional surface, see Rosenhead (1963) for further details.

Steady flow in the neighbourhood of a forward stagnation point on a general curved surface (Howarth, 1951)

The origin of the coordinates is taken to be the stagnation point. The coordinate system may be chosen so that the mainstream velocity is given by

$$\left(\frac{Ux}{L}, \frac{Vy}{L} \right)$$

(3.84)

where L is some typical length. The boundary equations are then

$$u \frac{\partial u}{\partial x} + v \frac{\partial u}{\partial y} + w \frac{\partial u}{\partial z} = \left(\frac{U}{L} \right)^2 x + v \frac{\partial^2 u}{\partial z^2},$$

(3.85)

$$u \frac{\partial v}{\partial x} + v \frac{\partial v}{\partial y} + w \frac{\partial v}{\partial z} = \left(\frac{V}{L} \right)^2 y + v \frac{\partial^2 v}{\partial z^2}$$

(3.86)

and

$$\frac{\partial u}{\partial x} + \frac{\partial v}{\partial y} + \frac{\partial w}{\partial z} = 0,$$

(3.87)

with the boundary conditions

$$u = v = w = 0 \qquad\qquad \text{on } z = 0,$$

and   (3.88)

$$u \to \frac{Ux}{L} \quad \text{and} \quad v \to \frac{Vy}{L} \quad \text{as } z \to \infty.$$

Letting $\eta = (U/\nu L)^{1/2} z$ we may write the velocity components in the form

$$u = \frac{Ux}{L} f'(\eta)$$

$$v = \left(\frac{Vy}{L}\right) g'(\eta) \qquad\qquad (3.89)$$

and

$$w = -\left(\frac{\nu}{LU}\right)^{1/2} [Uf(\eta) + Vg(\eta)].$$

Equations (3.85 and 3.86) then reduce to

$$f''' + \left(f + \frac{V}{U} g\right) f'' - f'^2 + 1 = 0, \qquad\qquad (3.90)$$

and

$$\frac{U}{V} g''' + \left(g + \frac{U}{V} f\right) g'' - g'^2 + 1 = 0, \qquad\qquad (3.91)$$

with the boundary conditions

$$f = g = f' = g' = 0 \qquad\qquad \text{on } \eta = 0$$

and   (3.92)

$$f' \to 1 \quad \text{and} \quad g' \to 1 \quad \text{as } \eta \to \infty.$$

When the radius of curvature of the surface is not large in comparison with the thickness of the boundary layer more terms must be included in the equations. For details of calculating the entry flow in circular pipes and flow around long, thin circular cylinders, where this occurs, see Rosenhead (1963) and Schlichting (1960).

Secondary flow

In the examples considered so far the direction of flow in the boundary layer corresponds with the direction of flow of the mainstream. However, under certain circumstances, the direction of flow in the boundary layer may be different from that in the mainstream due to some imposed pressure gradient. This deviation in direction from the external flow is generally referred to as a secondary flow.

Steady flow over sheared wings

We use a coordinate system in which x is the distance measured along the surface perpendicular to the generators, y is the distance parallel to the generators and z is the normal distance from the surface (see Figure 3.3).

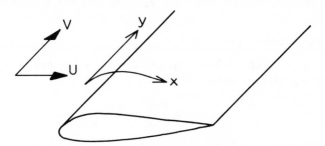

Figure 3.3: Flow over sheared wings.

The boundary layer equations for steady flow are

$$u \frac{\partial u}{\partial x} + w \frac{\partial u}{\partial z} = U \frac{dU}{dx} + v \frac{\partial^2 u}{\partial z^2}, \tag{3.93}$$

$$u \frac{\partial v}{\partial x} + w \frac{\partial v}{\partial z} = v \frac{\partial^2 v}{\partial z^2} \tag{3.94}$$

and

$$\frac{\partial u}{\partial x} + \frac{\partial w}{\partial z} = 0, \tag{3.95}$$

with the boundary conditions

$$u = v = w = 0 \qquad \text{on } z = 0$$

and
$$\tag{3.96}$$
$$u \to U \qquad \text{and} \qquad v \to V \qquad \text{as } z \to \infty.$$

It may be noted that equations (3.93) and (3.95) do not involve v and therefore (3.93) and (3.95) may be solved for u and w while equation (3.94) is also solved for v (Jones, 1947). The momentum equation for v is given by

$$\frac{d}{dx} \int_0^\infty u(V - v) \, dz = v \left(\frac{\partial v}{\partial z} \right)_{z=0}.$$

3.5 Flow in pipes

A particularly simple form of axisymmetric viscous flow, traditionally considered along with boundary layers, is the flow inside circular pipes for which the Reynolds number is defined as $U \, d\rho/\mu$, where U is the mean velocity, d is the diameter of the pipe, ρ is the density of the fluid and μ is viscosity. Generally flows at Reynolds numbers above 2,000 are *turbulent* and those at Reynolds numbers below 2,000 are *laminar*. Low R_e flow can be disturbed to become turbulent but this turbulence dies away if the pipe is long enough. Laminar flow can also be induced at very high R_e values if all distrubances to the fluid are eliminated, but it is unstable and if disturbed the flow becomes turbulent.

Shear stress in circular pipes

In circular pipes, with any kind of steady flow, the shear stress at a radius r is given by the equation of motion as

$$\tau = \frac{r}{2}\frac{dp^*}{dz}, \tag{3.97}$$

where z is the direction of the pipe, r is the radius and p^* is the piezometric pressure $p^* = p + \rho gh$.

Laminar flow

For this case an analytical expression relating discharge through the pipe Q, length of pipe L and piezometric pressure drop Δp^* can be obtained for circular pipes. The equation (3.97) is used in conjunction with Newton's law of viscosity to give

$$\tau = \mu\frac{du}{dr}, \tag{3.98}$$

where u is the velocity along the pipe. Integration over the pipe area [see IV, §6.3] gives the Hagen–Poiseuille pipe flow formula

$$Q = \frac{-\pi R^4}{8\mu L}\Delta p^*, \tag{3.99}$$

where R is the radius of the pipe and L is the length of pipe. The velocity profile as a function of radius is given by

$$u = \frac{1}{4\mu}\frac{dp^*}{dz}(R^2 - r^2), \tag{3.100}$$

a parabolic distribution. The maximum velocity is at the centre of the pipe and the mean velocity $= \frac{1}{2}$ maximum velocity. It can be shown that the velocity distribution for a pipe of general cross-section can be found by solving *Poisson's equation* [see IV, §8.1] in a domain of the same shape as the cross-section:

$$\nabla^2 u = \frac{1}{\mu}\frac{dp^*}{dz} \tag{3.101}$$

(see Lagerstrom and Chang, 1962). Here x and y are in the plane of the cross-section and the boundary condition is $u = 0$. Analytical solutions are available to Poisson's equation for many regular shapes [see IV, §8.2] and it can be solved *numerically* for irregular shapes [see III, §9.5].

Turbulent flow

Most real pipe flows occur in this regime, which is not amenable to a great deal of analysis. Results in this area are entirely experimental. Darcy's pipe flow law expresses the loss of head along a pipe as a function of length L, diameter d, mean velocity U, acceleration due to gravity g and the friction factor f:

$$h_f = \frac{fL}{d}\frac{U^2}{2g}, \qquad\qquad (3.102)$$

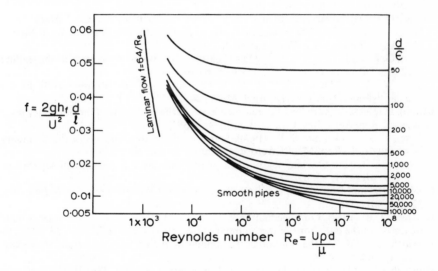

Roughness Table

Material	ϵ (m $\times$ 10^{-5})
Glass, Brass, Copper, Lead	Smooth
Steel, Wrought iron	4–6
Asphalted cast iron	12
Galvanized iron	15
Cast iron	26
Concrete	300
Riveted steel	910

Figure 3.4: Moody diagram.

where h_f is the head loss. This is for circular pipes of diameter d. For pipes of general shape the formula can be written as:

$$h_f = \frac{fL}{4m} \frac{U^2}{2g},$$

where m, the hydraulic mean depth, is given by the area of pipe/perimeter of pipe. The friction factor f depends upon the roughness of the pipe and is obtained from a Moody diagram (Figure 3.4), which gives f as a function of relative roughness and the Reynolds number.

R.B. & P.B.

References

Batchelor, G. K. (1954). The skin friction on infinite cylinders moving parallel to their length, *Quart. J. Mech.*, **7**, 179–192.

Blasius, H. (1908). Grenzschnichten in Flüssigkeiten mit Kleiner Reibung, *Z. Math. Phys.*, **56**, 1–37; (in English) The boundary layers in fluids with little friction, Tech. Memo. Nat. Adv. Comm. Aero., Washington, No. 1256.

Carrier, G. F., and Lin, C. C. (1958). On the nature of the boundary layer near the leading edge of a flat plate, *Quart. Appl. Math.*, **6**, 63–68.

Collins, W. M., and Dennis, S. C. R. (1973). The initial flow past an impulsively started circular cylinder, *Quart. Mech. Appl. Math.*, **26**, 53–75.

Crocco, L. (1939). A characteristic transformation of the equations of the boundary layer in gases, Rep. Aero. Res. Council Lond., No. 4582.

Falkner, V. M., and Skan, S. W. (1930). Some approximate solutions of the boundary layer equations. Rep. Memo. Aero Res. Council Lond., No. 1314.

Goldstein, S. (1948). On laminar boundary-layer flow near a position of separation, *Quart. J. Mech.*, **1**, 43–69.

Goldstein, S., and Rosenhead, L. (1936). Boundary layer growth, *Proc. Camb. Phil. Soc.*, **32**, 392–401.

Hartree, D. R. (1937). On an equation occurring in Falkner and Skan's approximate treatment of the equations of the boundary layer, *Proc. Camb. Phil. Soc.*, **33**, 223–239.

Howarth, L. (1938). On the solution of the laminar boundary layer equations, *Proc. Roy. Soc. (A)*, **164**, 547–579.

Howarth, L. (1951). The boundary layer in three dimensional flow. Part II: The flow near a stagnation point, *Phil. Mag.*, **42**(7), 1433–1440.

Jones, R. T. (1947). Effects of sweepback on bounary-layer and separation, Rep. Nat. Adv. Comm. Aero., Washington, No. 884.

Karman, Th. V., and Millikan, C. B. (1934). On the theory of laminar boundary layers involving separation, Rep. Nat. Adv Comm. Aero., Washington, No. 504.

Lagerstrom, P. A., and Chang, I. D. (1962). Flow at low Reynolds number, in *Handbook of Engineering Mechanics* (Ed. W. Hügge), Chap. 81, McGraw-Hill.

Mangler, W. (1945). Boundary layers on bodies of revolution in symmetrical flows, Ber Aerodyn, vers anst, Gottingen, no. 45/A/17.

Nayfeh, A. H. (1973). *Perturbation Methods*, Wiley.

Prandtl, L. (1904). Über Flüssigkeitsbewegung bei sehr Kleiner Reibung, Verh III int. Math. Kongr. Heidelberg 484–491; also available in English, Motion of fluids with very little viscosity, Tech. Memo. Nat. Adv. Comm. Aero., Washington, No. 452.

Proudman, I., and Johnson, K. (1962). Boundary layer growth near a rear stagnation point, *J. Fluid. Mech.*, **12**, 161–168.

Rayleigh, Lord (1911). On the motion of solid bodies through viscous liquids, *Phil. Mag.*, **21**(6), 697–711.

Robins, A. J., and Howarth, J. A. (1972). Boundary-layer development at a two-dimensional rear stagnation point, *J. Fluid Mech.*, **56**, 161–171.

Rosenhead, L. (Ed.) (1963). *Laminar Boundary Layers*, Oxford University Press.

Schlichting, H. (1960). *Boundary Layer Theory*, McGraw-Hill.

Smith, F. T. (1977). The laminar separation of an incompressible fluid streaming past a smooth surface, *Proc. Roy. Soc. (A)*, **356**, 443–463.

Stewartson, K. (1954). Further solutions of the Falkner–Skan equation, *Proc. Camb. Phil. Soc.*, **50**, 454–465.

Stewartson, K. (1974). Multistructured boundary layers on flat plates and related bodies, *Adv. Appl. Mech.*, **14**, 145–239.

Stewartson, K. (1981). D'Alembert's paradox, *Siam Rev.*, **23**(3), 308–343.

Van de Vooren, A. I., and Dijkstra, D. (1970). The Navier–Stokes solution for laminar flow past a semi-infinite flat plate, *J. Eng. Math.*, **4**, 9–27.

Van Dyke, M. D. (1969). Higher order boundary layer theory, *Ann. Rev. Fluid. Mech.*, **1**, 265–293.

Mathematical Methods in Engineering
Edited by G. A. O. Davies
© 1984, John Wiley & Sons, Ltd.

4

Compressible Flow

4.1 Compressibility, speed of sound and Mach number

In any gas flow there is no unique pressure–density relationship in general, but a measure of gas compressibility can be conveniently given by the speed of sound a. This is the propagation speed, relative to the gas, of pressure *waves* which are sufficiently weak that the associated entropy production can be regarded as negligible (which means, in turn, that gradients of temperature and velocity associated with the wave are very weak) and is given by [see IV, §5.3]

$$a^2 = \frac{\partial p}{\partial \rho}\bigg|_{\text{constant entropy}} . \tag{4.1}$$

For a steady gas stream at modest speeds the maximum pressure change is likely to be *of the order* of the dynamic head $\frac{1}{2}\rho U^2$ (see Section 2.4, 'the pressure equation'), so that from equation (4.1) the maximum fractional density change is given by [see IV, Definitions 2.3.2 and 2.3.3]

$$\frac{\Delta\rho}{\rho} = O\left(\frac{U^2}{a^2}\right) = O(M^2).$$

It can therefore be seen that compressibility effects, i.e. the appearance of significant density changes induced by the flow, can only become important once the local gas velocity U is significant compared with the speed of sound. This ratio, termed the *Mach number M*, is a parameter of immense importance in the theory of compressible flow.

It is usual to classify flow fields as *subsonic* or *supersonic*, depending upon whether the Mach number is less than or greater than one in the region of interest, although clearly both types are present in some problems such as transonic flows. The major distinction between these two types is the manner of propagation of disturbances, which can be conveniently illustrated by Figure 4.1. Consider a steady uniform flow with velocity U and speed of sound a passing a point P which is a source of infinitesimal strength acoustic disturbances. These disturbances radiate outwards at the speed of sound relative to the stream, and since in the illustration the velocity and speed of sound are

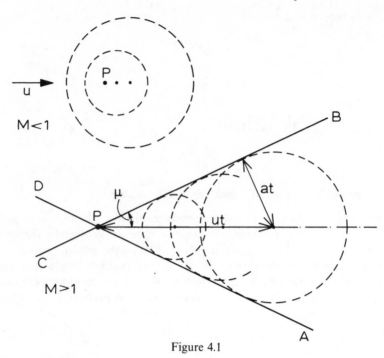

Figure 4.1

everywhere constant the wavefronts corresponding to disturbances emitted at particular instances are all spherical, the centres simply carried in the stream direction at the stream velocity. If the flow field is subsonic then clearly disturbances can propagate in all directions and eventually their effect can be felt everywhere. If the stream is supersonic, however, then the effect is felt only in a downstream cone-like region, which is the region of influence of point P. The limits of this region are the *Mach lines* PA and PB, inclined to the stream at the *Mach angle* μ where, from Figure 4.1,

$$\mu = \sin^{-1}\left(\frac{1}{M}\right). \tag{4.2}$$

Outside of this region P cannot influence the flow but, in turn, it is dependent on conditions in the region confined between the Mach lines PC and PD. In practice of course a flow field will be non-uniform, with variations in gas velocity and speed of sound so that the Mach lines resulting from infinitesimal strength disturbances will now follow curved paths, still inclined locally, however, at the Mach angle to streamlines. Finite strength disturbances resulting from shock waves (Section 4.5) are always inclined more steeply to streamlines than are the Mach lines.

This difference in the manner of propagation of disturbances is reflected in the governing equations of motion, which are *elliptic* for steady subsonic flows and *hyperbolic* for steady supersonic flows (see Section 4.6 and IV, §8.5).

4.2 Equation of state for a perfect gas

Once compressibility effects are important, an equation of state is required to specify the thermodynamic properties of the gas. Most analytical developments have been confined to perfect gases, to which we will *always* restrict our attention in this text, for which the equation of state is

$$\frac{p}{\rho} = RT, \tag{4.3}$$

where R is constant for any gas. This gas constant can also be expressed by

$$R = \frac{G}{m},$$

where G is the *universal gas constant* and m is the *molecular weight*, or by

$$R = C_p - C_v, \tag{4.4}$$

where C_p and C_v are constant for a perfect gas and are the *specific heats* at constant pressure and constant volume respectively. The *specific internal energy* i (energy per unit mass), which is a measure of the thermal energy, is given by

$$i = C_v T \tag{4.5}$$

or (see Section 4.3)

$$i = \frac{RT}{\gamma - 1} = \frac{a^2}{\gamma(\gamma - 1)}, \tag{4.6}$$

where γ is the ratio of specific heats C_p/C_v. The *specific enthalpy* h is defined by

$$h = i + \frac{p}{\rho} \tag{4.7}$$

for all substances or by

$$h = C_p T = \frac{\gamma R T}{\gamma - 1} = \frac{a^2}{\gamma - 1} \tag{4.8}$$

for a perfect gas.

4.3 The first law of thermodynamics, irreversibility, entropy and the second law of thermodynamics

It is well known that mechanical and thermal energy are equivalent. *The first law of thermodynamics* states this formally for an arbitrary *closed* system (fixed mass with no mass transfer across its boundary) moving with the flow between states 1 and 2 as

$$Q_{12} = E_2 - E_1 + W_{12}, \qquad (4.9)$$

where Q_{12} is the heat transfer to the system by radiation or conduction across the boundary, W_{12} is the work performed by the system on its surroundings, which in this text arises from the displacement of pressure forces at the system boundary, and $E_2 - E_1$ is the change of energy of the system which includes kinetic energy, internal energy i, chemical, gravitational, etc., although only the first two will be considered here. The first law therefore expresses the overall energy balance which must be satisfied and makes no statement about the directions in which the process may go. If there are no internal dissipative effects arising from viscous friction and heat transfer, i.e. from the effects of velocity and temperature gradients respectively in a real fluid, or if they are sufficiently small to be ignored, then the process is termed reversible. That is, the flow in equation (4.9) can be returned from state 2 to state 1, precisely reproducing throughout the fluid properties and the heat/work transfers in reverse of the original process. In practice the motion of a fluid particle can be considered reversible here, or nearly so, at sufficiently high Reynolds numbers outside boundary layers and wakes, and provided it does not cross a shock wave. Inside regions of dissipation kinetic energy or mechanical work are irreversibly converted to thermal energy and the process can proceed only in that one direction.

The second law of thermodynamics introduces the property *entropy*, *s* (specific entropy, defined per unit mass), which defines the possible direction of the process by the requirement that

$$T\,ds \geqslant dQ, \qquad (4.10)$$

where the equality refers to reversible processes only. Entropy is related to the other gas properties by

$$T\,ds = dh - \frac{dp}{\rho}, \qquad (4.11)$$

which on integrating between two end states 1 and 2 for a perfect gas gives

$$s_2 - s_1 = \frac{R}{\gamma - 1} \log_e \left(\frac{p_2}{p_1}\right)\left(\frac{\rho_1}{\rho_2}\right)^{\gamma}. \qquad (4.12)$$

For inviscid non-conducting or adiabatic flow the entropy of a fluid particle remains constant, therefore, and since we will neglect viscous boundary layers and wakes in this chapter this result will hold for all particle paths except in their passage through a shock wave. Mathematically, constant entropy for a moving particle is expressed by the total derivative, following a particle,

$$\frac{Ds}{Dt} = \frac{\partial s}{\partial t} + \mathbf{U} \cdot \mathbf{\nabla} s = 0, \qquad (4.13)$$

which includes the non-stationary term plus the term for the convection of the particle by the velocity field through a gradient of entropy. Equation (4.13) is termed an *isentropic motion* and for a perfect gas this provides.

$$\frac{\partial}{\partial t}\left(\frac{p}{\rho^{\gamma}}\right) + \mathbf{U}\cdot\mathbf{V}\left(\frac{p}{\rho^{\gamma}}\right) = 0, \tag{4.14}$$

i.e. p/ρ^{γ} holds constant for the particle. If in addition the entropy is constant throughout the flow field, i.e.

$$\mathbf{V}s = \mathbf{0} \tag{4.15}$$

the motion is termed *homentropic*. This applies to shock-free flow or to the flow downstream of a plane shock wave (see Section 4.5).

Finally, returning to equation (4.1) and using the homentropic statements above, we now see that the speed of sound in a perfect gas may be written as

$$a^2 = \frac{\gamma p}{\rho} = \gamma RT. \tag{4.16}$$

4.4 The equations of motion

Throughout this text we treat the gas in the region of interest as perfect, inviscid and non-conducting (except insofar as the latter two are implicit in the structure of a shock wave) with no sources or sinks of heat, mass or momentum, and with no body forces. The equations of motion will be derived in general *vector* form here. In practice [see IV, §17.3.1], a variety of coordinate systems are generally used for the solution of these equations (in addition to *rectangular cartesian* [see V, §2.1.4]) which often reflect some special symmetry or 'natural' system for the flow, such as *spherical* or *cylindrical polar* coordinates [see V, §2.1.5], streamline coordinates or the *characteristic* mesh for *hyperbolic* problems [see III, §§9.1 and 9.2].

Mass conservation or continuity equation

This is a statement that the net rate of loss of mass through the surface of a volume V fixed in space (see Figure 4.2a and b) equals the rate of depletion of mass within the volume. For the *steady* streamline flow in Figure 4.2(a), where the volume V is the segment of streamtube shown, the formulation is simply that

$$\rho U A = \text{constant}, \tag{4.17}$$

where ρ and U are uniform across the elemental area A of streamtube normal to the flow. Equation (4.17) therefore also holds for one-dimensional flow in ducts with area variations. This integral form of the continuity equation permits the appearance of flow discontinuities within the volume and is the form used later

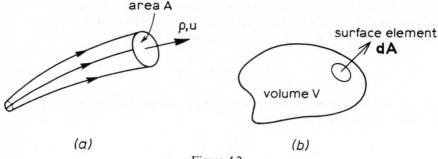

area A

ρ,u

surface element
dA

volume V

(a)   *(b)*

Figure 4.2

in the analysis of shock waves (Section 4.5). For unsteady flow the mass conservation is written for the general volume of Figure 4.2(b) as

$$\int_A \rho \mathbf{U} \cdot \mathbf{dA} = -\frac{\partial}{\partial t} \int_V \rho \, dV = -\int_V \frac{\partial \rho}{\partial t} \, dV \qquad (4.18)$$

or, by *Gauss' theorem* [see IV, Theorem 6.4.2],

$$\int_V \mathbf{V} \cdot (\rho \mathbf{U}) \, dV + \int_V \frac{\partial \rho}{\partial t} \, dV = 0,$$

so that at a point in space where all variables are differentiable

$$\text{div} (\rho \mathbf{U}) + \frac{\partial \rho}{\partial t} = 0. \qquad (4.19)$$

For steady two-dimensional motion

$$\frac{\partial}{\partial x} (\rho u) + \frac{\partial}{\partial y} (\rho v) = 0,$$

which is satisfied by the *stream function* ψ (see Section 2.4 also) where

$$\rho u = \frac{\partial \psi}{\partial y}, \qquad \rho v = -\frac{\partial \psi}{\partial x}. \qquad (4.20)$$

The momentum equation

The acceleration of a particle of fluid is expressed mathematically as

$$\frac{D\mathbf{U}}{Dt} = \frac{\partial \mathbf{U}}{\partial t} + (\mathbf{U} \cdot \mathbf{V})\mathbf{U},$$

giving the non-stationary term together with the effect of the convection of the fluid particle through a gradient of velocity.

Using Gauss' theorem the net force due to the pressure acting around the surface of an infinitesimal volume of fluid dV is

$$-\nabla p \, dV$$

so that the acceleration or momentum equation gives (cf. Section 2.1)

$$-\frac{1}{\rho}\nabla p = \frac{\partial \mathbf{U}}{\partial t} + (\mathbf{U}\cdot\nabla)\mathbf{U} \tag{4.21}$$

or

$$-\frac{1}{\rho}\nabla p = \frac{\partial \mathbf{U}}{\partial t} + \tfrac{1}{2}\nabla(U^2) + (\nabla \times \mathbf{U}) \times \mathbf{U}, \tag{4.22}$$

where $\nabla \times \mathbf{U}$ is the *vorticity*.

For the special, and important, case of irrotational motion

$$\nabla \times \mathbf{U} = 0 \tag{4.23}$$

so that the velocity field is satisfied by the *scalar potential* ϕ, where (equation 2.1)

$$\mathbf{U} = \nabla\phi. \tag{4.24}$$

In this case the momentum equation can be written as

$$-\frac{1}{\rho}\nabla p = \nabla\left(\frac{\partial \phi}{\partial t}\right) + \tfrac{1}{2}\nabla(U^2) \tag{4.25}$$

or on integrating between any two points at a given instant

$$\frac{\partial \phi}{\partial t} + \int \frac{dp}{\rho} + \tfrac{1}{2}U^2 = f(t), \tag{4.26}$$

which is the compressible form of the unsteady Bernoulli equation for inviscid irrotational flow given previously in Section 2.5. With the homentropic assumption, equation (4.26) gives further that

$$\frac{\partial \phi}{\partial t} + \frac{\gamma}{\gamma - 1}\frac{p}{\rho} + \tfrac{1}{2}U^2 = f(t). \tag{4.27}$$

In the case of steady, inviscid but rotational flow (therefore *not* homentropic), such as downstream of a curved shock wave, equation (4.22) gives that *along a streamline*

$$\int \frac{dp}{\rho} + \tfrac{1}{2}U^2 = \text{constant} \tag{4.28}$$

or that

$$\frac{\gamma}{\gamma - 1}\frac{p}{\rho} + \tfrac{1}{2}U^2 = \text{constant} \tag{4.29a}$$

since p/ρ^γ holds constant *along* the streamline. In contrast to the steady form of

(4.27), however, the constant of equation (4.29a) varies between streamlines and will be determined by the upstream shock wave.

Alternative forms of equation (4.29a), using the isentropic perfect gas relationship, are that *along a streamline*

$$p\left(1 + \frac{\gamma - 1}{2}M^2\right)^{\gamma/(\gamma-1)} = \text{constant} = p_0 \text{ (say)}, \tag{4.29b}$$

$$\rho\left(1 + \frac{\gamma - 1}{2}M^2\right)^{1/(\gamma-1)} = \text{constant} = \rho_0, \tag{4.29c}$$

and

$$T\left(1 + \frac{\gamma - 1}{2}M^2\right) = \text{constant} = T_0, \tag{4.29d}$$

where p_0, ρ_0 and T_0 are the total pressure, density and temperature respectively and are the appropriate values achieved if the streamline is isentropically brought to, or started from, rest. Equation (4.29d) is often referred to as *the energy equation*, since it can be derived in this integral form for steady streamline motion from energy considerations without recourse to the isentropic statement (see the next section), so that it permits the presence of shock waves. Equations (4.29a), (4.29b) and (4.29c) still refer specifically to isentropic (shock-free) streamline motion.

Energy equation

Referring to the fixed volume shown in Figure 4.2(b) the work done by the pressure forces at the volume boundary equals the rate of change of energy within the volume and the energy flux crossing its boundary. Thus

$$-\int_A p\mathbf{U}\cdot d\mathbf{A} = \int_V \frac{\partial}{\partial t}(\rho i + \tfrac{1}{2}\rho U^2)\,dV + \int_A (\rho i + \tfrac{1}{2}\rho U^2)\mathbf{U}\cdot d\mathbf{A}. \tag{4.30}$$

For the specific case of steady motion along the streamtube of Figure 4.2(a) (i.e. V is a segment of the streamtube) the continuity statement reduces equation (4.30) to

$$\frac{p}{\rho} + i + \tfrac{1}{2}U^2 = \text{constant}$$

or

$$h + \tfrac{1}{2}U^2 = \text{constant} = h_0 \text{ (total enthalpy)}, \tag{4.31}$$

which, utilizing equation (4.8), gives

$$T\left(1 + \frac{\gamma - 1}{2}M^2\right) = T_0,$$

which is again equation (4.29d). This integral derivation therefore does not exclude the presence of flow discontinuities, as remarked earlier.

More generally, using *Gauss' theorem* [see IV, Theorem 6.4.2] and considering the limit of infinitesimal volume (and excluding flow discontinuities), equation (4.30) gives

$$-\mathbf{V}\cdot(p\mathbf{U}) = \frac{\partial}{\partial t}(\rho i + \tfrac{1}{2}\rho U^2) + \mathbf{V}\cdot(\rho i + \tfrac{1}{2}\rho U^2)\mathbf{U}$$

or, using equation (4.19), for continuity

$$-\mathbf{V}\cdot(p\mathbf{U}) = \rho\frac{\partial}{\partial t}(i + \tfrac{1}{2}U^2) + \rho\mathbf{U}\cdot\mathbf{V}(i + \tfrac{1}{2}U^2)$$

$$= \rho\frac{D}{Dt}(i + \tfrac{1}{2}U^2). \tag{4.32}$$

On further manipulation this also gives

$$\frac{1}{\rho}\frac{\partial p}{\partial t} = \frac{Dh_0}{Dt}, \tag{4.33}$$

which shows that the total enthalpy h_0 (equals $h + \tfrac{1}{2}U^2$) of a particle is changed only by the non-stationary pressure term $\partial p/\partial t$. If the flow is steady equation (4.33) reverts to the statement of constant total enthalpy along a streamline again.

Provided that motion along a streamline in steady flow is shock free and hence isentropic, it is clear that the momentum and energy equations are equivalent. More generally, the unsteady forms of these equations are also equivalent in isentropic motion and are therefore not independent, so that often equation (4.13) (or 4.14) is written as an *alternative* form of the energy equation.

Steady streamline motion or one-dimensional duct flow

From the continuity equation (4.17) the streamtube area A for one-dimensional flow is given by

$$\rho U A = \rho M a A = \text{constant}$$

or

$$\rho M A T^{1/2} = \text{constant},$$

which holds irrespective of whether or not shock waves are present. For *isentropic* motion equations (4.29c) and (4.29d) means that this can be further rewritten as

$$AM\left(1 + \frac{\gamma - 1}{2}M^2\right)^{-(\gamma+1)/[2(\gamma-1)]} = \text{constant}. \tag{4.34}$$

In *differential* form [see IV, §3.2] this gives

$$\frac{dA}{A} + \frac{1 - M^2}{1 + [(\gamma - 1)/2]M^2} \frac{dM}{M} = 0, \tag{4.35}$$

which shows that:

(i) If the flow is subsonic a reduction in area accelerates it; an increase in area decelerates it.
(ii) At supersonic speeds the flow is accelerated by an increase in area, and *vice versa.*
(iii) It is only possible to accelerate flow steadily from subsonic to supersonic speeds by passing through a throat of minimum area ($dA = 0$) at which the flow is sonic.

If a nozzle is driven solely by reducing the pressure at the downstream end, once the flow becomes sonic at the throat it is no longer possible to further change the conditions there or upstream since acoustic disturbances cannot now propagate upstream. The mass flow rate is therefore a maximum irrespective of any subsequent downstream changes which occur, a condition which is referred to as choking. In a practical nozzle a shock wave may form in the supersonic portion, its position determined by the need for the resulting nozzle exit pressure to match correctly to the applied back pressure.

Crocco's equation, vorticity and circulation

If the momentum equation (4.22) is combined with the second law statement (equation 4.11), Crocco's equation results:

$$\frac{\partial \mathbf{U}}{\partial t} + \nabla h_0 = T\nabla S - (\nabla \times \mathbf{U}) \times \mathbf{U} \tag{4.36}$$

which relates the vorticity $\nabla \times \mathbf{U}$ to gradients of entropy and total enthalpy. In the steady flow through a curved shock wave, for example, $\nabla h_0 = 0$ but $\nabla s \neq 0$ so that the downstream flow is rotational. The circulation Γ about any closed line circuit is defined as [see IV, §9.4]

$$\Gamma = \oint \mathbf{U} \cdot \mathbf{dl}$$

where $\mathbf{dl}$ is a line element.

The rate of change of circulation of a closed circuit moving with the gas is given by

$$\frac{D}{Dt} \oint (\mathbf{U} \cdot \mathbf{dl}) = \oint \left[\frac{D\mathbf{U}}{Dt} \cdot \mathbf{dl} + \mathbf{U} \cdot d\left(\frac{D\mathbf{l}}{Dt} \right) \right]$$

$$= \oint \left[\frac{D\mathbf{U}}{Dt} \cdot \mathbf{dl} + \mathbf{U} \cdot d\mathbf{U} \right]$$

$$= \oint \left[-\frac{\nabla p}{\rho} \cdot \mathbf{dl} + d(\tfrac{1}{2} U^2) \right]$$

$$= \oint \left[-\frac{dp}{\rho} + d(\tfrac{1}{2} U^2) \right] \tag{4.37}$$

from the momentum equation.

The second term on the right-hand side of (4.37) vanishes for integration around a closed circuit. If the flow is homentropic (so that ρ is expressible as a function of p only) then the first integral term on the right-hand side is also zero and *Kelvin's theorem* results, i.e. the circulation is constant in a closed circuit moving with the fluid.

4.5 Shock waves

In the continuum analysis of an inviscid non-conducting gas it is possible for certain flow field *discontinuities* to occur [see IV, §5.3]. In the contact surface or vortex sheet discontinuities, the relative velocity normal to the discontinuity is zero (no flow across it) and the pressure across it is constant, but discontinuities in temperature, density, tangential velocity and entropy are possible. In contrast the shock wave discontinuity can only occur when the velocity component of the upstream flow normal to the shock wave is supersonic. In reality shock waves are very thin but have a complex structure, controlled by viscous stresses and thermal conduction arising from the large internal gradients of velocity and temperature, but from the viewpoint of calculating the changes in properties between the equilibrium states *either side* of the shock wave it can simply be regarded as an abrupt discontinuity in an inviscid adiabatic stream.

In general the shock wave may be curved, and it may be either moving or stationary. In practice it is best to consider the analysis from the viewpoint of an observer moving with the shock wave so that it appears stationary, as shown in Figure 4.3. Here we consider a shock wave normal to the flow direction, but it will be seen shortly that these results can be readily extended to the oblique shock wave case.

Because of the *discontinuous* behaviour, the equations of motion must be expressed in *integral* form and the isentropic assumption *cannot* be made. Across

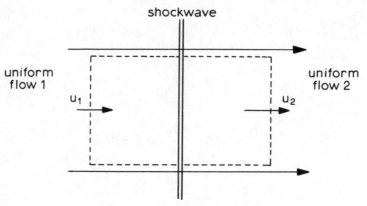

Figure 4.3: The normal shock wave.

the control volume these give for continuity (equation 4.17)

$$\rho U = \text{constant}, \tag{4.38}$$

for momentum normal to the shock wave

$$p + \rho U^2 = \text{constant},$$

which states that the applied pressure difference equals the change of momentum flux or

$$\rho U^2 \left(1 + \frac{1}{\gamma M^2}\right) = \text{constant}, \tag{4.39}$$

and for energy (equation 4.31)

$$\frac{a^2}{\gamma - 1} + \frac{U^2}{2} = \text{constant}$$

or

$$U^2 \left[\frac{2 + (\gamma - 1)M^2}{2(\gamma - 1)M^2}\right] = \text{constant}. \tag{4.40}$$

Eliminating U and ρ between equations (4.38), (4.39) and (4.40) and evaluating the resulting expression on either side of the shock wave gives the *quadratic* equation in M_2^2 [see IV, §2.1]:

$$[2\gamma M_1^2 - (\gamma - 1)]M_2^4 - (2\gamma M_1^4 + 2)M_2^2 + [(\gamma - 1)M_1^4 + 2M_1^2] = 0.$$

This has two roots for M_2^2 [see I, §14.5], either $M_2^2 = M_1^2$, which shows that a uniform unchanged flow is of course one possible solution, or

$$M_2^2 = \frac{(\gamma - 1)M_1^2 + 2}{2\gamma M_1^2 - (\gamma - 1)}, \tag{4.41}$$

which is the shock wave solution since it permits a change in Mach number. The corresponding changes in pressure and density are then given by

$$\frac{p_2}{p_1} = \frac{2\gamma}{\gamma + 1} M_1^2 - \frac{\gamma - 1}{\gamma + 1} \tag{4.42}$$

and

$$\frac{\rho_2}{\rho_1} = \frac{u_1}{u_2} = \frac{(\gamma + 1)M_1^2}{2 + (\gamma - 1)M_1^2}. \tag{4.43}$$

The temperature ratio T_2/T_1 is given by the equation of state (4.3) together with (4.42) and (4.43) and the entropy change can then be deduced from equation (4.12).

Equation (4.41) shows that M_2 is less than unity if M_1 is greater than unity (with $p_2 > p_1$, $T_2 > T_1$, $s_2 > s_1$) or $M_2 > 1$ for $M_1 < 1$ (with $p_2 < p_1$, $T_2 < T_1$, $s_2 < s_1$). The second of these two possibilities, i.e. a shock wave in a subsonic approach flow, violates the second law equation (4.10) because it gives an entropy decrease for an adiabatic process. It is therefore inadmissible and the shock wave can exist only for supersonic relative approach flow.

If the shock wave is inclined to the incident flow direction, as shown in Figure 4.4, there is then an abrupt deflection of the flow as well as compression. The analysis is very similar to that above for the normal shock wave, however, and most of the desired results can be deduced immediately if the flow is resolved into velocity components normal (U_{1N} and U_{2N}) and tangential (U_{1T} and U_{2T}) to the shock wave.

Considering the control volume shown by the dotted boundary in Figure 4.4, continuity gives

$$\rho_1 U_{1N} = \rho_2 U_{2N} = \dot{m} \tag{4.44}$$

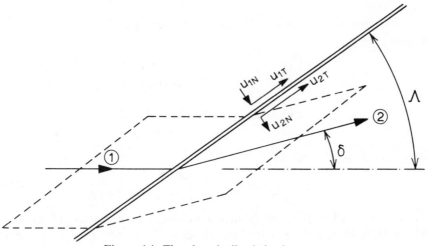

Figure 4.4: The plane inclined shock wave.

and momentum tangential to the shock wave gives

$$\dot{m}(U_{1T} - U_{2T}) = 0,$$

since the net pressure force acting on the control volume tangential to the shock is zero.

The mass flux $\dot{m}$ is non-zero so that

$$U_{1T} = U_{2T}$$

and the tangential velocity is then preserved for passage of a streamline through the shock wave. The momentum equation normal to the shock wave gives

$$p_1 + \rho_1 U_{1N}^2 = p_2 + \rho_2 U_{2N}^2, \tag{4.45}$$

whilst the energy equation is

$$h_1 + \tfrac{1}{2}(U_{1N}^2 + U_{1T}^2) = h_2 + \tfrac{1}{2}(U_{2N}^2 + U_{2T}^2)$$

or

$$h_1 + \tfrac{1}{2}U_{1N}^2 = h_2 + \tfrac{1}{2}U_{2N}^2. \tag{4.46}$$

The governing equations (4.44), (4.45) and (4.46) are in fact the same as the set (4.38), (4.39) and (4.40), and the results of equation (4.41), (4.42) and (4.43) hold provided that the Mach number M_1 for the normal shock wave is replaced for the oblique shock wave by the normal component of the approach Mach number $M_1 \sin \Lambda$.

The downstream *total* Mach number M_2 may be supersonic or subsonic depending upon the inclination of the shock wave Λ, and is given from equation (4.41) as

$$M_2^2 \sin^2 (\Lambda - \delta) = \frac{2 + (\gamma - 1)M_1^2 \sin^2 \Lambda}{2\gamma M_1^2 \sin^2 \Lambda - (\gamma - 1)}. \tag{4.47}$$

The relationship between the shock wave angle Λ and flow deflection δ is given simply by

$$\frac{U_{1N}}{\tan \Lambda} = U_{1T} = U_{2T} = \frac{U_{2N}}{\tan (\Lambda - \delta)}$$

or

$$\frac{\tan (\Lambda - \delta)}{\tan \Lambda} = \frac{(\gamma - 1)M_1^2 \sin^2 \Lambda + 2}{(\gamma + 1)M_1^2 \sin^2 \Lambda},$$

which may be written as

$$\tan \delta = 2 \cot \Lambda \frac{M_1^2 \sin^2 \Lambda - 1}{M_1^2(\gamma + \cos 2\Lambda) + 2}. \tag{4.48}$$

Data for shock waves are tabulated in several works (see, for example, NACA 1135, 1953, and Houghton and Brock, 1975) so only general comments upon shock wave behaviour will be given here.

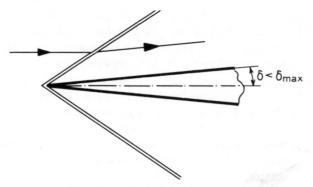

Figure 4.5: Plane attached shock wave on a wedge.

For given values of M_1 and γ equation (4.48) shows that there is always a *maximum* deflection δ_{max} through which a shock wave can turn the flow. For deflections less than this there are two possible solutions. The larger shock wave angle solution of these is referred to as the *strong* shock case and the smaller value of Λ as the *weak* shock case. In the limit of zero flow deflection these correspond to the normal shock wave and the Mach wave (see Figure 4.1) respectively. The *weak* shock wave is always the solution which occurs for a plane oblique attached shock wave on a two-dimensional wedge, as shown in Figure 4.5.

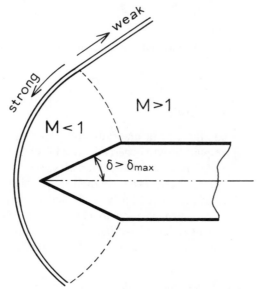

Figure 4.6: Shock detachment from a two-dimensional wedge.

If the wedge angle is greater than δ_{max} it is clearly not possible to turn the flow simply by a plane attached shock wave. A detached shock wave occurs instead, as shown in Figure 4.6, with an inner strong shock region and an outer weak shock region, with flow turning accomplished both through the shock wave and also in the post-shock flow field.

A related phenomenon is found for the intersection between two shock waves and in shock reflection from a surface. Regular intersection (or reflection) is possible, as shown in Figure 4.7(a), provided that the flow deflection δ required for the flow entering the second shock wave is less than the appropriate value of δ_{max} for this region. Once this is exceeded the triple-shock Mach intersection (or reflection) must occur (see Figure 4.7b). The precise conditions for transition from regular to Mach intersection are not understood, however, and there is an overlap region where both are possible theoretically.

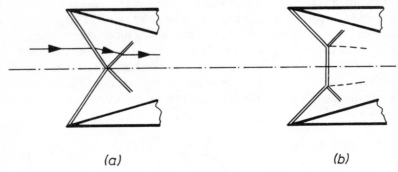

(a)   (b)

Figure 4.7: (a) Regular intersection, (b) Mach intersection.

4.6 The method of characteristics

In a steady supersonic flow, or in unsteady compressible flow at any Mach number, the governing equations are *hyperbolic* [see IV, §8.5] and can be solved, usually *numerically*, by the method of *characteristics* [see III, §9.2]. This is an important technique because it covers a wide range of problems in compressible aerodynamics, it can be implemented fairly easily and it provides more or less 'exact' solutions which can also be used as checks upon more approximate techniques. The method will be illustrated here by several problems in unsteady and steady compressible flow with *two independent variables*.

One-dimensional unsteady homentropic flow

The governing *quasi-linear homogeneous* equations [see IV, §8.6.0] of continuity and momentum are

$$\rho \frac{\partial u}{\partial x} + u \frac{\partial \rho}{\partial x} = -\frac{\partial \rho}{\partial t} \tag{4.49}$$

and

$$\frac{\partial u}{\partial t} + u \frac{\partial u}{\partial x} = -\frac{1}{\rho} \frac{\partial p}{\partial x}. \tag{4.50}$$

The homentropic statement $p/\rho^\gamma = $ constant, together with the speed of sound $a^2 = \gamma p/\rho$, enables (4.49) and (4.50) to be rewritten as

$$a \frac{\partial u}{\partial x} + \frac{2u}{\gamma - 1} \frac{\partial a}{\partial x} + \frac{2}{\gamma - 1} \frac{\partial a}{\partial t} = 0 \tag{4.51}$$

and

$$u \frac{\partial u}{\partial x} + \frac{\partial u}{\partial t} + \frac{2a}{\gamma - 1} \frac{\partial a}{\partial x} = 0. \tag{4.52}$$

Consider a curve C in the t–x plane shown in Figure 4.8. Along this curve we have [see IV, §5.4]

$$\frac{\partial u}{\partial x} dx + \frac{\partial u}{\partial t} dt = du \tag{4.53}$$

and

$$\frac{\partial a}{\partial x} dx + \frac{\partial a}{\partial t} dt = da, \tag{4.54}$$

so that using equations (4.51) to (4.54) it is possible to write explicitly each derivative as a function of (u, a, du, da, dx, dt). Thus, for example,

$$\frac{\partial u}{\partial x}\bigg|_{\text{along C}} = \frac{\dfrac{2a}{\gamma - 1} \dfrac{da}{dt} - \left(u - \dfrac{dx}{dt}\right) \dfrac{du}{dt}}{\left(\dfrac{dx}{dt}\right)^2 - 2u \left(\dfrac{dx}{dt}\right) + (u^2 - a^2)}, \tag{4.55}$$

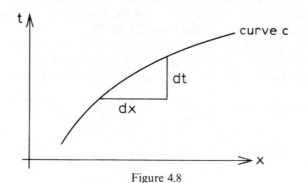

Figure 4.8

with similar expressions for the other derivatives although it is only necessary to study one here.

The expression becomes indeterminate for orientations of C such that the denominator becomes zero (when the numerator must also be zero). This condition occurs here if

$$\left(\frac{dx}{dt}\right)^2 - 2u\frac{dx}{dt} + (u^2 - a^2) = 0,$$

which has two real solutions for dx/dt provided that $a^2 > 0$. These real solutions mean that the equations are *hyperbolic* [see III, §9.1], giving the two families of characteristics

$$\frac{dx}{dt} = u \pm a. \tag{4.56}$$

The *compatability relationship* along these two characteristics are then obtained by setting the numerator of (4.55) to zero, giving the simple algebraic results that

$$\left.\begin{array}{l} d\left(u + \dfrac{2a}{\gamma - 1}\right) = 0 \\[2mm] \text{or } u + \dfrac{2a}{\gamma - 1} = \text{constant} \end{array}\right\} \text{along } \frac{dx}{dt} = u + a \ (C^+ \text{ characteristic}) \tag{4.57}$$

and

$$\left.\begin{array}{l} d\left(u - \dfrac{2a}{\gamma - 1}\right) = 0 \\[2mm] \text{or } u - \dfrac{2a}{\gamma - 1} = \text{constant} \end{array}\right\} \text{along } \frac{dx}{dt} = u - a \ (C^- \text{ characteristic}). \tag{4.58}$$

The quantities $u \pm 2a/(\gamma - 1)$ are referred to as the *Riemann invariants*; they are constant along the appropriate characteristic but it should be noted that within a given family of characteristics (C^+, say) the constant in general will change from one characteristic to the next [see III, §9.2].

The general solution technique for equations (4.57) and (4.58) is illustrated in Figure 4.9. If data for u and a are known at both A and B, then at the intersection C between the C^+ characteristic from A and C^- from B we have

$$u_A + \frac{2a_A}{\gamma - 1} = u_C + \frac{2a_C}{\gamma - 1}$$

and

$$u_B - \frac{2a_B}{\gamma - 1} = u_C - \frac{2a_C}{\gamma - 1},$$

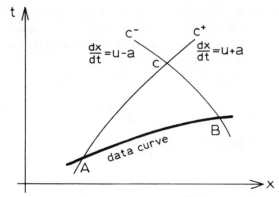

Figure 4.9: Characteristics for unsteady homentropic
flow.

which simply requires the solution of a pair of *simultaneous equations* for u_C and
a_C [see III, §3.3.1]. The position of C is not yet known, however, but in a
practical computation the initial data curve AB is subdivided into sufficiently
small intervals that the characteristic slope can be regarded as constant between
neighbouring mesh points. The array of characteristic mesh points is then
constructed as the computation advances in the direction of increasing time, as
shown in Figure 4.10. The local characteristic slope $u \pm a$ can be evaluated by
taking average values for u and a over each element of characteristic between
adjacent mesh points.

In compressive regions there may be a coalescence or intersection of
characteristics of the same family. In reality this corresponds to *shock wave
formation* which violates the constant entropy assumption required here so that
the method fails.

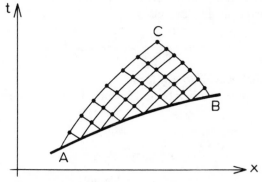

Figure 4.10: Sub-division of characteristic mesh.

A simple unsteady flow is illustrated by looking at the expansion flow (say) on the left of the piston in Figure 4.11 which is being accelerated along the duct from rest at $t = 0$. Here all C^+ characteristics have exactly the same constant along them since they all originate at $t = 0$ in the same region of undisturbed flow. This is referred to as a *simple wave flow,* and here

$$u + \frac{2a}{\gamma - 1} = \text{constant} \tag{4.59}$$

throughout the flow field behind the piston.

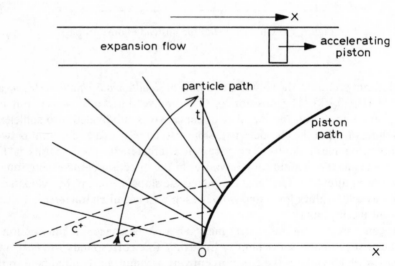

Figure 4.11: Simple wave flow.

In a region where all C^- characteristics have the same constant, in front of the piston in this particular illustration,

$$u - \frac{2a}{\gamma - 1} = \text{constant}, \tag{4.60}$$

provided that shock waves do not form.

Once the speed of sound has been determined then the pressure is given by

$$\frac{p}{\rho^\gamma} = \text{constant}$$

or

$$\frac{p}{a^{2\gamma/(\gamma - 1)}} = \text{constant}. \tag{4.61}$$

Unsteady one-dimensional isentropic (non-homentropic) flow

This corresponds (say) to the flow field behind a propagating shock wave of varying strength, although it should be noted at once that the characteristics, and the relationships along them, cannot be continued across the shock wave. Since the entropy varies between particles we must now use the isentropic statement of equations (4.13) or (4.14) so that the governing equations of continuity (4.49), momentum (4.50) and energy (or entropy) give, after some manipulation,

$$a\frac{\partial u}{\partial x} + \frac{2u}{\gamma - 1}\frac{\partial a}{\partial x} + \frac{2}{\gamma - 1}\frac{\partial a}{\partial t} = 0, \tag{4.62}$$

$$\frac{\partial u}{\partial t} + u\frac{\partial u}{\partial x} + \frac{2a}{\gamma - 1}\frac{\partial a}{\partial x} = \frac{a^2}{\gamma(\gamma - 1)}\frac{\partial}{\partial x}\left[\ln\left(\frac{p}{\rho^\gamma}\right)\right]$$

$$= \frac{a^2}{\gamma R}\frac{\partial s}{\partial x} \tag{4.63}$$

and

$$\frac{\partial s}{\partial t} + u\frac{\partial s}{\partial x} = 0. \tag{4.64}$$

Following the procedure of the previous section for homentropic flow these equations can be shown to be hyperbolic but now with *three characteristics* which, together with the compatability relationships along them, are given by

$$s = \text{constant along } \frac{dx}{dt} = u \text{ (particle path)}, \tag{4.65}$$

$$d\left(u + \frac{2a}{\gamma - 1}\right) = \frac{a}{\gamma R}ds \text{ along } \frac{dx}{dt} = u + a \text{ } (C^+ \text{ characteristic)}, \tag{4.66}$$

$$d\left(u - \frac{2a}{\gamma - 1}\right) = -\frac{a}{\gamma R}ds \text{ along } \frac{dx}{dt} = u - a \text{ } (C^- \text{ characteristic)}. \tag{4.67}$$

This therefore introduces the *particle path* as an extra characteristic, whilst along the wave paths the compatability relationships are given as ordinary differential equations rather than the simpler algebraic relationships found previously. The solution procedure differs somewhat from that illustrated earlier in Figure 4.9, since the values of the flow variables at position C now depend upon its location; this must therefore be found as an integral part of the solution procedure. Figure 4.12 illustrates the basic mesh construction which can be used to advance the computation by an increment which is sufficiently small that characteristics are again assumed straight between mesh points.

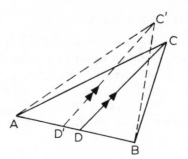

Figure 4.12: Basic computation
mesh for unsteady one-
dimensional flow (isentropic, non-
homentropic).

Equations (4.65), (4.66) and (4.67) are written in *difference form* [cf. III, §9.3] as

$$s_C = s_D, \tag{4.68}$$

$$(u_C - u_A) + \frac{2}{\gamma - 1}(a_C - a_A) = \frac{(a_C + a_A)(s_C - s_A)}{2\gamma R}, \tag{4.69}$$

$$(u_C - u_B) - \frac{2}{\gamma - 1}(a_C - a_B) = -\frac{(a_C + a_B)(s_C - s_B)}{2\gamma R}, \tag{4.70}$$

and are then solved numerically as follows:

(i) From known data at points A and B calculate the slope of the C^+ and C^- characteristics and hence make a first estimate of their intersection position at C.

(ii) From the known velocity distribution along AB, assumed linear, determine the particle path passing through position C, which hence gives position D.

(iii) With s_C known (equals s_D) solve (4.69) and (4.70) for u_C and a_C.

(iv) The calculation can be refined *iteratively*, since a better estimate of the characteristic slopes and particle path can be made using the values at point C as well as on the initial data line. This determines new positions C', D', and so on.

Two-dimensional steady homentropic irrotational supersonic flow

This implies, from Crocco's equation (4.36), that h_0 is constant everywhere. The governing equations of continuity, x and y momentum and irrotationality are

$$\frac{\partial}{\partial x}(\rho u) + \frac{\partial}{\partial y}(\rho v) = 0, \tag{4.71}$$

$$u\frac{\partial u}{\partial x} + v\frac{\partial u}{\partial y} = -\frac{1}{\rho}\frac{\partial p}{\partial x} = -\frac{a^2}{\rho}\frac{\partial \rho}{\partial x}, \tag{4.72}$$

$$v\frac{\partial v}{\partial y} + u\frac{\partial v}{\partial x} = -\frac{1}{\rho}\frac{\partial p}{\partial y} = -\frac{a^2}{\rho}\frac{\partial \rho}{\partial y} \tag{4.73}$$

and

$$\frac{\partial u}{\partial y} - \frac{\partial v}{\partial x} = 0. \tag{4.74}$$

Eliminating ρ between (4.71), (4.72) and (4.73) gives

$$(u^2 - a^2)\frac{\partial u}{\partial x} + (v^2 - a^2)\frac{\partial v}{\partial y} + uv\left(\frac{\partial u}{\partial y} + \frac{\partial v}{\partial x}\right) = 0. \tag{4.75}$$

Again following the earlier procedure, equation (4.74) and (4.75) are found to be hyperbolic provided that $u^2 + v^2 > a^2$, i.e. if the flow is supersonic, with the two sets of characteristics given by the *Mach lines* inclined to the left and right of the streamlines respectively. The compatability relationship can then be derived for this basic cartesian system (see, for example, Zucrow and Hoffman, 1977) but it is more convenient here to *transform* equations (4.74) and (4.75) to a streamline *coordinate* system, as shown in Figure 4.13. Thus,

$$\frac{\partial U}{\partial n} - U\frac{\partial \theta}{\partial t} = 0, \tag{4.76}$$

$$(M^2 - 1)\frac{\partial U}{\partial t} - U\frac{\partial \theta}{\partial n} = 0, \tag{4.77}$$

where U is the total velocity along the streamline $(=(u^2 + v^2)^{1/2})$, t and n are measured along and normal to the streamline and θ is the streamline inclination to the datum direction.

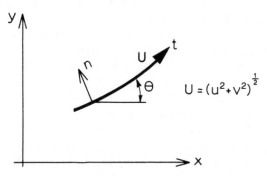

Figure 4.13: Streamline coordinates for steady two-dimensional flow.

The characteristics and their compatability relationships are then readily shown to be

$$(M^2 - 1)^{1/2} \frac{dU}{U} - d\theta = 0 \qquad \text{along } \frac{dn}{dt} = \tan \mu$$

$$\text{(left-running Mach line)} \quad (4.78)$$

and

$$(M^2 - 1)^{1/2} \frac{dU}{U} + d\theta = 0 \qquad \text{along } \frac{dn}{dt} = -\tan \mu$$

$$\text{(right-running Mach line).} \quad (4.79)$$

From the energy equation (4.31),

$$U^2 \left[\frac{1}{2} + \frac{1}{(\gamma - 1)M^2} \right] = \text{constant},$$

so that equations (4.78) and (4.79) can be integrated to give

$$v \mp \theta = \text{constant} \qquad \text{along } \frac{dn}{dt} = \pm \tan \mu. \qquad (4.80)$$

The Prandtl–Meyer angle

$$v = \int (M^2 - 1)^{1/2} \frac{dU}{U}$$

is a function of Mach number only, and evaluating, the *integral* [cf. IV, Example 4.3.11] gives

$$v(M) = \sqrt{\frac{\gamma + 1}{\gamma - 1}} \tan^{-1} \sqrt{\left(\frac{\gamma - 1}{\gamma + 1} \right) (M^2 - 1)} - \tan^{-1} \sqrt{M^2 - 1}. \quad (4.81)$$

The 'constant' of equation (4.80) only applies along the appropriate characteristic. Within a given family, say left-running Mach lines, $v - \theta$ varies generally from one characteristic to the next.

The general solution procedure is similar to that outlined earlier. Tabulated values of v as a function of Mach number, or Mach number as functions of v, can be found in various works such as NACA 1135 (1953) or Houghton and Brock (1975), or can be rapidly calculated for a computer solution. Generally the *inversion* of v [see IV, §5.12], i.e. the expression of M as $M(v)$, must be done numerically or taken from tables, but for the specific values of $\gamma = 5/3$ and $5/4$ it can be expressed analytically (Probstein, 1957).

As in the simple wave solutions for unsteady homentropic flows, simple wave solutions are of particular value in steady supersonic flow. In Figure 4.14 the flow is turned in each case from a uniform flow 1 to a uniform flow 2 by *significant* characteristics of one family only (these are left running here).

All *right-running* characteristics crossing the non-uniform fan-like region have the same constant, i.e.

$$v + \theta = \text{constant}$$

for turning by this *left-running* simple wave. In Figures 4.14(a) and (b) θ_2 is negative, so that v_2 increases, M_2 increases, μ_2 and p_2 decrease and the fan of characteristics diverge. In Figure 4.14(c) the flow is compressive and decelerates. Eventually the fan of characteristics will coalesce some distance from the surface to form a shock wave, in which case the method fails.

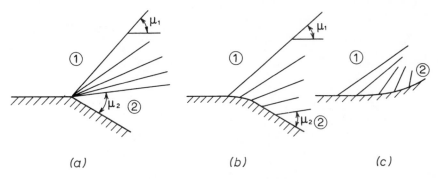

(a) (b) (c)

Figure 4.14: Simple wave flows.

Steady rotational supersonic flow (with $\nabla h_0 = 0$)

If the steady supersonic flow is rotational, such as in the region downstream of a curved shock wave, the equations of motion must be rewritten unless entropy variations can be taken as very weak. Along a streamline equation (4.77) is still correct, since it essentially incorporates only the isentropic statement that

$$\frac{\partial p}{\partial t} = a^2 \frac{\partial \rho}{\partial t}$$

and the vorticity is given from Crocco's equation by

$$\frac{\partial U}{\partial n} - U \frac{\partial \theta}{\partial t} = \frac{T}{U} \frac{\partial s}{\partial n}. \tag{4.82}$$

Equations (4.77) and (4.82) and the isentropic statement $\partial s/\partial t = 0$ are again found to be hyperbolic provided that $U^2 > a^2$. The streamline becomes an extra characteristic in addition to the two Mach lines and the construction of the numerical solution is then similar to that outlined in Figure 4.12.

Further comments on the method of characteristics

So far the numerical procedure has dealt only with interior flow field points. In practice the solution must also satisfy certain *boundary conditions*, such as those arising at a solid surface or at a shock wave whose location is essentially part of the required solution. In either case one characteristic (e.g. the left-running wave shown in the steady flow of Figure 4.15) will impinge upon the boundary mesh point from the flow field interior, providing one relationship between the variables, and another is provided by the boundary condition itself (e.g. at a solid surface $U = U_{surface}$ or $\theta = \theta_{surface}$ for unsteady and steady flows respectively together with $s_{surface} = $ constant always if the flow is isentropic; or

Figure 4.15: Solid surface and shock wave boundary conditions.

appropriate post-shock wave conditions will allow the shock wave strength to be determined [see III, §9.2]). The solution must also commence at some *specific data curve*, e.g. a transonic solution for a nozzle throat region to start the calculation of the downstream supersonic flow or a cone solution for the region downstream of the sharp nose of a supersonic slender body.

The method already outlined in the earlier sections has been confined to *homogeneous* equations in *two independent variables*. In some problems, such as *unsteady spherical flow* or *steady axially symmetric* flows (spherical and cylindrical polar coordinates respectively [see V, §2.1.5]) *non-homogeneous terms* are introduced (i.e. terms in the flow variables but not the derivatives). The characteristic equations can still be readily derived, although as the complexity of the governing equations increases it may be appropriate to use a more formal mathematical procedure to *eliminate* the derivatives between arrays of simultaneous equations. In other problems extensive analytical solutions have been developed for weak (but still non-linear) waves (see, for example, Kluwick, 1981). The method of characteristics can also be extended to flows in three independent variables, although the simplicity associated with reducing the governing equations to ordinary differential equations for the two-dimensional problem studied here is then lost. A description of this is beyond the scope of the present work, but is discussed in detail in several other texts (e.g. Ferri, 1955; Zucrow and Hoffman, 1977).

4.7   Conical and pseudo-steady similarity

Because of the complexity of the compressible flow equations in general, particularly when combined with the additional non-linearity imposed by the presence of shock waves, similarity-type solutions are often sought. These enable the number of independent variables to be reduced or may allow two apparently dissimilar flows to be related.

Conical similarity

Consider the circular-section cone at zero incidence to a steady supersonic stream shown in Figure 4.16. If the flow field is treated as inviscid there is no

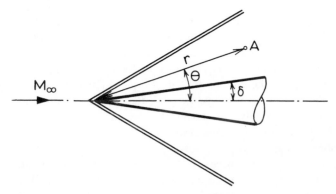

Figure 4.16: Circular cone at zero incidence to supersonic stream.

viscous length scale to describe the flow. Nor is there a characteristic physical length scale if the flow field between the shock wave and the body is everywhere supersonic, since any finite truncation of the trailing edge cannot be felt upstream of the first trailing edge Mach disturbance. From the viewpoint of position A, therefore, the cone is effectively infinite. In the absence of any such characteristic length scales the flow properties cannot depend upon the distance r from the vertex and can only vary with θ. This constancy of properties along rays through the vertex is referred to as a conical flow or *conical similarity*. For the circular-section cone the simplification is significant, since the governing equations can now be reduced simply to a non-linear ordinary differential equation in θ which may be solved numerically [see III, §8.2] (see also, for example, Taylor and Maccoll, 1933).

The same arguments may be extended to *any conical shape* (as shown generally in Figure 4.17 and mathematically defined by $\theta_{body} = f(\phi)$), provided

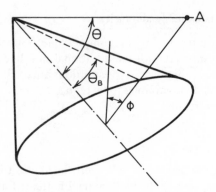

Figure 4.17: General conical body.

that there are no viscous effects or upstream influences from the trailing edge and provided that detachment of the shock wave from the vertex does not occur.

Again flow properties are constant along rays through the vertex so that the governing partial differential equations now have only ϕ and θ as independent variables. In general these flows are still complex. The governing equations will be *hyperbolic* or *elliptic* or *mixed* [see IV, §8.5] depending upon the region of interest, since essentially the governing Mach number now becomes the component in the conical plane, embedded shock waves may occur and if the shock strength varies entropy or vorticity *singularities* [see IV, §9.8] can arise in the *conical cross-plane*. There are, however, a range of solutions, usually numerical, for both exact and approximate forms of the conical equations. These include the circular cone (see, for example, Jones, 1969; Kopal, 1947; Sims, 1964), elliptic-section cones and various geometries with sharp leading edges such as the flat-plate delta wing (South and Klunker, 1969; Voskresenkii, 1965).

Pseudo-stationary flows

A classical problem in unsteady gas dynamics is the one-dimensional shock tube shown in Figure 4.18, where a diaphragm separating a low pressure gas (1) and high pressure gas (4) is ruptured at time $t = 0$. If the subsequent motion is treated as one-dimensional and inviscid, and if during time t no waves have reached the ends of the tube so that it can in fact be regarded as effectively infinite, the only length scale to which x can be referred is one constructed from the elapsed time t and a reference velocity (say the ambient speed of sound in gas 1). Thus x and t are not independent and the problem can be reduced to the solution of *an ordinary differential equation* in (say) $\tilde{x} = x/a_1 t$ [see IV, Chapter 7]. The solution is in fact straightforward here and all the necessary elements have been obtained earlier in the text. Regions 1 and 2 are both uniform and must satisfy the jump condition across the shock wave which lies between them.

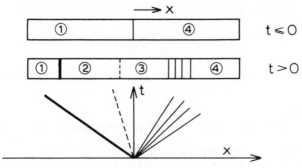

Figure 4.18: Shock tube.

Regions 3 and 4 are also uniform, but are separated by an unsteady isentropic simple wave. Between regions 2 and 3 is a contact surface, across which are jumps in density, temperature and entropy but across which the velocities and pressures are equal: $u_2 = u_3$, $p_2 = p_3$. It is these velocity and pressure conditions which provide the required matching between the shock wave and simple wave solutions (see, for example, Liepmann and Roshko, 1963).

Other pseudo-steady flows can readily be envisaged, where again there is no relevant length scale other than the combination of some reference speed and the time elapsed during the process, such as the diffraction of a plane shock wave by an edge as shown in Figure 4.19. The resulting flow field is highly complex, but the governing equations in r, θ, t are reducible to equations in $\tilde{r}$, θ only, where $\tilde{r} = r/at$.

Another important class of unsteady flows for which *similarity solutions or parameters* have been developed is in the general field of blast propagation (see, for example, Baker, 1973). A classical solution (Taylor, 1950) is that for the spherically propagating blast from an intense point release of energy E, where

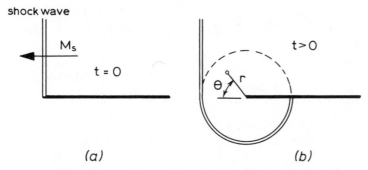

Figure 4.19: Diffraction of a plane shock wave by an isolated edge in two dimensions.

the independent variables r and t are combined as the dimensionless distance

$$\tilde{r} = \frac{r}{(Et^2/\rho_0)^{1/5}},$$

using arguments of *dimensional similarity*, where ρ_0 is the ambient density.

4.8 Small perturbation solutions, higher approximations and iterative solutions

Often solutions to the full equations are sought as *small perturbations* from some known solution. For thin streamline bodies, a natural scheme is to take the resultant flow field as a mild perturbation of the uniform approach stream. Together with the use of a velocity potential for irrotational flow, this has been developed into a powerful technique which will be considered next.

Small perturbation potential solutions for irrotational steady flow

In two-dimensional steady flow the equations of momentum and continuity can be combined to give equation (4.75) (or, alternatively, see equation 4.101). Provided that the flow is shock free or that any shock waves are extremely weak, a *velocity potential* (see equation 4.24) can be introduced for irrotational flow. Here we put

$$u = U_\infty + u' = U_\infty + \frac{\partial \phi}{\partial x}, \qquad v = v' = \frac{\partial \phi}{\partial y}, \qquad (4.83)$$

where u' and v' are the velocity perturbations from a uniform approach stream and ϕ is the perturbation velocity potential. Equation (4.75) now gives

$$\left[(U_\infty^2 - a_\infty^2) + (\gamma + 1)U_\infty \frac{\partial \phi}{\partial x} + \left(\frac{\gamma + 1}{2} \right) \left(\frac{\partial \phi}{\partial x} \right)^2 \right] \frac{\partial^2 \phi}{\partial x^2}$$

$$+ \left[-a_\infty^2 + (\gamma - 1)U_\infty \frac{\partial \phi}{\partial x} + \left(\frac{\partial \phi}{\partial y} \right)^2 + \left(\frac{\gamma - 1}{2} \right) \left(\frac{\partial \phi}{\partial x} \right)^2 \right] \frac{\partial^2 \phi}{\partial y^2}$$

$$+ 2 \left(\frac{\partial \phi}{\partial x} + U_\infty \right) \frac{\partial \phi}{\partial y} \frac{\partial^2 \phi}{\partial x \partial y} = 0, \qquad (4.84)$$

where the local speed of sound is expressed by

$$a^2 = a_\infty^2 - \frac{\gamma - 1}{2} \left[2U_\infty \frac{\partial \phi}{\partial x} + \left(\frac{\partial \phi}{\partial x} \right)^2 + \left(\frac{\partial \phi}{\partial y} \right)^2 \right]. \qquad (4.85)$$

These equations are still exact for inviscid irrotational flow, but can be simplified considerably if the solution is now restricted to small perturbations only. Thus if all *third-order terms* are neglected, as well as all *second-order* terms except specifically that which includes $\partial^2\phi/\partial x^2$, we obtain

$$(1 - M_\infty^2)\frac{\partial^2\phi}{\partial x^2} + \frac{\partial^2\phi}{\partial y^2} = (\gamma + 1)\frac{M_\infty^2}{U_\infty}\frac{\partial\phi}{\partial x}\frac{\partial^2\phi}{\partial x^2} \tag{4.86}$$

and

$$a^2 = a_\infty^2 - (\gamma - 1)U_\infty\frac{\partial\phi}{\partial x}. \tag{4.87}$$

Equation (4.86) is valid for subsonic, transonic and supersonic flow. It is not valid for hypersonic flow ($M_\infty \gg 1$) since some of the neglected higher order terms must then be included. The term on the right-hand side of (4.86) actually needs to be retained only for transonic flows ($U_\infty \approx a_\infty$) since then it is of the same order as the term in $\partial^2\phi/\partial x^2$ on the left-hand side. For subsonic and supersonic flow, with M_∞ not too close to unity, the right-hand side is set to zero to give a *linear equation* in the perturbation potential [see IV, (8.1.5)]. Subsonic and supersonic solutions are therefore easier to obtain, and considerably more extensive, than transonic solutions.

Finally, the energy equation (4.87) or Bernoulli's equation gives the pressure coefficient C_p as

$$C_p = \frac{p - p_\infty}{\frac{1}{2}\rho U_\infty^2} = -\frac{2u'}{U_\infty} = -\frac{2}{U_\infty}\frac{\partial\phi}{\partial x}. \tag{4.88}$$

Supersonic flow

With $M_\infty^2 > 1$ equation (4.86) gives the *wave equation*, with the *general solution* [see IV, §8.4]

$$\phi = F_1(x - \sqrt{M_\infty^2 - 1}\,y) + F_2(x + \sqrt{M_\infty^2 - 1}\,y),$$

where the *two families of characteristics* are inclined to the free stream direction at the Mach angle

$$\frac{dy}{dx} = \pm(M_\infty^2 - 1)^{-1/2} = \pm\tan\mu_\infty.$$

All characteristics of one family are therefore parallel in this approximation, so there is no steepening of compression waves or spreading of expansion waves as found in solutions of the full equations.

The compatability relationships are easily found, but can also be seen readily from the *linearization* of the characteristic results of equations (4.78) and (4.79). These give

$$(M_\infty^2 - 1)^{1/2}\frac{d(u')}{U_\infty} = d\theta, \qquad \text{along } \frac{dy}{dx} = \tan\mu_\infty,$$

or, using equation (4.88),

$$d\left(C_p + \frac{2\theta}{\sqrt{M_\infty^2 - 1}} \right) = 0. \tag{4.89}$$

Along $dy/dx = -\tan\mu_\infty$,

$$d\left(C_p - \frac{2\theta}{\sqrt{M_\infty^2 - 1}} \right) = 0. \tag{4.90}$$

The solution of these can be readily implemented using a *marching procedure* (Smith, 1978; Courant and Hilbert, 1962) as outlined in Section 4.7, noting that the characteristic mesh is specified simply by the value of M_∞.

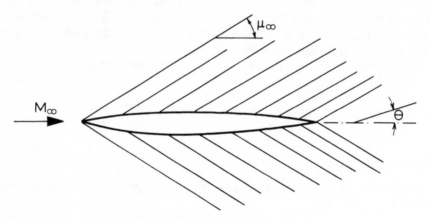

Figure 4.20: Linearized supersonic simple wave flow.

For simple wave flows, such as the upper and lower surface of the aerofoil in Figure 4.20 where significant waves are left running and right running respectively, equations (4.89) and (4.90) give

$$C_{pu} = \frac{2\theta_u}{\sqrt{M_\infty^2 - 1}} \qquad \text{(upper surface)}, \tag{4.91}$$

$$C_{pL} = -\frac{2\theta_L}{\sqrt{M_\infty^2 - 1}} \qquad \text{(lower surface)}, \tag{4.92}$$

where θ_u, θ_L are the local surface slopes relative to the approach stream. These are the Ackeret or linearized supersonic aerofoil results.

Subsonic flows

With $M_\infty^2 < 1$ and outside the transonic region, equation (4.86) gives the *elliptic equation* [see III, §9.5, and IV, §8.5]

$$(1 - M_\infty^2)\frac{\partial^2 \phi}{\partial x^2} + \frac{\partial^2 \phi}{\partial y^2} = 0, \tag{4.93}$$

which must be solved subject to the body surface boundary condition.

$$\left\{\frac{v'}{(U_\infty + u')}\right\}_{\text{body}} = \left(\frac{dy_{\text{body}}}{dx}\right) \simeq \frac{v'}{U_\infty} \simeq \frac{1}{U_\infty}\left(\frac{\partial\phi}{\partial y}\right)_{\text{body}}. \tag{4.94}$$

Within linear theory accuracy for *two-dimensional* bodies the above can be rewritten as

$$\frac{dy_{\text{body}}}{dx} \simeq \frac{1}{U_\infty}\left(\frac{\partial\phi}{\partial y}\right)_{y=0}. \tag{4.95}$$

Using the transformations

$$\bar{x} = x,$$

$$\bar{y} = y(1 - M_\infty^2)^{1/2},$$

$$\bar{y}_{\text{body}} = y_{\text{body}}, \tag{4.96}$$

$$\frac{\bar{\phi}}{\bar{U}_\infty} = \frac{\phi}{U_\infty}(1 - M_\infty^2)^{1/2},$$

equation (4.93) gives *Laplace's equation* for $\bar{\phi}$ [see IV, §8.2], i.e.

$$\frac{\partial^2 \bar{\phi}}{d\bar{x}^2} + \frac{\partial^2 \bar{\phi}}{\partial\bar{y}^2} = 0,$$

together with

$$\frac{1}{\bar{U}_\infty}\left(\frac{\partial\bar{\phi}}{\partial\bar{y}}\right)_{\bar{y}=0} = \frac{d\bar{y}_{\text{body}}}{d\bar{x}}.$$

These are the equations for incompressible flow ($M = 0$) about the *same* aerofoil at the *same* incidence (since neither x nor y_{body} were scaled); therefore all quantities with an overbar can be equated with the equivalent incompressible result.

It should be noted that it is permissible here to apply different scalings to y and y_{body}, since the body surface boundary condition is evaluated on $y = 0$ and not on the actual body surface. In general flow field scalings, this decoupling cannot be assumed, of course; in particular, in the application of the small perturbation potential theory to slender bodies (e.g. bodies of revolution) or any three-dimensional bodies with both lateral dimensions (y and z) which are small compared with its chord, the body and flow field must be properly matched (see Ward, 1955).

The pressure coefficient in the compressible flow is given by

$$C_p = -\frac{2}{U_\infty}\frac{\partial \phi}{\partial x} = -\frac{2}{\bar{U}_\infty(1 - M_\infty^2)^{1/2}}\frac{\partial \bar{\phi}}{\partial \bar{x}} = \frac{\bar{C}_p}{(1 - M_\infty^2)^{1/2}}$$

or   (4.97)

$$C_{p\,\text{compressible}} = \frac{C_{p\,\text{incompressible}}}{(1 - M_\infty^2)^{1/2}}$$

This is a form of the Prandtl–Glauert rule, stating here that the pressure coefficient on a thin two-dimensional aerofoil in subsonic flow is $(1 - M_\infty^2)^{-1/2}$ times the pressure coefficient at the same position on the same aerofoil at the same incidence in incompressible flow. Incompressible results may therefore be readily related to subsonic compressible flows within the accuracy of the small perturbation approximation, whether they be deduced from subsonic thin aerofoil theory, from solution of the exact incompressible equations or from experiment.

Detailed discussions of linearized subsonic and supersonic flows for two-dimensional, three-dimensional, and slender bodies may be found in several texts (e.g. Ward, 1955). Singularity distribution methods are used extensively for supersonic wing design (e.g. Carlson and Miller, 1974).

Two-dimensional transonic flow

If the free-stream Mach number is close to or equal to unity then the non-linear right-hand side of equation (4.86) must be retained. Subject to certain scaling requirements, given below, one transonic flow can be related to another by considering a scaling procedure similar to that outlined in the previous section. In particular, if we have two transonic flows 1 and 2, then the equations

$$x_2 = x_1,$$

$$y_2(1 - M_2^2)^{1/2} = y_1(1 - M_1^2)^{1/2},$$

$$\frac{\phi_2}{U_2}\frac{M_2^2(1 + \gamma_2)}{1 - M_2^2} = \frac{\phi_1}{U_1}\frac{M_1^2(1 + \gamma_1)}{1 - M_1^2},$$ (4.98)

$$\frac{y_{2B}M_2^2(1 + \gamma_2)}{(1 - M_2^2)^{3/2}} = \frac{y_{1B}M_1^2(1 + \gamma_1)}{(1 - M_1^2)^{3/2}}$$

transform the transonic equations for flow 1 to the transonic equations for flow 2. The body surface boundary conditions are still given by (4.94), but it should be noted, however, that now it is no longer possible to keep the body coordinate (y_{2B}, y_{1B}) unchanged, and since this implies that both incidence and thickness distributions must scale correspondingly, the similarity requirement becomes that

$$\frac{\tau M^2 (1 + \gamma)}{(1 - M^2)^{3/2}}$$

be the same for both aerofoils, where τ is a characteristic thickness parameter, say maximum thickness chord. This transformation is referred to as an *affine transformation*, where one geometry is simply a linear stretching of the other. The transonic similarity parameter is usually expressed as

$$x = \frac{1 - M^2}{[\tau(\gamma + 1)M^2]^{2/3}}, \tag{4.99}$$

and provided that this is invariant between two affine aerofoils the pressure coefficients at equivalent points are then related by

$$\frac{C_p[(\gamma + 1)M^2]^{1/3}}{\tau^{2/3}} = \text{constant}. \tag{4.100}$$

Equations (4.99) and (4.100) are clearly useful for correlating experimental data and solutions to the transonic equations. Analytical solution of equation (4.86) is difficult because of the non-linear terms (see Guderley, 1962). In Murman and Cole's (1971) *implicit numerical solution scheme* [cf. III, §9.3] for small disturbance transonic flow, each mesh point is computed and tested to determine whether the flow is locally subsonic or supersonic and the appropriate *elliptic-type* or *hyperbolic-type differencing scheme* is then selected for that point, [see III, §§9.5 and 9.3] so that domains of influence and dependence are properly treated. For a discussion of full transonic solution methods (either full potential or Euler) see, for example, Yoshihara (1972) and Roe (1982).

Other methods

For steady potential flows the continuity equation gives

$$\mathbf{V} \cdot (\rho \mathbf{U}) = \mathbf{V} \cdot (\rho \mathbf{V} \phi) = 0$$

which, using $\nabla p = a^2 \nabla \rho$ for homentropic motion and the steady irrotational form of the momentum equation (4.22), provides

$$a^2 \nabla^2 \phi = \mathbf{V} \phi \cdot \mathbf{V}(\tfrac{1}{2} U^2), \tag{4.101}$$

where

$$a^2 = \frac{\gamma - 1}{2}(U_\infty^2 - U^2) + a_\infty^2.$$

The *Rayleigh–Janzen* solution for fully subsonic flow expands ϕ as a series in M_∞^2:

$$\phi = U_\infty(\phi_0 + M_\infty^2 \phi_1 + M_\infty^4 \phi_2 + \cdots). \tag{4.102}$$

Substitution of (4.102) into (4.101) and equating like powers of M_∞^2 gives

$$\nabla^2 \phi_0 = 0,$$

$$\nabla^2 \phi_1 = f_1(\phi_0), \tag{4.103}$$

i.e. the solution of *Laplace's equation* for ϕ_0 which is the incompressible solution and *Poisson's equation* for ϕ_1, etc. [see IV, §8.2]. This can be accomplished if the body shape is one for which *Green's function* is known [see IV, Definition 8.2.1] but in practice this is very laborious (see Lighthill, 1960).

An *iterative solution technique* to equation (4.101), again of course for subsonic shock-free flow, is obtained by taking the solution for ϕ_0 as the basic incompressible solution, evaluating the non-linear right-hand side of (4.101) with this first estimate and then solving the resulting Poisson equation, and so on [see IV, §8.5]. Again as an analytical technique this is extremely laborious and extremely difficult for all but the simplest geometries and limited number of iterations. It can, however, be used as the basis of an *iterative numerical scheme* for solution of the governing equation [see III, §9.5]. Sell's (1967) method for subcritical aerofoils in fact iterates upon the stream function equation rather than the velocity potential, but the principle of the solution of Poisson's equation where the right-hand side is evaluated on the basis of a previous iteration is the same.

The Karman–Tsien approximation (which is extensively discussed in various references such as Kuo and Sears, 1955) is a *hodograph solution* of the subsonic equations, simplified by the approximate gas relationship $p = a/\rho + b$, where the constants a and b are chosen to give the correct pressure–density derivative at the free stream conditions. It leads to the correction formula relating the pressure coefficient at a point on an aerofoil in compressible flow C_p to its incompressible value (strictly a slight distortion of the aerofoil shape is also involved) by

$$C_p = \frac{C_{p_i}}{\sqrt{1 - M_\infty^2} + [M_\infty^2/(1 + \sqrt{1 - M_\infty^2})]\, C_{p_i}/2}.$$

R.H.

References

Baker, W. E. (1973). *Explosions in Air*, University of Texas Press.

Carlson, H. W., and Miller, D. S. (1974). Numerical methods for the design and analysis of wings at supersonic speeds, NASA Rep. No. TN D-7713.

Courant, R., and Hilbert, D. (1962). *Methods of Mathematical Physics*, Interscience.

Ferri, A. (1955). *The Method of Characteristics, General Theory of High Speed Aerodynamics*, Vol. 6 of *The Series in High Speed Aerodynamics and Jet Propulsion*, Princeton University Press and Oxford University Press.

Guderley, K. G. (1962). *Theory of Transonic Flow*, Pergamon.

Houghton, E. L., and Brock, A. E. (1975). *Tables for the Compressible Flow of Dry Air*, Edward Arnold.

Jones, D. J. (1969). Tables of inviscid supersonic flow about circular cones at incidence, $\gamma = 1.4$, AGARDograph No. 137 (Parts I and II).

Kluwick, A. (1981). The analytical method of characteristics, *Progress in Aerospace Sciences*, Vol. 19, pp. 197–313, Pergamon.

Kopal, Z. (1947). *Tables for the Supersonic Flow around Cones*, prepared by the Staff of the Computing Section, under the direction of Zdenek Kopal, Technical Report No. 1, Department of Electrical Engineering, Centre of Analysis, Massachusetts Institute of Technology, Cambridge.

Kuo, Y. H., and Sears, W. R. (1955). *Plane Subsonic and Transonic Potential Flows, General Theory of High Speed Aerodynamics*, Vol. 6 of *The Series in High Speed Aerodynamics and Jet Propulsion*, Princetown University Press and Oxford University Press.

Liepmann, H. W., and Roshko, A. (1963). *Elements of Gasdynamics*, Wiley.

Lighthill, M. J. (1960). *Higher Approximations in Aerodynamic Theory*, Princetown University Press.

Murman, E. M., and Cole, J. D. (1971). Calculation of Plane Steady Transonic Flows, *AIAA J.*, **9**, 114–121.

NACA (1953). Equations, tables and charts for compressible flow, Report No. 1135.

Probstein, R. F. (1957). Inversion of the Prandtl–Meyer relation for specific heat ratios of 5/3 and 5/4, *J. Aero. Sci.*, **24**, 316–317, 632.

Roe, P. L. (Ed.) (1982). *Numerical Methods in Aeronautical Fluid Dynamics*, Academic Press.

Sells, C. C. L. (1967). Plane subcritical flow past a lifting aerofoil, Roy. Aircraft Estab. Tech. Rep. No. 67146.

Sims, J. L. (1964). Tables for supersonic flow around right circular cones at zero angle of attack, NASA Rep. No. SP-3004.

Smith, G. D. (1978). *Numerical Solution of Partial Differential Equations: Finite Difference Methods*, 2nd edn, Clarendon Press.

South, J., and Klunker, E. B. (1969). Methods for calculating non-linear conical flows, NASA Rep. No. SP-228.

Taylor, G. I. (1950). The formation of a blast wave from a very intense explosion, *Proc. Roy. Soc. (A)*, **201**, 159–174.

Taylor, G. I., and Maccoll, J. W. (1933). The air pressure on a cone moving at high speeds, *Proc. Roy. Soc. (A)*, **139**, 278–297.

Voskresenkii, G. P. (1965). Numerical solution of the problem of a supersonic gas flow past an arbitrary surface of a delta wing in the compression region, *Mekhanika Zhidkosti i Gaza*, **4**, 134.

Ward, G. N. (1955). *Linearised Theory of Steady High-Speed Flow*, Cambridge University Press.

Yoshihara, H. (1972). Plane inviscid transonic airfoil theory, AGARDograph No. 156.

Zucrow, M. J., and Hoffman, J. D. (1977). *Gas Dynamics*, Vol. II, *Multi-Dimensional Flow*, Wiley.

5

Free Surfaces and Waves

5.1 Hydrostatics

When a fluid is at rest the *pressure* at a point (measured as a force per unit area) is the same in every direction. When the only external force is due to gravity and the density ρ varies with depth, the difference between the pressures at two points A and B is

$$p_A - p_B = \int_{z_A}^{z_B} g\rho \, dz, \tag{5.1}$$

where z is the vertical coordinate.

If the density ρ is constant

$$p_A - p_B = g\rho h,$$

where h is the difference in height between A and B. 'The pressure due to a *head* of k metres of fluid' means the pressure at a depth of k metres below the effective surface.

The *thrust* of a fluid on a surface is equivalent to a single force at the *centre of pressure*. If the pressure intensity p is constant the force in a direction Ox, say, is $p \times$ projected area of the surface at right angles to Ox. The centre of pressure is then at the *centroid* of the projected area.

If the pressure intensity is proportional to depth in the fluid, i.e. $p = \rho g h$, then the force on a vertical plane surface of area A is

$$F = \rho g \int yh \, dh, \tag{5.2}$$

where y is the breadth of the surface at depth h [see IV, §4.2], i.e.

$$F = \rho g \times \text{the } \textit{first moment of area} \text{ about the surface of the fluid}$$

or

$$F = \rho g A \bar{h}, \tag{5.3}$$

where $\bar{h}$ is the depth of the centroid G of the plane surface.

The depth of the centre of pressure $\bar{k}$ is obtained by taking moments about the surface of the fluid:

$$F\bar{k} = \rho g \int y h^2 \, dh \qquad (5.4)$$

$$= \rho g \times \text{the second moment of area about the surface of the fluid}$$

$$= \rho g A k^2, \qquad (5.5)$$

where k is the *radius of gyration* of the plane surface about a coplanar axis in the surface of the fluid. Hence

$$\bar{k} = \frac{k^2}{\bar{h}}. \qquad (5.6)$$

If I_G is the second moment of area of the surface about a coplanar axis through its centroid G then by the *theorem of parallel axes*

$$Ak^2 = A\bar{h}^2 + I_G$$

and hence

$$\bar{k} = \bar{h} + \frac{I_G}{A\bar{h}}. \qquad (5.7)$$

Thus the centre of pressure lies below the centre of gravity.

First and second moments of area can be obtained exactly for various simple shapes (Lamb, 1932).

The *principle of Archimedes* explains the effect of *buoyancy*. When a solid is wholly or partly immersed in a heavy fluid at rest the resultant thrust of the fluid on the solid is equal and opposite to the weight of fluid displaced by the solid and acts in a vertical line through the centre of gravity of the fluid displaced (the *centre of buoyancy*).

The *stability* of a *floating body* depends on the position of the *metacentre* which is defined as follows. If H is the centre of buoyancy when the body is in

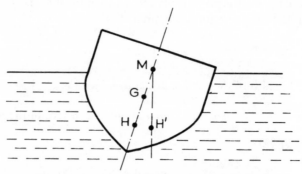

Figure 5.1: The metacentre of a floating body.

equilibrium and H′ is the centre of buoyancy after a small rotation about an axis perpendicular to the plane through H, H′ and the centre of gravity of the body G, which does not alter the weight of fluid displaced, then the vertical through H′ meets HG at the *metacentre* M (see Figure 5.1). The equilibrium is stable if M is above G and unstable if M is below G. The distance GM is called the *metacentric height*. See Francis (1958) for ways of determining this.

5.2 Open channel flow

The form of open channel flow most often of interest to engineers is the flow of water in rivers and canals with the free surface exposed to air at atmospheric pressure. Usually the water is moving under the action of gravity due to the bed slope (assumed to be small) and opposed by the friction at the surface of the bed and the sides of the channel.

Many useful results can be obtained by considering a one-dimensional flow governed by the *equation of continuity*:

$$\frac{\partial A}{\partial t} + \frac{\partial Q}{\partial x} = 0, \tag{5.8}$$

where A = wetted cross-section area,
           Q = discharge.

and the equation of momentum:

$$\frac{\partial Q}{\partial t} + \frac{\partial}{\partial x}\left(\frac{\alpha Q^2}{A}\right) + gA\left(\frac{\partial h}{\partial x} + \frac{\partial z}{\partial x}\right) + \text{friction term} = 0, \tag{5.9}$$

where h = water depth,
           z = height of the bed above datum,
           x = distance downstream,
           t = time,
           g = acceleration due to gravity,
           α = correction factor to allow for the variation of velocity across the section (for uniform velocity distribution $\alpha = 1$, which is assumed in elementary theory)

Assuming a steady-state flow, a rectangular cross-section and $\alpha = 1$, we have $Q = \text{constant} = bhu$, where b is the breadth of the channel and u is the velocity (assumed constant across a cross-section). Then equation (5.9) becomes

$$\frac{d}{dx}\left(\frac{u^2}{2g} + h + z\right) + \frac{1}{gA}\,(\text{friction terms}) = 0. \tag{5.10}$$

The form of equation (5.10) introduces the idea of the *energy* at a cross-section

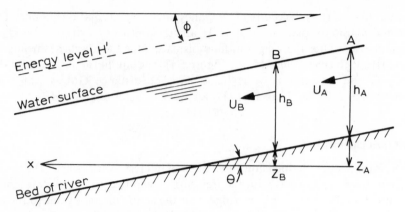

Figure 5.2: Vertical cross-section through the open channel.

expressed as a head in length units (Figure 5.2):

$$H' = \frac{u^2}{2g} + h + z. \tag{5.11}$$

The *specific energy* H is the energy related to the level of the bed; thus

$$H = \frac{u^2}{2g} + h. \tag{5.12}$$

If there is a steady flow q (volume per unit time per unit width of stream) so that $q = uh = Q/b$, then

$$H = \frac{q^2}{2gh^2} + h. \tag{5.13}$$

Figure 5.3 shows the *specific energy curve* of H against h. For a fixed q, a given $H = H_1$ can correspond either to:

(i) Two values of h:
   The smaller h_F which gives *fast flow*.
   The larger h_S which gives *slow flow*.

$$\text{For the fast flow the } \textit{Froude number } \frac{u}{\sqrt{gh}} > 1, \tag{5.14}$$

and the flow is *supercritical*.

$$\text{For the slow flow the } \textit{Froude number } \frac{u}{\sqrt{gh}} < 1, \tag{5.15}$$

and the flow is *subcritical*.

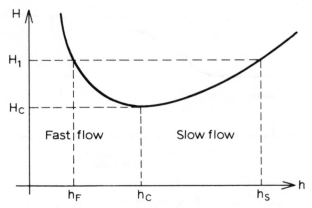

Figure 5.3: Specific energy curve.

(ii) When $H = H_c$, there is just one value of h:

$$h = \left(\frac{q^2}{g}\right)^{1/3} = h_c, \text{ the critical depth.} \qquad (5.16)$$

Then the *Froude number* $\dfrac{u}{\sqrt{gh}} = \dfrac{u_c}{\sqrt{gh_c}} = 1,$ $\qquad (5.17)$

and the flow is *critical*. Thus for a given value of q, H is a minimum.

(iii) When $H < H_c$, there is no corresponding value of h. If the flow remains steady we have

$$H'_A = H'_B + \text{loss of energy due to bottom friction, etc.,} \qquad (5.18)$$

where H'_A, H'_B are the energies at the cross-section through A and B in Figure 5.2. If the loss of energy due to friction, etc., between A and B is negligible then the *energy level* is horizontal and $\phi = 0$. Then the equation

$$H'_A = H'_B \qquad (5.19)$$

can be used to assess the effect of a change in the cross-section of the bed. Francis (1958) gives a number of examples to illustrate this.

Again assuming a steady-state flow, $\alpha = 1$, and a uniform breadth of channel so that $q = uh = Q/b$ is constant, the momentum equation (5.9) may be written in the form:

$$\frac{d}{dx}\left(\frac{q^2}{h} + \tfrac{1}{2}gh^2\right) + gh\frac{dz}{dx} + \frac{\tau_0 P}{b} = 0, \qquad (5.20)$$

where τ_0 is the boundary shear stress due to the surface roughness at the bed and P is the wetted perimeter.

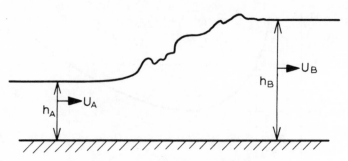

Figure 5.4: An example of a hydraulic jump.

A case of particular interest where there is a sudden decrease in energy not due to friction is the *hydraulic jump*. This occurs, for example, when the fast flow from under a sluice or at the bottom of a spillway changes to slow flow at a greater depth (Figure 5.4). This may be due to some resistance downstream or to a flattening of the bedslope. There is then an abrupt transition from supercritical to subcritical flow. From equation (5.20), if there is zero bedslope and friction and the depth changes from h_A to h_B we have

$$\frac{q^2}{h_A} + \tfrac{1}{2}gh_A^2 = \frac{q^2}{h_B} + \tfrac{1}{2}gh_B^2. \tag{5.21}$$

Equation (5.21) may be rewritten as

$$(h_B - h_A)\left[h_B^2 + h_A h_B - \frac{2q^2}{gh_A}\right] = 0. \tag{5.22}$$

Thus if there is to be a change of depth we must have

$$h_B = \frac{h_A}{2}\left(-1 + \sqrt{1 + \frac{8q^2}{h_A^3 g}}\right). \tag{5.23}$$

(Evidently $h_B > h_A$ only if $q^2/h_A^3 g = u_A^2/h_A g > 1$, i.e. the flow upstream is supercritical.) The loss in energy is, then, from equation (5.11),

$$\Delta H' = \left(\frac{u_A^2}{2g} + h_A\right) - \left(\frac{u_B^2}{2g} + h_B\right). \tag{5.24}$$

The *hydraulic jump* is often used to decelerate flow and reduce scouring.

In practice friction forces normally reduce the energy as the flow proceeds so that the energy level as shown in Figure 5.2 is sloping at an angle ϕ to the horizontal. Taking account of the force applied to the fluid, equation (5.21) becomes

$$(\tfrac{1}{2}\rho gh_A^2 + \rho qu_A) - (\tfrac{1}{2}\rho gh_B^2 + \rho qu_B) = F', \tag{5.25}$$

where F' is the force applied to the fluid between cross-sections A and B.

Writing

$$F = \tfrac{1}{2}\rho g h^2 + \rho q u \tag{5.26}$$

equation (5.25) may be written

$$F_A - F_B = F' \tag{5.27}$$

and F_A, F_B can be taken to mean the force at cross-sections A, B of the river. F is a minimum for $h = h_c = (q^2/g)^{1/3}$.

Figure 5.5 shows $F/\rho g h_c^2$ plotted against h_c/h. The curve is similar to that in

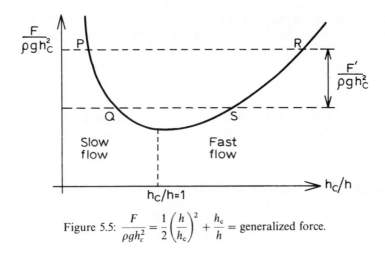

Figure 5.5: $\dfrac{F}{\rho g h_c^2} = \dfrac{1}{2}\left(\dfrac{h}{h_c}\right)^2 + \dfrac{h_c}{h} =$ generalized force.

Figure 5.3 but the regions corresponding to slow and fast flow are now reversed. The effect of a resistance F' to the flow is to move from P to Q in slow flow thus increasing h_c/h or from R to S in fast flow thus decreasing h_c/h.

If there is a loss of energy due to friction forces and

$$m = \text{hydraulic mean radius} = \frac{\text{cross-sectional area}}{\text{wetted perimeter}},$$

then the *Chezy formula* for the mean velocity gives

$$\bar{u} = Cm^{1/2}i^{1/2} \tag{5.28}$$

and the *Manning formula* gives

$$\bar{u} = Mm^{2/3}i^{1/2}, \tag{5.29}$$

where C, M are roughness coefficients (Sellin, 1969, gives values of these) and $i = dH'/dx$, the slope of the energy level.

If τ_0 is the shear force acting on the fluid due to the surface roughness at the bed then

$$\tau_0 = \rho g m \frac{dH'}{dx} = \rho g m i. \tag{5.30}$$

In terms of the Chezy constant, C,

$$\frac{dH'}{dx} = \frac{\bar{u}^2}{C^2 m} = \frac{q^2}{C^2 h^2 m}. \tag{5.31}$$

If dH'/dx is constant, with C, m, q constant, then $h = h_0$ (the *normal depth*), $i = dH'/dx = dz/dx = s$ (the *bed slope*) and $q = q_0$.

If the stream is wide enough to take $h_0 \approx m_0$ and measurements of q_0, h_0 are made on a length of river at normal depth, where $s = i$, then C can be found from

$$C = \frac{q_0}{\sqrt{h_0^3 s}}. \tag{5.32}$$

Hence on a comparable stretch of river where h is changing,

$$\frac{dh}{dx} = s \frac{1 - (h_0/h)^3}{1 - (h_c/h)^3}, \tag{5.33}$$

where h_c is given by equation (5.16). Thus [see IV, §4.2]

$$h = \int_0^l s \frac{1 - (h_0/h)^3}{1 - (h_c/h)^3} \, dx \tag{5.34}$$

and this equation can be solved *numerically* [see III, §10.6.1] to give the distances l_1, $l_2, \ldots$ at which the depth is some given d_1, $d_2, \ldots$.

The corresponding equation with Manning's formula is

$$h = \int_0^l s \frac{1 - (h_0/h)^{10/3}}{1 - (h_c/h)^3} \, dx. \tag{5.35}$$

See Francis (1958), Henderson (1966) and Sellin (1969) for a further discussion of methods of dealing with channels which are not wide or of more complicated cross-section. Problems with varying shapes of channel and unsteady flow (see Henderson, 1966) can be solved by small-scale models or numerically as above.

5.3 Surface water waves

Surface water waves are complicated phenomena but formulae useful to the engineer can be derived by making various simplifying assumptions such as assuming smooth low regular waves and approximating to allow for *deep* or *shallow* water. Higher order, and hence more accurate, formulae are referred to where necessary.

Surface waves are generally generated by wind blowing across the water for a distance known as the *fetch*. There are empirical formulae such as

$$\text{Wave height in metres} = 0.0169\sqrt{U^2F} \qquad (5.36)$$

and

$$\text{Wave period in seconds} = 0.50\sqrt[4]{U^2F}, \qquad (5.37)$$

where U = windspeed in knots (1 knot = 0.516 m/s),
F = fetch length in nautical miles (1 nautical mile = 1,858 m) (see Ippen, 1966).

The term *fully developed* (or *fully arisen*) *sea* refers to the condition when no more wave energy can be added to the wave system even though the wind continues and the fetch increases. As waves move out of the generating area the longer waves travel faster; these can persist for a long distance as *swell* with long crests perpendicular to the direction of travel. As the waves travel they gradually decay; an empirical law states that the waves lose one third of their height each time they travel a distance in nautical miles equal to their length in feet. These wind-generated waves are mainly in the period range 1 to 20 seconds. For a further discussion see Ippen (1966) and Wiegel (1964).

The basic theory (Wiegel, 1964) assumes that the motion is such that there is a velocity potential ϕ satisfying Laplace's equation in two dimensions:

$$\frac{\partial^2\phi}{\partial x^2} + \frac{\partial^2\phi}{\partial y^2} = 0 \qquad (5.38)$$

[see Section 2.1 and also IV, §9.13] and we have *Bernoulli's equation*

$$\frac{p}{\rho} = -gy + \frac{\partial\phi}{\partial t} - \frac{1}{2}\left[\left(\frac{\partial\phi}{\partial x}\right)^2 + \left(\frac{\partial\phi}{\partial y}\right)^2\right], \qquad (5.39)$$

where p = pressure,
ρ = density,
$-y$ = depth,
x, y are horizontal and vertical coordinates,
t = time.

There are also the boundary conditions on the surface and the sea bottom (see Figure 5.6) and various theories can be developed depending on what quantities are regarded as small enough to be neglected. The simplest theory is called *Stokes first-order* or *Airy* wave theory which gives sinusoidal waves. These are smooth low waves corresponding to two-dimensional oscillatory motion in water of uniform depth h where the amplitude H of the wave is small enough for squares of velocities to be neglected (i.e. the theory is linearized). This means that

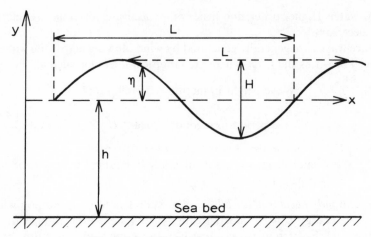

Figure 5.6: A Stokes first-order wave.

the wave steepness and $L^2 H/2h^3$ must be 'small'. The potential function satisfying Laplace's equation [see IV, §8.2] for the linearized solution is [see IV, §§2.12 and 2.13]

$$\phi = \frac{HL}{2T} \frac{\cosh\left[2\pi(y+h)/L\right]}{\sinh\left(2\pi h/L\right)} \sin 2\pi(x/L - t/T), \qquad (5.40)$$

where $L = 2\pi/k$ is the wave length,
             $T = 2\pi/\sigma$ is the wave period,
             $H =$ wave height trough to crest (see Figure 5.6),
             $h =$ depth of water.

The *wave speed*

$$C = \frac{L}{T} = \frac{\sigma}{k} = \frac{gT}{2\pi} \tanh\left(\frac{2\pi h}{L}\right) = \left(\frac{gL}{2\pi} \tanh\frac{2\pi h}{L}\right)^{1/2} \qquad (5.41)$$

since $2\pi L/T^2 = g \tanh(2\pi h/L)$. The surface elevation is

$$\eta = \frac{H}{2} \cos 2\pi\left(\frac{x}{L} - \frac{t}{T}\right). \qquad (5.42)$$

By this linearized theory the particles move in closed orbits, i.e. there is no mass transport. The displacements of the particles are as follows:

Horizontally: $X = -\dfrac{H}{2} \dfrac{\cosh\left[2\pi(\bar{y}+h)/L\right]}{\sinh\left(2\pi h/L\right)} \sin\left[2\pi\left(\dfrac{\bar{x}}{L} - \dfrac{t}{T}\right)\right],$ (5.43)

Vertically: $Y = \dfrac{H}{2} \dfrac{\sinh\left[2\pi(\bar{y}+h)/L\right]}{\sinh\left(2\pi h/L\right)} \cos\left[2\pi\left(\dfrac{\bar{x}}{L} - \dfrac{t}{T}\right)\right],$ (5.44) where $(\bar{x}, \bar{y})$ is the mean position ($\bar{y} = -h$ corresponds to the bottom of the sea).

The *particle orbits* are elliptical; in deeper water they are more nearly circular (Figure 5.7). The water is almost still at a depth equal to half the wave length (i.e. the motion here is about 4 per cent of that at the surface).

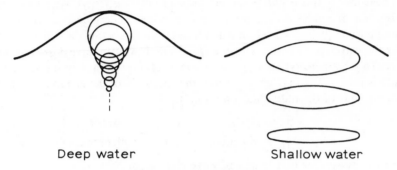

Deep water   Shallow water

Figure 5.7: Particle orbits.

The *particle velocities* and *accelerations* are needed in calculating wave forces on structures by Morison's equation (equations 2.22 and 5.81). They are obtained by differentiating the particle displacements with respect to time. The horizontal velocity is

$$\frac{\partial X}{\partial t} = u = \frac{\pi H}{T} \frac{\cosh \left[2\pi(\bar{y} + h)/L\right]}{\sinh \left(2\pi h/L\right)} \cos 2\pi \left(\frac{\bar{x}}{L} - \frac{t}{T}\right). \tag{5.45}$$

The vertical velocity is

$$\frac{\partial Y}{\partial t} = v = \frac{\pi H}{T} \frac{\sinh \left[2\pi(\bar{y} + h)/L\right]}{\sinh \left(2\pi h/L\right)} \sin 2\pi \left(\frac{\bar{x}}{L} - \frac{t}{T}\right). \tag{5.46}$$

Differentiating again, the horizontal acceleration is

$$\frac{\partial u}{\partial t} = \frac{2\pi^2 H}{T^2} \frac{\cosh \left[2\pi(\bar{y} + h)/L\right]}{\sinh \left(2\pi h/L\right)} \sin 2\pi \left(\frac{\bar{x}}{L} - \frac{t}{T}\right) \tag{5.47}$$

and the vertical acceleration is

$$\frac{\partial v}{\partial t} = -\frac{2\pi^2 H}{T^2} \frac{\sinh \left[2\pi(\bar{y} + h)/L\right]}{\sinh \left(2\pi h/L\right)} \cos 2\pi \left(\frac{\bar{x}}{L} - \frac{t}{T}\right). \tag{5.48}$$

The *subsurface pressure* is

$$p = -\rho g y + \rho \frac{g H}{2} \frac{\cosh \left[2\pi(y + h)/L\right]}{\cosh \left(2\pi h/L\right)} \cos 2\pi \left(\frac{x}{L} - \frac{t}{T}\right) \tag{5.49}$$

(where ρ = density). The second term on the right of equation (5.49) is called the *dynamic pressure*.

The total *energy* in the wave per unit length of crest is

$$E = \rho\,\frac{gH^2L}{8} \tag{5.50}$$

(in linear theory this is half potential and half kinetic energy). Waves of greater energy can, of course, do greater damage than lower energy waves. Over 99 per cent of the energy in a wave is in a depth less than $L/2$ below the still water level.

As the wave speed varies with the wave length we can have a *group-velocity* effect. This happens when a group of waves of approximately the same wave length is proceeding in deep water and the speed C_G of the group as a whole is less than that of the individual waves in it:

$$C_G = \frac{C}{2}\left[1 + \frac{2kh}{\sinh{(2kh)}}\right] = \frac{C}{2}\left[1 + \frac{4\pi h/L}{\sinh{(4\pi h/L)}}\right]. \tag{5.51}$$

The *mean power* per unit length of wave crest is

$$P = \frac{C_G E}{CT}. \tag{5.52}$$

For shallow water, where $h < 0.04L$, a useful approximation is [see IV, §2.13] $\tanh{(2\pi h/L)} \approx \sinh{(2\pi h/L)} \approx 2\pi h/L$ (error < 3 per cent) and the formula for wave speed is then

$$C_S = (gh)^{1/2}. \tag{5.53}$$

The particle orbits are now very flat and, from equation (5.51), $C_G = C$. For deep water, where $h > 0.5L$, a useful approximation is $\tanh{(2\pi h/L)} \approx 1$ (error < 4 per cent) and the wave speed is given by

$$C_D = \frac{gT}{2\pi}. \tag{5.54}$$

From equation (5.41), then,

$$L = gT^2/2\pi. \tag{5.55}$$

The particle orbits are now circular with radius

$$r = \frac{H}{2}\,e^{2\pi \bar{y}/L}. \tag{5.56}$$

The ratio of wave speed in water of general depth to that in deep water is

$$\frac{C}{C_D} = \tanh{\frac{2\pi h}{L}}. \tag{5.57}$$

The ratio of wave length in water of general depth to that in deep water is

$$\frac{L}{L_D} = \tanh\frac{2\pi h}{L}.$$ (5.58)

Standing waves occur when the water surface moves up and down periodically but the wave stays in the same position. When this effect is due to the reflection of a progressive wave train from a vertical wall it is often called *clapotis*; the wave height is then double that of the original wave train. The surface elevation is

$$\eta = \frac{H}{2}\sin\left(\frac{2\pi x}{L}\right)\sin\left(\frac{2\pi t}{T}\right),$$ (5.59)

where

$$\frac{2\pi L}{T^2} = g\tanh\left(\frac{2\pi h}{L}\right).$$ (5.60)

The *particle motions* are given by:

Horizontally: $X = \dfrac{H}{2}\dfrac{\cosh\left[2\pi(\bar{y}+h)/L\right]}{\sinh\left(2\pi h/L\right)}\cos\left(\dfrac{2\pi\bar{x}}{L}\right)\sin\left(\dfrac{2\pi t}{T}\right),$ (5.61)

Vertically: $Y = \dfrac{H}{2}\dfrac{\sinh\left[2\pi(\bar{y}+h)/L\right.}{\sinh\left(2\pi h/L\right)}\sin\left(\dfrac{2\pi\bar{x}}{L}\right)\sin\left(\dfrac{2\pi t}{T}\right).$ (5.62)

The particles move in straight lines up and down vertically beneath the troughs and crests and they move horizontally beneath the nodes. The velocities and accelerations can be obtained as before by differentiating with respect to the time t. For deep water ($h/L > 0.5$), $\tanh(2\pi h/L) \approx 1$ (error < 4 per cent) and we have for the particle motions:

Horizontally: $X = \dfrac{H}{2}e^{2\pi\bar{y}/L}\cos\left(\dfrac{2\pi\bar{x}}{L}\right)\sin\left(\dfrac{2\pi t}{T}\right),$ (5.63)

Vertically: $Y = \dfrac{H}{2}e^{2\pi\bar{y}/L}\sin\left(\dfrac{2\pi\bar{x}}{L}\right)\sin\left(\dfrac{2\pi t}{T}\right).$ (5.64)

For water in a *canal* of length l and vertical sides the wave lengths of the standing waves in the natural modes of oscillation are

$$L = \frac{2l}{m}, \quad m = 1, 2, 3, \ldots.$$ (5.65)

The *subsurface pressure* at a point (x, y) is

$$p = -\rho g y + \rho\frac{gH}{2}\frac{\cosh\left[2\pi(y+h)/L\right]}{\cosh\left(2\pi h/L\right)}\sin\left(\frac{2\pi x}{L}\right)\sin\left(\frac{2\pi t}{T}\right).$$ (5.66)

The *energy* of a standing wave per unit length of crest is

$$E = \tfrac{1}{16}g\rho H^2 L. \tag{5.67}$$

(This is one half of the corresponding quantity for a progressive wave of height H; see equation 5.50.)

For the foregoing linear theory to be valid the wave steepness and also $L^2 H/h^3$ must be 'small'. Strictly speaking, these assumptions should only be made for long period waves such as swell, tsunamis, seiches, and tides. However, to a first approximation they can give useful results for shorter, steeper waves. More accurate results may be obtained from the higher order *Stokes waves* (Wiegel, 1964). Other types of wave theory are those of the Boussinesq *solitary* wave, *cnoidal* waves, and Gerstner's *trochoidal* waves. Figure 5.8 shows their ranges of application. See Muir Wood (1969) and Wiegel (1964) for details of these.

The theoretical limiting steepness of a wave is when $H/L \approx 1/7$ or when the crest angle equals 120°. Miche's equation is

$$\left(\frac{H}{L}\right)_{\text{MAX}} = 0.142 \tanh\left(\frac{2\pi h}{L}\right). \tag{5.68}$$

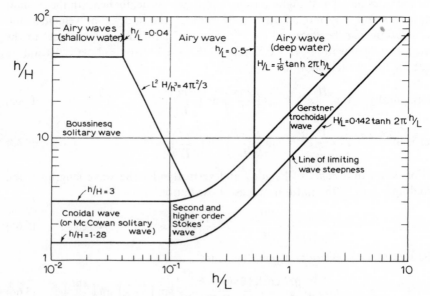

Figure 5.8: Wave types (by permission of A. M. Muir Wood).

5.4 Reflection, refraction and diffraction

With the linear theory of progressive waves the wave speed C given by equation (5.41) depends on the ratio h/L of water depth to wave length. When $h/L < 0.5$ approximately the effect of *refraction* becomes noticeable, i.e. the part of the wave entering shallower water slows down and the wave crest line is bent (Figure 5.9). Snell's law of refraction gives

$$\frac{C_1}{\sin \alpha_1} = \frac{C_2}{\sin \alpha_2}. \tag{5.69}$$

Where the waves approach an abrupt change to deeper water there can be total *reflection* for

$$\frac{C_2}{C_1} \sin \alpha_1 > 1. \tag{5.70}$$

Refraction also occurs when a wave travels obliquely into a current (see Muir Wood, 1969, or Wiegel, 1964, for a further discussion).

The above theory is of course neglecting the density of the air above the water. When we have two otherwise unlimited superposed fluids of densities ρ_1 and ρ_2, waves of length L at the boundary have a wave speed C given by

$$C^2 = \frac{gL}{2\pi} \frac{\rho_1 - \rho_2}{\rho_1 + \rho_2}. \tag{5.71}$$

If the upper and lower fluids are of depths h_2, h_1 respectively, and h_2/L is large

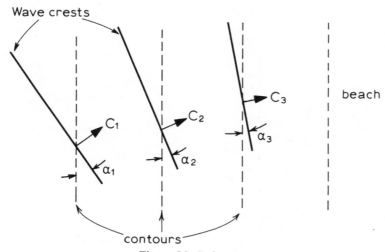

Figure 5.9: Refraction.

and h_1/L small, then approximately

$$C^2 = gh_1 \left(1 - \frac{\rho_2}{\rho_1} \right) \tag{5.72}$$

(see Lamb, 1932).

Reflection

Reflection is most evident from a vertical (or nearly vertical) wall. Water wave reflection in general follows the same rule as for sound and light waves, i.e.

$$\text{Angle of incidence } \alpha_i = \text{angle of reflection } \alpha_r \tag{5.73}$$

(see Figure 5.10). However, if $\alpha_1 > 45°$ a wave called the *Mach stem* (Wiegel, 1964) appears travelling along the face of the wall and of height greater than the incident wave height. For $\alpha_i > 70°$ there is only the Mach stem and no reflected wave. For $\alpha_i = 0$ the reflection produces the *standing wave* (equation 5.59).

When $0 < \alpha < 45°$ the incident and reflected waves produce a pattern of short-crested waves. With the linear theory the surface elevation is then

$$\eta = \frac{H}{2} \sin\left[\frac{2\pi}{L}(x - Ct) \right] \cos\left(\frac{2\pi z}{L'} \right), \tag{5.74}$$

where L = wave length as before,
             L' = crest length,

and the z axis is horizontal.

The *wave speed* C is given by

$$C^2 = \frac{gL}{2\pi} \sqrt{1 + \left(\frac{L}{L'}\right)^2} \tanh\left[\frac{2\pi h}{L} \sqrt{1 + \left(\frac{L}{L'}\right)^2} \right] \geqslant \frac{gL}{2\pi} \tanh\left(\frac{2\pi h}{L} \right) \tag{5.75}$$

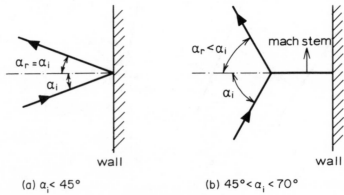

(a) $\alpha_i < 45°$ (b) $45° < \alpha_i < 70°$

Figure 5.10: Reflection.

Thus, comparing with equation (5.41), the short-crested wave travels faster than the long-crested wave of the same length L in the same depth of water h.

For reflection from a sloping wall or steep impermeable beach the ratio of height of reflected wave H_r to height of incident wave H_i,

$$\frac{H_r}{H_i} = K_r, \tag{5.76}$$

is called *the reflection coefficient*. If β is the angle between the wall and the horizontal, then $K_r \approx 0$ for $\beta < 15°$ because of energy loss by bottom friction.

$$K_r \text{ varies from 0.8 to 1.0 for } \beta > 20°. \tag{5.77}$$

Diffraction

When waves strike a barrier with a small aperture those waves which pass through the aperture do not continue in a narrow beam but spread laterally so that all the water beyond the barrier is affected. This effect is called diffraction. In general when surface waves are incident upon any body in the water both diffraction and reflection occur. To find the resulting wave pattern, the wave equation must be solved in an exterior domain in such a way that the *Sommerfeld radiation condition*

$$\left. \text{Lim} \right|_{r \to \infty} r^{(n-1)/2} \left(\frac{\partial \phi}{\partial r} - ik\phi \right) = 0 \tag{5.78}$$

(n dimensions) must be satisfied. This can be done by a number of *numerical methods* [see III, §9.2] such as the boundary integral (surface element) method and pseudo differential operators. Some analytical solutions are also available as follows.

(a) *Waves incident upon a semi-infinite breakwater.* For the geometry shown in Figure 5.11 the wave elevations can be obtained using Fresnel integrals as follows. For an incident wave of unit amplitude, the amplitudes in the subregions are:

Region 1:

$$\eta = f(r_1) \exp\left[-ikr \cos(\theta - \theta_0)\right] + f(r_2) \exp\left[-ikr \cos(\theta + \theta_0)\right],$$

Region 2:

$$\eta = \exp\left[-ikr \cos(\theta - \theta_0)\right]$$
$$- g(r_1) \exp\left[-ikr \cos(\theta - \theta_0)\right] + f(r_2) \exp\left[-ikr \cos(\theta + \theta_0)\right],$$

Region 3:

$$\eta = \exp\left[-ikr \cos(\theta - \theta_0)\right] + \exp\left[-ikr \cos(\theta + \theta_0)\right]$$
$$- g(r_1) \exp\left(-ikr \cos(\theta - \theta_0)\right] + g(r_2) \exp\left[-ikr \cos(\theta + \theta_0)\right],$$

where $f(r) = \dfrac{1+i}{2}\left[\dfrac{1-i}{2} + C(r) - S(r)\right],$

$g(r) = \tfrac{1}{2}\{1 - C(r) - S(r) + i[S(r) - C(r)]\},$

$r_1 = 2\sqrt{\dfrac{kr}{\pi}}\,\sin\,[\tfrac{1}{2}(\theta_0 - \theta)],$

$r_2 = 2\sqrt{\dfrac{kr}{\pi}}\,\sin\,[\tfrac{1}{2}(\theta_0 + \theta)].$

$C(r)$ and $S(r)$ are the Fresnel cosine and sine integrals:

$$C(r) = \int_r^\infty \cos^2 u\,du \qquad \text{and} \qquad S(r) = \int_r^\infty \sin^2 u\,du.$$

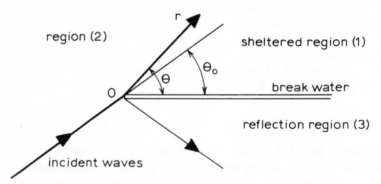

Figure 5.11: Waves incident upon semi-infinite breakwater.

(b) *Waves incident upon a circular cylinder, of radius a, axis vertical.* The surface elevation, η, is given, for incident waves of unit amplitude, by (see MacCamey and Fuchs, 1952):

$$\eta = \sum_{n=0}^{\infty} \varepsilon_n i^n \cos n\theta \left[J_n(kr) - \dfrac{J'(ka)}{H'(ka)} H_n(kr) \right], \qquad (5.79)$$

where J_n and H_n are *Bessel* [see IV, §10.4] and *Hankel* [see IV, (10.4.12)] functions of order n and ε_n is the *Jacobi symbol* (i.e. $\varepsilon_0 = 1$, $\varepsilon_n = 2$ for $n \geqslant 1$).
   The *diffraction coefficient* is

$$K = \frac{\text{wave height in diffraction area}}{\text{wave height outside diffraction area}} \qquad (5.80)$$

Wiegel (1964) gives a useful diagram showing diffraction coefficients as a function of incident wave angle for a semi-infinite rigid breakwater.

Wave forces on structures generally can be described as due to *diffraction, drag,* and *inertia.* The diffraction effects are mainly significant when the submerged part of the structure has dimensions which are comparable with the wave length. Morison's equation is widely used by engineers to compute the forces due to drag and inertia (Morison *et al.,* 1950). It should only be applied when the wave length *L* is about five times greater than the characteristic dimension *D* of the structure because it assumes that the wave motion is not affected by the structure. For larger structures a more complicated theory is necessary because diffraction should be included as well.

Morison's equation gives for the force on a length *ds* of the structure

$$dF = |\tfrac{1}{2}C_D\rho DU|U| + C_m\rho A\dot{U}| \, ds \tag{5.81}$$

and

$$\text{Total force} = \int_0^\eta dF, \tag{5.82}$$

where η = water level,
          ρ = density,
          U = velocity of water in direction of *F*,
          $\dot{U}$ = acceleration of water in direction of *F*,
          D = width of section perpendicular to *F*,
          A = cross-sectional area,
          C_D = drag coefficient,
          C_M = inertia coefficient

and

          $D/L < 0.2$.

The particle velocities and accelerations are to be found from the appropriate wave theory (e.g. equations 5.45 and 5.48). Wiegel discusses the assumptions on which Morison's equation depends and gives values of C_M and C_D for various shapes. Hogben (1976) and Hogben and Standing (1974) give further discussion on this topic.

                                                                        W.L.W.

References

Francis, J. R. D. (1958). *A Textbook of Fluid Mechanics for Engineering Students,* Edward Arnold.

Henderson, F. M. (1966). *Open Channel Flow,* Macmillan.

Hogben, N. (1976). Wave loads on structures, Paper presented at the Behaviour of Offshore Structures conference at the Norwegian Institute of Technology.

Hogben, N., and Standing, R. G. (1974). Wave loads on large bodies, Paper No. 26, Int. Symp. on Dynamics of Marine Vehicles and Structures in Waves, Inst. Mech. Engrs.

Ippen, A. T. (Ed.) (1966). *Estuary and Coastline Hydrodynamics,* McGraw-Hill.

Lamb, H. (1932). *Hydrodynamics,* Cambridge University Press.

MacCamey, R. C., and Fuchs, R. A. (1952). *Wave Forces on Piles: A Diffraction Theory*, Vol. 3, p. 334, Inst. Eng. Research; Waves Investigation Lab., Berkeley, California.

Morison, J. R., O'Brien, M. P., Johnson, J. W., and Schaaf, S. A. (1950). The forces exerted by surface waves on piles, *Petroleum Trans. Am. J. Mech. Engrs*, **189**.

Muir Wood, A. M. (1969). *Coastal Hydraulics*, Macmillan.

Sellin, R. H. J. (1969). *Flow in Channels*, Macmillan.

Wiegel, R. L. (1964). *Oceanographical Engineering*, Prentice-Hall.

Mathematical Methods in Engineering
Edited by G. A. O. Davies
© 1984, John Wiley & Sons, Ltd.

6

Hydrology

6.1 Introduction

The current chapter is limited to the development of basic classical theory depicting surface run-off, stream flow and atmospheric circulation. The other major feature, subsurface flow, is covered in Chapter 7. The utilization of numerical methods for predicting surface flow and atmospheric circulation has not been as widespread as that in groundwater hydrology. This is largely due to the complex three-dimensional nature of the coupled equations depicting such flow. In surface run-off this is characterized by the pseudo three-dimensional non-linear nature of the governing equations and the complexity of the ground topology and roughness. The roughness is usually simplified via a Chezy (5.28) or Manning (5.29) type of equation which proves to be permissible owing to the pseudo three-dimensional nature of the equations. This concept is amplified in Section 6.2. However, the application of numerical methods, particularly relating to streamflow, has been found to be quite useful and the relative importance of some parameters has been investigated.

Similar inhibitive features are associated with the equations which can be used to predict atmospheric circulation. Usually, however, the flow to be modelled is fully three dimensional and also incorporates the effect of temperature variations. A simplifying assumption which has been utilized in most texts to date is that the flow is laminar as opposed to the usually turbulent nature of surface run-off. This makes the equations slightly less non-linear and avoids the introduction of equations depicting turbulence, kinetic energy, and dissipation.

6.2 Surface run-off and stream flow

If the rate of precipitation or snowmelt exceeds the combined effect of evaporation and infiltration, water begins to accumulate on the surface of the land mass. This *static* local storage appears in the form of pools or is retained locally, against the effect of gravity, by surface tension. When the gravitational effect becomes dominant the water moves, contributing to surface run-off which appears as a thin sheet flow over the land surface. Topographic features tend to concentrate this sheet flow into discrete primarily one-dimensional flow in the form of streams. Many such discrete concentrations contribute to the network which usually results in a larger stream or river which eventually discharges into a lake, estuary, or ocean. In each type of flow situation, during the passage of water over the land surface, the equations depicting the flow can be considered to be identical in form.

Classical mathematical concepts

In the case of three-dimensional motion the velocity field can be specified by a velocity vector,

$$\mathbf{U} = \mathbf{i}u + \mathbf{j}v + \mathbf{k}w, \tag{6.1}$$

where u, v and w are the three orthogonal components of velocity in the x, y and z cartesian coordinate directions respectively [see IV, §§17.2.2 and 17.3.1]. In such a system the main variables, for a compressible fluid, would be the three velocity components, the local pressure p and density ρ. The five equations required for solution are the equation of continuity (conservation of mass), the three equations of motion (conservation of momentum) and the equation of state, $p = f(\rho)$.

Continuity

For an elemental volume the mass balance for compressible non-steady flow results in an equation of the form (see Schlichting, 1960, and also Section 2.1)

$$\frac{\partial \rho}{\partial t} + \operatorname{div}(\rho \mathbf{U}) = 0, \tag{6.2}$$

where the divergence $\operatorname{div} \mathbf{U}$ is the same as $\nabla \cdot \mathbf{U}$. If the fluid is considered incompressible this reduces to the simplified form

$$\operatorname{div} \mathbf{U} = 0, \tag{6.3}$$

which is the form used in the current text.

Momentum

Utilizing Newton's second law of motion [see IV, §17.1], the following version of the equation in Section 2.1 can be derived:

$$\rho \frac{D\mathbf{U}}{Dt} = \mathbf{F} + \mathbf{P}, \tag{6.4}$$

in which **F** denotes a body force, usually a gravitational force per unit volume ρg where g is the acceleration due to gravity, and **P** denotes a boundary force per unit volume. The convective acceleration is

$$\frac{D\mathbf{U}}{Dt} = \frac{\partial \mathbf{U}}{\partial t} + (\mathbf{U} \cdot \nabla)\mathbf{U}. \tag{6.5}$$

The body forces can be regarded as known external forces whereas the surface forces are dependent on the rate of strain, the velocity field, of the fluid. The system of surface forces determines the local state of stress which, via Stokes' law of friction, can be empirically related to the rate of strain term.

We now expand the Newtonian law of (2.1) in the following cartesian form:

$$\begin{bmatrix} \sigma_{xx} & \tau_{xy} & \tau_{xz} \\ \tau_{xy} & \sigma_{yy} & \tau_{yz} \\ \tau_{xz} & \tau_{yz} & \sigma_{zz} \end{bmatrix} = \begin{bmatrix} -P & 0 & 0 \\ 0 & -P & 0 \\ 0 & 0 & -P \end{bmatrix} + \mu \begin{bmatrix} \dfrac{\partial u}{\partial x} & \dfrac{\partial u}{\partial y} & \dfrac{\partial u}{\partial z} \\[2mm] \dfrac{\partial v}{\partial x} & \dfrac{\partial v}{\partial y} & \dfrac{\partial v}{\partial z} \\[2mm] \dfrac{\partial w}{\partial x} & \dfrac{\partial w}{\partial y} & \dfrac{\partial w}{\partial z} \end{bmatrix}$$

$$+ \mu \begin{bmatrix} \dfrac{\partial u}{\partial x} & \dfrac{\partial v}{\partial x} & \dfrac{\partial w}{\partial x} \\[2mm] \dfrac{\partial u}{\partial y} & \dfrac{\partial v}{\partial y} & \dfrac{\partial w}{\partial y} \\[2mm] \dfrac{\partial u}{\partial z} & \dfrac{\partial v}{\partial z} & \dfrac{\partial w}{\partial z} \end{bmatrix} - \tfrac{2}{3}\mu \begin{bmatrix} \text{div } \mathbf{U} & 0 & 0 \\ 0 & \text{div } \mathbf{U} & 0 \\ 0 & 0 & \text{div } \mathbf{U} \end{bmatrix}, \tag{6.6}$$

in which σ denotes a normal stress and τ a shear (Figure 6.1). Subtracting the pressure from the normal stress we can define a new variable,

$$\sigma_{xx} = -p + \sigma'_{xx}; \qquad \sigma_{yy} = -p + \sigma'_{yy}; \qquad \sigma_{zz} = -p + \sigma'_{zz} \tag{6.7}$$

so that the frictional terms of the stress components become

$$\sigma'_{xx} = \mu\left(2\frac{\partial u}{\partial x} - \frac{2}{3}\text{div } \mathbf{U}\right); \qquad \tau_{xy} = \mu\left(\frac{\partial u}{\partial y} + \frac{\partial v}{\partial x}\right),$$

$$\sigma'_{yy} = \mu\left(2\frac{\partial v}{\partial y} - \frac{2}{3}\text{div } \mathbf{U}\right); \qquad \tau_{yz} = \mu\left(\frac{\partial v}{\partial z} + \frac{\partial w}{\partial y}\right), \tag{6.8}$$

$$\sigma'_{zz} = \mu\left(2\frac{\partial w}{\partial z} - \frac{2}{3}\text{div } \mathbf{U}\right); \qquad \tau_{zx} = \mu\left(\frac{\partial w}{\partial x} + \frac{\partial u}{\partial z}\right),$$

in which μ denotes the molecular viscosity of the fluid.

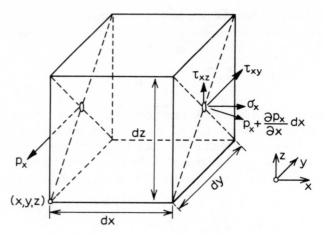

Figure 6.1: Stress tensor on an element of fluid.

The surface forces per unit volume can be written

$$\mathbf{P}_{xx} = \mathbf{i}\left(\frac{\partial \sigma_{xx}}{\partial x} + \frac{\partial \tau_{xy}}{\partial y} + \frac{\partial \tau_{xz}}{\partial z}\right)$$

$$\mathbf{P}_{yy} = \mathbf{j}\left(\frac{\partial \tau_{xy}}{\partial x} + \frac{\partial \sigma_{yy}}{\partial y} + \frac{\partial \tau_{yz}}{\partial z}\right) \qquad (6.9)$$

$$\mathbf{P}_{zz} = \mathbf{k}\left(\frac{\partial \tau_{xz}}{\partial x} + \frac{\partial \tau_{yz}}{\partial y} + \frac{\partial \sigma_{zz}}{\partial z}\right)$$

(face yz) (face zx) (face xy)

such that (6.4) becomes

$$\rho\frac{Du}{Dt} = F_{xx} + \left(\frac{\partial \sigma_{xx}}{\partial x} + \frac{\partial \tau_{xy}}{\partial y} + \frac{\partial \tau_{xz}}{\partial z}\right)$$

$$\rho\frac{Dv}{Dt} = F_{yy} + \left(\frac{\partial \tau_{xy}}{\partial x} + \frac{\partial \sigma_{yy}}{\partial y} + \frac{\partial \tau_{yz}}{\partial z}\right) \qquad (6.10)$$

$$\rho\frac{Dw}{Dt} = F_{zz} + \left(\frac{\partial \tau_{xz}}{\partial x} + \frac{\partial \tau_{yz}}{\partial y} + \frac{\partial \sigma_{zz}}{\partial z}\right)$$

Utilizing equations (6.6) to (6.9), equation (6.10) produces the three scalar components of the previous vector equation in (2.1), thus:

$$\rho\frac{Du}{Dt} = F_{xx} - \frac{\partial p}{\partial x} + \frac{\partial}{\partial x}\left[\mu\left(2\frac{\partial u}{\partial x} - \frac{2}{3}\text{div } \mathbf{U}\right)\right] + \frac{\partial}{\partial y}\left[\mu\left(\frac{\partial u}{\partial y} + \frac{\partial v}{\partial x}\right)\right]$$

$$+ \frac{\partial}{\partial z}\left[\mu\left(\frac{\partial w}{\partial x} + \frac{\partial u}{\partial z}\right)\right],$$

$$\rho\frac{Dv}{Dt} = F_{yy} - \frac{\partial p}{\partial y} + \frac{\partial}{\partial y}\left[\mu\left(2\frac{\partial v}{\partial y} - \frac{2}{3}\text{div } \mathbf{U}\right)\right] + \frac{\partial}{\partial z}\left[\mu\left(\frac{\partial v}{\partial z} + \frac{\partial w}{\partial y}\right)\right]$$

$$+ \frac{\partial}{\partial x}\left[\mu\left(\frac{\partial u}{\partial y} + \frac{\partial v}{\partial x}\right)\right], \tag{6.11}$$

and

$$\rho\frac{Dw}{Dt} = F_{zz} - \frac{\partial p}{\partial z} + \frac{\partial}{\partial z}\left[\mu\left(2\frac{\partial w}{\partial z} - \frac{2}{3}\text{div } \mathbf{U}\right)\right] + \frac{\partial}{\partial x}\left[\mu\left(\frac{\partial w}{\partial x} + \frac{\partial u}{\partial z}\right)\right]$$

$$+ \frac{\partial u}{\partial y}\left[\mu\left(\frac{\partial v}{\partial z} + \frac{\partial w}{\partial y}\right)\right].$$

For an incompressible fluid, div $\mathbf{U} = 0$, the complete set of equations are

$$\frac{D\mathbf{U}}{Dt} = \mathbf{F} - \text{grad } p + \mu\nabla^2\mathbf{U} \tag{6.12}$$

and the equation of continuity is

$$\text{div } \mathbf{U} = 0; \tag{6.13}$$

these form the basis of the equation utilized in the current text for depicting overland and stream flow.

A generalized mathematical model, based on hydrodynamic principles derived previously, can be readily obtained to describe surface run-off from a watershed. The required flow equations are based on the Navier–Stokes equations and continuity equations. These are used in a vertically averaged form and the resulting set of non-linear time-dependent partial differential equations will be presented in a form suitable for numerical analysis.

In the analysis presented in the current chapter, a three-dimensional cartesian coordinate system is used and the vertically averaged velocity is taken parallel to the ground at discrete points in space (Figure 6.2).

Governing hydrodynamic equations

The equations depicting conservation of mass and momentum are derived for incompressible Newtonian fluids and an idealized physical model is shown in Figure 6.2. Referring to an elemental control volume in the chosen fixed

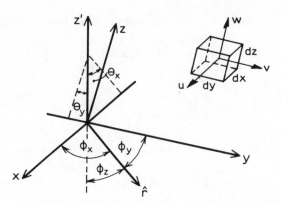

Figure 6.2: Definition sketch of coordinate system.

Cartesian coordinate system, the equation of conservation of mass of an incompressible fluid is, from equation (6.13),

$$\frac{\partial u}{\partial x} + \frac{\partial v}{\partial y} + \frac{\partial w}{\partial z} = 0. \tag{6.14}$$

The z axis is taken to be normal to the local watershed surface which defines the xy plane. For integration normal to the watershed surface the boundary conditions at the fluid free surface and ground must be defined.

Boundary conditions

(i) The kinematic boundary condition at the free surface, $z = d(x, y)$, is

$$\left.\left(\frac{\partial d}{\partial t} + u\frac{\partial d}{\partial x} + v\frac{\partial d}{\partial y} - w\right)\right|_{z=d} = r\cos\theta_z, \tag{6.15}$$

in which r is the rainfall intensity, θ_z the angle that the gravitational direction makes with the z axis, and d the local depth of flow.

(ii) The kinematic condition at the ground surface, $z = 0$, is

$$w|_{z=0} = i\cos\theta_z, \tag{6.16}$$

in which i is the rate of infiltration into the ground. Note that both the infiltration and rainfall rate are measured vertically and a positive infiltration depicts flow into the underlying aquifer.

Applying both boundary conditions to (6.14) results in, upon integration,

$$\frac{\partial h}{\partial t} + \frac{\partial}{\partial x}(\bar{u}h) + \frac{\partial}{\partial y}(\bar{v}h) = (r - i), \tag{6.17}$$

in which h is the local depth of flow, measured normal to the xy plane, and $\bar{u}$ and $\bar{v}$ are the vertically averaged velocities in the x and y directions respectively.

The expanded form of the Navier–Stokes equations (2.2) for an incompressible Newtonian fluid in a gravitational field are

$$\frac{\partial u}{\partial t} + u\frac{\partial u}{\partial x} + v\frac{\partial u}{\partial y} + w\frac{\partial u}{\partial z} = g\sin\theta_x - \frac{1}{\rho}\frac{\partial p}{\partial x} \rightarrow \nu\nabla^2 u, \qquad (6.18)$$

$$(F_{xx})$$

$$\frac{\partial v}{\partial t} + u\frac{\partial v}{\partial x} + v\frac{\partial v}{\partial y} + w\frac{\partial v}{\partial z} = g\sin\theta_y - \frac{1}{\rho}\frac{\partial p}{\partial y} + \nu\nabla^2 v \qquad (6.19)$$

$$(F_{yy})$$

and

$$\frac{\partial w}{\partial t} + u\frac{\partial w}{\partial x} + v\frac{\partial w}{\partial y} + w\frac{\partial w}{\partial z} = g\sin\theta_z - \frac{1}{\rho}\frac{\partial p}{\partial z} + \nu\nabla^2 w, \qquad (6.20)$$

$$(F_{zz})$$

in which θ_x, θ_y and θ_z are the angles which the gravitational direction makes with x, y and z axes respectively.

In order to facilitate the vertical integration of equations (6.18) to (6.20), the following boundary conditions and assumptions are employed.

(a) *Assumptions*
  (i) The boundary shear at the ground is the only force resisting flow and some terms, such as the expressions representing internal stress, in the Navier–Stokes equations can be ignored.
 (ii) For shallow water flow w can be considered to be very small in comparison with u and v. For simplicity, therefore, this component can be assumed to be zero everywhere within the flow domain except at the free surface. Integrating (6.20) and incorporating this assumption leads to

$$p(x, y, 0; t) = \rho g d \cos\theta_z + B_r\rho r\Lambda\cos(\theta_z + \phi_z) \qquad (6.21)$$

in which B_r is the momentum correction factor for the distribution of the terminal velocity of descent of the droplets, Λ is the mean terminal velocity of the rain droplets and ϕ_z is the angle between the velocity vector representing the terminal velocity and the vertical direction (Figure 6.2). Note that θ_z is taken to be positive when in the first quadrant and negative when in the second.
(iii) If the total local pressure is assumed to be composed of the hydrostatic pressure, $\rho g(d - z)\cos\theta_z$ and a dynamic pressure $\rho g h^*(x, y, z; t)$, arising from the raindrop impact with the water surface, the following term for pressure can be assumed:

$$p(x, y, z; t) = \rho g[(d - z)\cos\theta_z + h^*], \qquad (6.22)$$

in which h^* is used to denote the excess pressure caused by raindrop impact (Eagleson, 1970).

(b) *Boundary conditions*

(i) The kinematic condition at the free surface

$$z = d(x, y; t)$$

is that depicted in equation (6.15). Ignoring the surface tension effects, to a first approximation, the dynamic condition at the free surface is

$$p(x, y, d; t) = 0. \tag{6.23}$$

The momentum flux per unit mass influx over a unit surface area can be approximated by

$$w(x, y, d; t)r \cos \theta_z = -B_r r \Lambda \cos \theta_z \cos (\theta_z + \phi_z). \tag{6.24}$$

In order to satisfy the dynamic condition, equation (6.23), at the free surface, $z = d$, equation (6.22) results in

$$h^*(x, y, d; t) = 0. \tag{6.25}$$

(ii) At the ground surface, the momentum flux due to infiltration

$$w(x, y, 0; t)i \cos \theta_z \tag{6.26}$$

can usually be ignored since the average velocity of infiltration is so small that such values are negligible.

At the ground surface, $z = 0$, equation (6.8) and (6.9) results in

$$h^*(x, y, 0; t) = \frac{B_r}{g} r \Lambda \cos \theta_z \cos (\theta_z + \phi_z). \tag{6.27}$$

It is readily seen that the dynamic pressure varies from zero at the free surface to that represented by equation (6.27) at the ground surface. The variation over the flow depth is, however, unknown and this is usually taken to be linear with depth. However, in order to conform with the basic concept that the total head, T_h, remains constant over a vertical cross-section, i.e.

$$T_h = z \cos \theta_z + \frac{p}{\rho g} + \alpha \frac{v}{2g} = \text{constant}, \tag{6.28}$$

where α is the energy correction factor, it is necessary to assume that h^* is uniformly distributed over the cross-section except at the free surface. The overpressure distribution is then represented by

$$h^*(x, y, z; t) = 0 \qquad \text{for } z = d \tag{6.29}$$

and

$$h^*(x, y, z; t) = \left(\frac{B_r}{g}\right) r\Lambda \cos \theta_z \cos (\theta_z + \phi_z)$$

$$\text{for } 0 \leqslant z \leqslant d. \tag{6.30}$$

Based on the above approximations and assumptions and using the *Leibnitz rule of differentiation* under the integral sign [see IV, Theorem 4.7.3], the Navier–Stokes equations can be integrated to yield a vertically averaged form:

$$\frac{\partial}{\partial t} (h\bar{u}) + \frac{\partial}{\partial x} (B_x h\bar{u}^2) + \frac{\partial}{\partial y} (B_{xy} h\bar{u}\bar{v}) - B_r r\Lambda \cos \phi_x$$

$$= gh \sin \theta_x - g(h \cos^2 \theta_z + h^*) \frac{\partial h}{\partial x} - \frac{\tau_{ox}}{\rho \cos \theta_z} \tag{6.31}$$

and

$$\frac{\partial}{\partial t} (h\bar{v}) + \frac{\partial}{\partial x} (B_{xy} h\bar{u}\bar{v}) + \frac{\partial}{\partial y} (B_y h\bar{v}^2) - B_r r\Lambda \cos \phi_y$$

$$= gh \sin \theta_y - g(h \cos^2 \theta_z + h^*) \frac{\partial h}{\partial y} - \frac{\tau_{oy}}{\rho \cos \theta_z}, \tag{6.32}$$

in which τ_{ox} and τ_{oy} are boundary shears in the negative direction of x and y respectively; B_x, B_y and B_{xy} are the momentum correction factors which are defined by

$$B_x \bar{u}^2 = \frac{1}{d} \int_0^d u^2 \, dz, \tag{6.33}$$

$$B_{xy} \bar{u}\bar{v} = \frac{1}{d} \int_0^d uv \, dz, \tag{6.34}$$

$$B_y \bar{v}^2 = \frac{1}{d} \int_0^d v^2 \, dz, \tag{6.35}$$

$$B_r r\Lambda \cos \phi_x \cos \theta_z = \int_0^d uq_s \, dz, \tag{6.36}$$

$$B_r r\Lambda \cos \phi_y \cos \theta_z = \int_0^d vq_s \, dz, \tag{6.37}$$

in which q_s is a source or sink or combination of both.

6.3 Simplified dynamic model

Some simplifications to the generalized continuity and momentum equations are found to be acceptable and do not have a marked influence, for practical purposes, on the results obtained. The usual assumptions which are invoked are:

(i) Momentum correction factors are unity.
(ii) The pressure distribution is hydrostatic.
(iii) The overpressure due to precipitation impact is zero.
(iv) Bed slopes are small such that $\cos \theta \cong 1$ and $\sin \theta \cong \theta$.
(v) The fluid is Newtonian and incompressible.
(vi) The rainfall is vertical and uniform.

Taking the vertically integrated forms of both the continuity and momentum equations we have

Continuity:
$$\frac{\partial h}{\partial t} + \bar{u}\frac{\partial h}{\partial x} + h\frac{\partial \bar{u}}{\partial x} + \bar{v}\frac{\partial h}{\partial y} + h\frac{\partial \bar{v}}{\partial y} = r - i - e \qquad (6.38)$$

and the corresponding

Momentum:

$$\frac{\partial \bar{u}}{\partial t} + \bar{u}\frac{\partial \bar{u}}{\partial x} + \bar{v}\frac{\partial \bar{u}}{\partial y} + g\frac{\partial h}{\partial x} = g(S_x - S_{fx}) + (r - i - e)\frac{\bar{u}}{h} + \frac{\Lambda r}{h} \qquad (6.39)$$

and

$$\frac{\partial \bar{v}}{\partial t} + \bar{u}\frac{\partial \bar{v}}{\partial y} + \bar{v}\frac{\partial \bar{v}}{\partial y} + g\frac{\partial h}{\partial y} = g(S_y - S_{fy}) + (r - i - e)\frac{\bar{v}}{h} + \frac{\Lambda r}{h}, \qquad (6.40)$$

in which $\tau_{ox}/\rho g h = S_{fx}$; $\tau_{oy}/\rho g h = S_{fy}$.

It is immediately apparent that the friction slopes must be defined in terms of the basic variables $\bar{u}$, $\bar{v}$ and h if a solution to equations (6.39) and (6.40) is to be obtained. It has been suggested (Taylor, Al-Mashidani and Davis, 1974) that the following form of relationship could be used:

$$S_{fx} = \frac{f}{8gh} \bar{u}(\bar{u}^2 + \bar{v}^2)^{1/2}, \qquad (6.41)$$

$$S_{fy} = \frac{f}{8gh} \bar{v}(\bar{u}^2 + \bar{v}^2)^{1/2}, \qquad (6.42)$$

in which f is a resistance coefficient which again requires further definition. For laminar flow the relationship is usually written (Eagleson, 1970) as

$$f = \frac{k}{R_e}, \qquad (6.43)$$

in which k is a dimensionless resistance parameter and R_e the Reynolds number defined by

$$R_e = \frac{[(\bar{u}h)^2 + (\bar{v}h)^2]^{1/2}}{v} \qquad (6.44)$$

where v is the fluid kinematic viscosity. If the Chezy formula is employed, then

$$f = \frac{8g}{C^2} \tag{6.45}$$

and, when the Manning relationship is used,

$$f = \frac{8gn^{1/2}}{C^2}, \tag{6.46}$$

in which C and n are the Chezy and Manning coefficients respectively.

The relationship represented by equation (6.45) has been found to be fairly representative for the resistance of sheet flow up to a Reynolds number of approximately 1,000 (Yoon and Wenzel, 1971) in which k is a function of rainfall intensity and slope. For Reynolds numbers greater than 900 the effect of rainfall intensity can be considered to be a constant contribution and for $R_e > 2,000$ the effect can be ignored (Shen and Li, 1973). It has also been concluded that uncertainties in the selection of the friction factor do not result in large errors in the evaluation of flow depth and velocity. Indeed, the conclusion is that if the friction factor is selected within a certain percentage error, e, the resulting error in the evaluation of the boundary shear and flow depth is approximately $e/3$. Turning to some detailed comments regarding the choice of friction factors for two-dimensional flow (Chow and Ben-Zui, 1973), the following relationships have been employed:

$$S_{fx} = \frac{1}{8gh}(f + f')\bar{u}(\bar{u}^2 + \bar{v}^2)^{1/2} \tag{6.47}$$

and

$$S_{fy} = \frac{1}{8gh}(f + f')\bar{v}(\bar{u}^2 + \bar{v}^2)^{1/2}, \tag{6.48}$$

in which f' is an apparent resistance coefficient due to the overpressure and other impeding effects related to rainfall impact. This results in a modified Darcy–Weisbach friction factor coefficient,

$$f^* = f' + f,$$

leading to a slightly modified form of (6.26) and (6.27), as presented in Harbaugh and Chow (1967) and defined as a *conceptual watershed roughness*.

Without taking non-linearities such as the free-surface effect and transient nature of the flow into account, the Darcy–Weisbach friction coefficient for laminar flow is defined by equation (6.28) with $k = 24$.

For turbulent flow on smooth surfaces,

$$(f)^{-1/2} = 2 \log R_e(f)^{1/2} + 0.404 \tag{6.49}$$

has been advocated, and for such flow over rough surfaces

$$(f)^{-1/2} = 2\log\frac{2R_e}{k} + 1.74, \tag{6.50}$$

in which k is the roughness size of the surface and R_e a Reynolds number now defined by

$$R_e = \frac{(q^2 + p^2)^{1/2}}{v}, \tag{6.51}$$

in which $q = \bar{u}h$ and $p = \bar{v}h$.

To account for the effects of non-linearity, curvature of the flow surface, the transient nature of the flow and other inaccuracies due to the simplifying assumptions, it has been suggested by Chow and Ben-Zui (1973) that k be increased by 5 per cent.

Rewriting (6.39) as

$$f^* = \frac{k}{R_e} \tag{6.52}$$

and utilizing (6.51), equations (6.41) and (6.42) can be re-represented by

$$f_{fx} = \frac{vk}{8gh^2}\bar{u} \tag{6.53}$$

and

$$f_{fy} = \frac{vk}{8gh^2}\bar{v}. \tag{6.54}$$

In view of the complexity of surface flow over watersheds it is doubtful whether any of the formulae defining the friction factor would be strictly valid for the spatially varying unsteady flow. These have been proven to be quite realistic for laboratory models and may be taken as reasonably realistic for practical purposes.

The equations have now been defined which can be used to predict stream flow and overland flow as either a combined or separate system. *Finite difference* (Chow and Ben-Zui, 1973) and *finite element* (Al-Mashidani and Taylor, 1974a, 1974b) techniques have been utilized to predict such flow and a comparison with experimental results has been effected [see III, Chapter 9]. Whichever numerical technique is employed, both are subject to similar input and boundary conditions. If such a flow is considered as part of the hydrologic cycle the boundary conditions would consist of point flows on the boundary to the domain under consideration and/or spatially varying distributed input within the domain. They could consist of either historical or synthetic rainfall, infiltration, stream flow and abstraction or recharge/dispersion associated with water resource systems or disposal. Rainfall can be regarded as a time-dependent spatially distributed inflow and streams as temporal point inflows on the boundary, whilst infiltration is usually a spatially distributed outflow into the groundwater zone. The associated groundwater flow has been treated in Chapter 7 and will not be discussed here.

Irrespective of whether the overland flow is divided into stream and sheet flow the basic equations derived are applicable.

6.4 Simplified theory: the kinematic wave

Kinematic flow over planes and through channels arises whenever a balance between gravitational and frictional forces is present. The existence of such a balance implies that the derivatives of the velocity term in the momentum equations are negligible in comparison to gravitational and frictional effects. Although this may seem a gross assumption, the validity of the kinematic wave approximation has been adequately demonstrated by Woolhiser and Ligget (1967) with experimental data used as a basis for evaluating the necessary parameters (Morgali, 1970, and Schreiber and Bender, 1972). Indeed, converging overland flow has also been analysed, utilizing the simplified theory by Woolhiser *et al.* (1971).

The momentum equations, (6.39) and (6.40), are reduced to

$$S_{ox} = S_{fx} \tag{6.55}$$

and, assuming zero infiltration, the equation of continuity becomes

$$\frac{\partial h}{\partial t} + \frac{\partial q}{\partial x} = r, \tag{6.56}$$

in which, again, $q = \bar{u}h$.

A relationship which can be used to relate the mean velocity and the local water depth h is

$$\bar{u} = \alpha h^m \tag{6.57}$$

or

$$q = \alpha h^{m+1}, \tag{6.58}$$

in which α is a function of the slope and roughness of a plane representing the surface of the catchment and m is a constant (see Henderson, 1966). Equations (6.56) and (6.57) or (6.58) are the required set of equations when the kinematic theory is employed. This quasi-steady-state approach was first developed by Lighthill and Whitham (1955) and although the normal solution to these equations is not trivial, it is not as difficult as a solution to the complete equations. Furthermore, analytic techniques can be used to obtain solutions in many cases.

The kinematic wave theory can be solved, for all Froude numbers, using *the method of characteristics* (Eagleson, 1970) [see III, §9.2] or *finite element methods* (Al-Mashidani, 1974a, 1974b; see also Chapter 15).

An extension to the current theory is that of representing flow over a cascade (Figure 6.3). For present purposes this can be represented by a series of discrete flow planes each with its own slope and roughness characteristics. An upland watershed can then be represented as a series of such cascades, and discharges into a channel. Due to the discontinuities, represented by an abrupt change of surface slope, roughness or width of successive planes, discontinuities in depth occurred resulting in a *kinematic shock*. The physical manifestation of the shock is a rapid rise or near vertical front in the outflow hydrograph. This shock phenomenon was detected when both the method of characteristics and finite element techniques were used. A general *shock* criterion for the Pth plane is (Kibler and Woolhiser, 1972)

$$\frac{W_{p-1}\alpha_{p-1}}{W_p\alpha_p} = P_s > 1, \tag{6.59}$$

in which W_p is the width of the Pth plane and α_p denotes the constant of the Pth plane. Equation (6.59) defines the *shock parameter* P_s and establishes that a shock formation will occur on plane P for $P_s > 1$.

In their numerical analysis, Kibler and Woolhiser (1972) considered the solutions obtained by the method of characteristics as a standard for comparison purposes with other solution techniques.

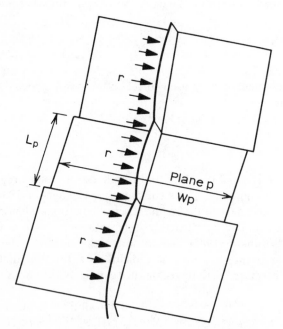

Figure 6.3: Typical cascade.

6.5 Atmospheric circulation

> *The air moves like a river and*
> *carries the clouds with it.*
>   *Leonardo Da Vinci*

The penetration of solar energy through the earth's atmosphere to the surface is effected by an energy redistribution through circulation of the atmosphere and oceans. The basic equations for describing conservation of mass and momentum flow were presented in Section 6.2. An additional energy equation is required, however, before an attempt can be made to construct a computer-based numerical model of circulation. This will be outlined later.

Due to the spatial and temporal magnitudes required for depicting overall gaseous movement in the atmosphere, only simplified models have been attempted. These are simplified both with respect to the complexity of the equations and spatial topography and extent. However, with the advent of increasing observations of atmospheric movement, particularly via satellites, the more pertinent parameters or terms may be isolated and included in the numerical model. Equally important, the less influential parameters or terms can be excluded, making the analyst's task more tractable.

Heat transfer

The energy balance for a fluid in motion involves the heat added via external processes together with that due to friction. In an incompressible fluid the energy balance is determined by considering the internal energy, heat conduction, heat convection and the generation of heat due to friction. For a compressible fluid the additional factor is the work done during expansion or compression. Radiation, for present purposes, is neglected.

Considering an elemental volume of fluid (Figure 6.4), the quantity of heat, δQ, added to the volume externally and through friction is

$$\delta Q = \Delta W C_v \, \delta T + p \, \delta(\Delta V), \tag{6.60}$$

in which the first term on the right-hand side depicts internal energy and the second that due to expansion. In the above equation, C_v denotes the specific heat at constant volume, ΔW the weight of the elemental volume, T the local temperature and p the pressure.

The total change in heat can be written, in general terms, as

$$Q = Q_c + Q_f, \tag{6.61}$$

where Q_c is associated with conduction and Q_f with friction.

Invoking *Fourier's equations*, the heat flux crossing an area A is assumed to be proportional to the gradient of temperature normal to that surface, such that

$$\frac{dQ_c}{dt} = q = -k'\frac{\partial T}{\partial n},\qquad(6.62)$$

in which A is the surface area, q the flux and k' the material conductivity.

Considering the element of fluid shown in Figure 6.4, the quantity of heat introduced into the volume in time t due to conduction is

$$\delta Q_c = \delta t\,\Delta V\left[\frac{\partial}{\partial x}\left(k'\frac{\partial T}{\partial x}\right) + \frac{\partial}{\partial y}\left(k'\frac{\partial T}{\partial y}\right) + \frac{\partial}{\partial z}\left(k'\frac{\partial T}{\partial z}\right)\right]\qquad(6.63)$$

or, in the limit $\delta t \to 0$,

$$\frac{dQ_c}{dt} = \Delta V\left[\frac{\partial}{\partial x}\left(k'\frac{\partial T}{\partial x}\right) + \frac{\partial}{\partial y}\left(k'\frac{\partial T}{\partial y}\right) + \frac{\partial}{\partial z}\left(k'\frac{\partial T}{\partial z}\right)\right].\qquad(6.64)$$

Considering the heat generation due to friction which can be obtained by the part of the frictional forces converted into mechanical energy and partly dissipated as heat, the total work per unit time due to the normal and shearing stresses and due to friction can be written as

$$\Delta V\left[\frac{\partial}{\partial x}\left(\sigma_x u + \tau_{xy} v + \tau_{xz} w\right) + \frac{\partial}{\partial y}\left(\tau_{yx} u + \sigma_y v + \tau_{yz} w\right) + \frac{\partial}{\partial z}\left(\tau_{zx} u + \tau_{zy} v + \sigma_z w\right)\right].$$
$$(6.65)$$

The amount of work performed in unit time due to the change in the elemental volume is

$$-p\,\frac{D}{Dt}\,\frac{\delta(\Delta V)}{\Delta V},\qquad(6.66)$$

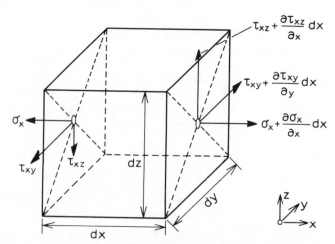

Figure 6.4: Frictional stresses on fluid element.

which may be rewritten in the form

$$-p\frac{D}{dt}\delta(\Delta V) = p\,\Delta V\left(\frac{\partial u}{\partial x} + \frac{\partial v}{\partial y} + \frac{\partial w}{\partial z}\right). \tag{6.67}$$

Combining equations (6.65) and (6.67), the quantity of mechanical energy transformed into heat through friction can be derived as

$$\delta Q_f = \delta t\,\Delta V\left[\left(\sigma_x\frac{\partial u}{\partial x} + \tau_{xy}\frac{\partial v}{\partial x} + \tau_{xz}\frac{\partial w}{\partial x}\right) + \left(\tau_{yx}\frac{\partial u}{\partial y} + \sigma_y\frac{\partial v}{\partial y} + \tau_{yz}\frac{\partial w}{\partial y}\right)\right.$$
$$\left. + \left(\tau_{zx}\frac{\partial u}{\partial z} + \tau_{zy}\frac{\partial v}{\partial z} + \sigma_z\frac{\partial w}{\partial z}\right)\right] + \delta t\,\Delta V\,p\left(\frac{\partial u}{\partial x} + \frac{\partial v}{\partial y} + \frac{\partial w}{\partial z}\right). \tag{6.68}$$

Invoking Stokes' hypothesis for the normal and shearing stress we have

$$\delta Q_f = \delta t\,\Delta V\mu\phi, \tag{6.69}$$

in which ϕ denotes a *dissipation function* which can now be defined by

$$\phi = 2\left[\left(\frac{\partial u}{\partial x}\right)^2 + \left(\frac{\partial v}{\partial y}\right)^2 + \left(\frac{\partial w}{\partial z}\right)^2\right] + \left(\frac{\partial v}{\partial x} + \frac{\partial u}{\partial y}\right)^2 + \left(\frac{\partial w}{\partial y} + \frac{\partial v}{\partial z}\right)^2$$
$$+ \left(\frac{\partial u}{\partial z} + \frac{\partial w}{\partial x}\right)^2 - \frac{2}{3}\left(\frac{\partial u}{\partial x} + \frac{\partial v}{\partial y} + \frac{\partial w}{\partial z}\right)^2. \tag{6.70}$$

Utilizing equations (6.61), (6.63) and (6.69), we can derive a generalized expression:

$$\rho C_v\frac{DT}{Dt} + p\frac{D}{Dt}\frac{\delta(\Delta V)}{\Delta V} = \frac{\partial}{\partial x}\left(k'\frac{\partial T}{\partial x}\right) + \frac{\partial}{\partial y}\left(k'\frac{\partial T}{\partial y}\right) + \frac{\partial}{\partial z}\left(k'\frac{\partial T}{\partial z}\right) + \mu\phi. \tag{6.71}$$

The work due to compression can be simplified by the following approach:

$$\frac{D}{Dt}\frac{\delta(\Delta V)}{\Delta V} = \text{div }\mathbf{w} = \left(\frac{\partial u}{\partial x} + \frac{\partial v}{\partial y} + \frac{\partial w}{\partial z}\right), \tag{6.72}$$

such that the work per unit volume per unit time can be written as

$$p\left(\frac{\partial u}{\partial x} + \frac{\partial v}{\partial y} + \frac{\partial w}{\partial z}\right). \tag{6.73}$$

For an incompressible fluid this term vanishes. The usual form in which this expression is included is

$$\left(\frac{\partial u}{\partial x} + \frac{\partial v}{\partial y} + \frac{\partial w}{\partial z}\right) = -\frac{1}{\rho}\left(\frac{\partial\rho}{\partial t} + u\frac{\partial\rho}{\partial x} + v\frac{\partial\rho}{\partial y} + w\frac{\partial\rho}{\partial z}\right) = -\frac{1}{\rho}\frac{D\rho}{Dt} \tag{6.74}$$

and

$$p\left(\frac{\partial u}{\partial x} + \frac{\partial v}{\partial y} + \frac{\partial w}{\partial z}\right) = -\frac{p}{\rho}\frac{D\rho}{\partial t} = -\frac{Dp}{Dt} + \frac{p}{\rho}\frac{D\rho}{Dt}. \tag{6.75}$$

Incorporating (6.75) into (6.71) we have

$$\rho \left[C_v \frac{DT}{Dt} + \frac{D}{Dt} \left(\frac{p}{\rho g} \right) \right] = \frac{Dp}{Dt} + \frac{\partial}{\partial x} \left(k' \frac{\partial T}{\partial x} \right) + \frac{\partial}{\partial y} \left(k' \frac{\partial T}{\partial y} \right) + \frac{\partial}{\partial z} \left(k' \frac{\partial T}{\partial z} \right) + \mu \phi.$$
$$(6.76)$$

Employing the thermodynamic equation for a perfect gas,

$$C_p \, \delta T = C_v \, \delta T + \delta \left(\frac{p}{\rho g} \right), \qquad (6.77)$$

where C_p denotes the specific heat at constant pressure. A generalized form normally used in the study of atmospheric convection can then be written as

$$\rho \frac{D}{Dt} (C_p T) = \frac{Dp}{Dt} + \frac{\partial}{\partial x} \left(k' \frac{\partial T}{\partial x} \right) + \frac{\partial}{\partial y} \left(k' \frac{\partial T}{\partial y} \right) + \frac{\partial}{\partial z} \left(k' \frac{\partial T}{\partial z} \right) + \mu \phi. \quad (6.78)$$

For an incompressible fluid, $Dp/Dt = 0$, $C_p = C_v = C$ and we have

$$\rho c \frac{DT}{Dt} = \frac{\partial}{\partial x} \left(k' \frac{\partial T}{\partial x} \right) + \frac{\partial}{\partial y} \left(k' \frac{\partial T}{\partial y} \right) + \frac{\partial}{\partial z} \left(k' \frac{\partial T}{\partial z} \right) + \mu \phi \qquad (6.79)$$

and if k' is independent of temperature and there is no dissipation,

$$\rho c \frac{DT}{Dt} = k' \left(\frac{\partial^2 T}{\partial x^2} + \frac{\partial^2 T}{\partial y^2} + \frac{\partial^2 T}{\partial z^2} \right). \qquad (6.80)$$

The above equations, combined with the momentum and continuity equations where the body forces now include buoyancy terms due to changes in local temperature, represent the equations necessary for the prediction of atmospheric circulation.

Recent developments in the utilization of such equations associated with the finite element method of Chapter 15 are readily available (Upson *et al.*, 1981). Although the study of such phenomena is in its infancy, the application of such numerical prediction models, with the advent of larger core and faster computers, is rapidly becoming a reality.

C.T.

References

Al-Mashidani, G., and Taylor, C. (1974a). Finite element solution of the shallow water equations—Surface run-off, *Proc. Int. Conf. on Finite Elements in Flow Problems, Swansea.*

Al-Mashidani, G., and Taylor, C. (1974b). A numerical technique for simulating direct run-off, *Proc. Int. Coll. on Field Simulation.*

Chow, V. T., and Ben-Zui, A. (1973). Hydrodynamic modelling of two-dimensional watershed flow, *ASCE J. Hyd. Div.*, **99** (HY5), 772–792.

Eagleson, P. S. (1970). *Dynamic Hydrology*, McGraw-Hill.

Harbaugh, T. E., and Chow, V. T. (1967). A study of the roughness of conceptual river systems of watersheds, *Proc. Twelfth Congress IAHR*, **1**, 9–17.

Henderson, F. M. (1966). *Open Channel Flow*, Macmillan.

Kibler, D. F., and Woolhiser, D. A. (1972). Mathematical properties of the kinematic cascade, *J. Hydrology*, **15**, 131–147.

Lighthill, M. J., and Whitham, G. B. (1955). On kinematic waves, 1. Flood movement in long rivers, *Proc. Roy. Soc.*, **1955**, 281–316.

Morgali, J. R. (1970). Laminar and turbulent overland flow hydrographs. *ASCE J. Hyd. Div.*, **96** (HV2), 441–460.

Schlichting, H. (1960). *Boundary Layer Theory* (Trans. J. Kestin), McGraw-Hill.

Schreiber, D. L., and Bender, D. L. (1972). Obtaining overland flow resistance by optimization, *ASCE J. Hyd. Div.*, **98** (HY3), 429–446.

Shen, H. W., and Li, R. M. (1973). Rainfall effect of sheet flow over a smooth surface, *ASCE J. Hyd. Div.*, **99** (HY5), 772–792.

Taylor, C., Al-Mashidani, G., and Davis, J. M. (1974). A finite element approach to watershed run-off. *J. Hydrology*, **21**, 231–246.

Upson, C. D., Gresho, P. M., Sani, R. L., Chan, S. T., and Lee, R. L. (1981). A thermal convection simulation in three dimensions by a modified finite element method, Paper presented at Second Int. Symp. on Num. Meth. in Heat Transfer, University of Maryland.

Woolhiser, D. A., Holland, M. E., Smith, G. L., and Smith, R. E. (1971). Experimental investigation of converging overland flow, *Trans. ASAE*, **14**(4), 684–687.

Woolhiser, D. A., and Ligget, J. A. (1967). Unsteady one dimensional flow over a plane— The rising hodograph, *Water Resources Res.*, **3** (3, third quarter), 753–771.

Yoon, Y. N., and Wenzel, H. G. (1971). Mechanics of sheet flow under simulated rainfall. *ASCE J. Hyd. Div.*, **97** (HY9), 1367–1386.

Mathematical Methods in Engineering
Edited by G. A. O. Davies
© 1984, John Wiley & Sons, Ltd.

7

Flow and Deformation in Soils

7.1 Introduction and basic concepts

Soil mechanics has in many ways lagged behind other branches of engineering, despite the fact that the ground, be it soil or rock, is an integral part of the design of any civil engineering structure. The neglect of soil behaviour is mainly due to the fact that, as a material, soil is somewhat different from most engineering materials. Apart from its nature—neither a true solid nor a liquid—soil occurs in a wide variety of types, it is to some extent non-homogeneous at any site and it exhibits anisotropic properties with regard to strength, compressiblity and flow of water through its pores.

A soil mass is a *multiphase* material consisting of a complex array of grains of various shapes and sizes, mixed with water, which fills partially or completely the void spaces. In the field, soils lying below the water table are more or less completely saturated, i.e. all air is excluded from the voids, though some air will be dissolved in the water phase. The soil grains are discrete particles ranging in size from very small particles of the order of 10^{-3} mm through medium grain sizes to coarser gravel and boulders of size > 10 mm. The finer-grained soils—the clay fraction—are composed of flake- or rod-shaped mineral particles (montmorillonite, illite, kaolinite, etc.) which exhibit the characteristics of chemical bonding with films of adsorbed water, and due to the small particle size pellicular water effects are correspondingly more important. Because of the cohesional effects at grain-to-grain contact points the structure of the solid phase may be very loose, and this may lead to high compressibility. Hydrodynamic expulsion of the water, however, is impeded by the small sinuous void network, so that consolidation is a slow process. In contrast to the behaviour of clays, silts, sand and coarser-grained fractions exhibit little or no cohesional bonding—unless due to some cementing agent—and thus these soil types are generally described as cohesionless. The individual particles are generally less plate-like in shape—deriving mainly from quartz—being more rounded with a smoother surface texture in transported weathered soils, such as fluvial or aeolian (wind-borne) deposits. Some soils and crushed rock are angular and rougher.

The physical behaviour of a soil is characterized more than anything else by its grain *size distribution* (Figure 7.1). This representation (or alternatively that of a *histogram*) enables one to see in a quantifiable way what is the predominant grain size fraction and if the soil is poorly (mainly of uniform grain size) or well graded (with a wide range of grain sizes). Well-graded soils are likely to have higher dry densities. If one visualizes a system of spheres—on the one hand of uniform size and on the other hand of different sizes—this is readily apparent. It is obvious that the smaller-sized spheres can fill the voids between the larger ones. In this way a more stable system of discrete particles is obtained. In other words, the solid phase is stiffer both with respect to compression and distortion. Due to the much greater surface area of clay particles their influence in a sand–silt–clay mixture is much greater than the weight proportion would indicate.

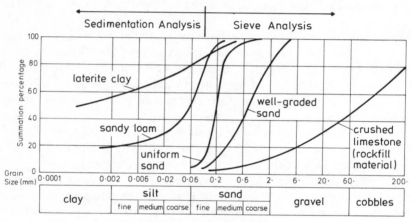

Figure 7.1: Typical grading curves.

Although the science of soil mechanics developed relatively late—around 1925 marks the true beginnings with the work of Terzaghi (1925)—it was able to build on the classical theories of *elasticity, plasticity* (see Chapters 10 and 11) and *viscous fluid flow*. The basic assumption in these theories is that the material is a continuous medium, so that in the mathematical formulation the abstract idea of a *continuum* is used. From the microscopic viewpoint all solid substances are composed of discrete particles (hence particulate mechanics), and this is very evident in naturally occurring soils. However, as a working hypothesis the concept of statistical averages, i.e. bulk or macroscopic quantities, is introduced. Quantities such as stress and strain are related to a unit volume of material, not only for total stress analysis, in which the overall properties of the material are considered, but also in effective stress analysis, in which the stresses acting in the various phases are considered. Experience shows that the concept of macroscopic stresses is adequate in most practical situations as the dimensions of a typical structural element (foundation pad, pile, etc.) are usually large in comparison with the mean grain size of the material in contact with the structure. The material in this chapter relates specifically to *soils*, but it is clear that many of the concepts can be applied to other *porous media* as well.

Concept of effective stress

If u_w and u_g are the pressures in the liquid and gas phases of a small cube of soil, the forces acting normal to a cross-section of area A are equal to the total force P, viz.

$$P = A_s\sigma_s + A_wu_w + A_gu_g, \tag{7.1}$$

where σ_s is the average microscopic stress in the solid phase and A_s, A_w and A_g are the average elemental areas for the solid, liquid and gas phases respectively. Then σ_s is actually a meaningless quantity for a particulate system, since the stress varies considerably within grains and at grain boundaries. A more useful quantity is the intergranular pressure which will now be obtained for a saturated system ($A_g = 0$). If A_c is the average contact area, the force transmitted through the grain boundaries is (see Figure 7.2)

$$P_c = P - (A - A_c)u_w. \tag{7.2}$$

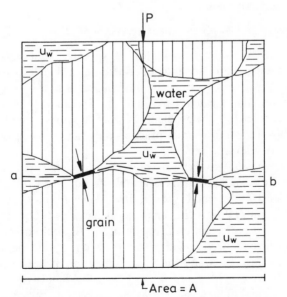

Figure 7.2: Two-dimensional representation of inter-granular stresses in a soil element (Note that the average contact area in the plane normal to the force P is the total projected area of the individual contact points along **ab**.)

Dividing by A gives an expression for the macroscopic value of the intergranular pressure

$$\bar{\sigma} = \sigma - (1 - a_c)u_w, \tag{7.3}$$

where a_c is the contact area ratio and σ is the total external stress. Experimental evidence suggests that for soils a_c is very small and for all practical purposes the intergranular pressure

$$\bar{\sigma} \cong \sigma' = \sigma - u_w, \tag{7.4}$$

where σ' is *the effective stress* normal to the section through the mixture. In soil mechanics the pressure in the pore fluid is usually denoted by the symbol u without a suffix. The *volumetric* or *octahedral effective stress* is given by

$$\sigma'_{oct} = \tfrac{1}{3}(\sigma_{xx} + \sigma_{yy} + \sigma_{zz}) - u, \tag{7.5}$$

where σ_{xx}, σ_{yy} and σ_{zz} are the total stresses in the three orthogonal coordinate directions. σ'_{oct} is the normal effective stress on the octahedral plane and controls both the volumetric and shear behaviour of the soil (see also Appendix B at the end of this chapter).

7.2 Steady groundwater flow

Darcy's law, seepage forces and critical hydraulic gradient

The *steady* flow of water in soils (called *seepage*) requires that the grain structure of the soil is stationary and not undergoing a change in volume (i.e. it is effectively rigid). Modern theories of multiphase flow involving interphasal displacements are not treated here, but the special case of a non-linear deformable grain skeleton (the problem of *consolidation*) is treated in Section 7.3 on *transient* flow. In soil mechanics practice, only the problems of seepage and consolidation are relevant.

A useful background to the flow of water through a porous medium is supplied by the *hydraulics* problem of the flow of a liquid in a tube, from which it is known that two fundamental types of flow—*laminar* and *turbulent*—may be distinguished. Laminar flow is characterized by particles of liquid flowing along their own definite paths without interfering with each other, whereas in turbulent flow the paths are irregular and intersecting (see Chapter 2).

The uniaxial flow of water in a soil may be investigated using the permeameter shown diagrammatically in Figure 7.3. Soil is packed at uniform density into a cylindrical tube and a *hydraulic gradient* is applied. The hydraulic gradient i is defined as the head loss between two points on a flowline (or streamline) which are a unit distance apart. The hydraulic head h consists of the hydrostatic water pressure relative to some datum—for convenience this is usually taken as the exit or tail water level in an engineering application—plus the excess water pressure u, viz.

$$h = \frac{u}{\gamma_w} + z_0, \tag{7.6}$$

where γ_w is the unit weight of water and z_0 the height relative to the datum. In soil mechanics the velocity head can be safely neglected. With reference to Figure 7.3, the hydraulic gradient is the difference between the piezometer readings for points A and B divided by the length AB, where AB tends to zero, that is [see IV, §3.1]

$$i = -\lim_{\delta s \to 0} \left(\frac{\delta h}{\delta s}\right) = -\frac{dh}{ds}. \tag{7.7}$$

The negative sign indicates that the hydraulic head decreases in the direction of flow. The head loss results mainly from work done in overcoming the drag of the soil to seepage through its pores. The soil structure experiences a force or drag. Referring to Figure 7.3, it may be seen from equilibrium considerations for the whole specimen that the seepage force equals the difference between the water pressures at the ends of the specimen minus the force required to raise the water

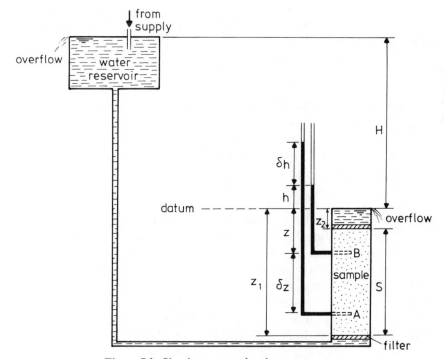

Figure 7.3: Simple constant-head permeameter.

through the height S, i.e.

$$F_s = A\gamma_w(H + Z_1 - Z_2) - A\gamma_w S. \tag{7.8}$$

Since $(Z_1 - Z_2)$ equals S, the seepage force per unit volume of soil specimen J is given by

$$J = \frac{F_s}{AS} = \gamma_w \frac{H}{S} = i\gamma_w. \tag{7.9}$$

If the drag on the soil particles is increased by increasing the driving head H an unstable condition (called the quick condition, boiling or piping if the condition develops on the downstream side of an earth dam) may result in which the effective stress in the soil is reduced to zero. This happens when the upward seepage force equals the effective weight of the specimen, viz.

$$i\gamma_w A = (\gamma_{sat} - \gamma_w)A = \gamma'A.$$

The *critical* hydraulic gradient i_c is

$$i_c = \frac{\gamma'}{\gamma_w}. \tag{7.10}$$

The effective density may be expressed in terms of other soil constants, e.g. the void ratio e and the specific gravity of the grains G_s (see Appendix A). The following expression then results:

$$i_c = \frac{G_s - 1}{1 + e}. \tag{7.11}$$

For a typical sand $e \simeq 0.6$, $G_s = 2.65$ so for these values the critical hydraulic gradient would be a little in excess of unity.

In 1856 Darcy carried out experiments in which he showed that the rate of discharge v_a is proportional to the hydraulic gradient. *Darcy's law* is

$$v_a = \frac{Q}{At} = ki, \tag{7.12}$$

where Q is the volume discharged in time t and A is the cross-sectional area. The constant of proportionality k is called the *coefficient of permeability*. The quantity v_a is in fact the superficial velocity of the flowing water and thus k has the units of velocity. The actual velocity of the water through the sinuous system of pores is greater than v_a. If n is the porosity of the soil, the average velocity in the pores is v_a/n. This quantity, however, is not of practical interest.

The pores of granular soils are relatively fine and as a result flow is almost always laminar, except perhaps in coarse sands and gravels. In cohesive soils some of the water is adsorbed on the surface of the clay platelets. The permeability is then related to the amount of 'free' water and depends very much on the mineralogical characteristics. For clays the permeability is a few orders of magnitude smaller than for granular soils. A rough comparison of values of the coefficient of permeability for various soil types is given in Table 7.1.

Various factors besides the mean grain size affect the permeability of a soil. These include the void ratio, shape and structural arrangement (fabric) of the grains, pore air content, chemical composition, and in porous media other than natural soils the properties of the pore fluid. In other applications (e.g. petroleum engineering) the permeability of the medium alone, K, is used. The following relation then applies:

$$k = \frac{\gamma_f K}{\mu}, \tag{7.13}$$

where μ is the dynamic viscosity of the pore fluid, and γ_f its unit weight.

Table 7.1   Approximate range of values of the coefficient of permeability in soils

Soil type	Gravel	Sand	Silt	Clay
Representative size (mm)	4	0.6	0.008	0.001
Coefficient of permeability k (cm/s)	1	10^{-2}	5×10^{-6}	10^{-8}

Three-dimensional steady-state flow

Darcy's law, which was discussed in the last section, may be extended to three dimensions, viz.

$$v_x = k_x i_x,$$
$$v_y = k_y i_y, \tag{7.14}$$
$$v_z = k_z i_z,$$

where for general anisotropic conditions the permeabilities k_x, k_y, k_z in the three orthogonal directions x, y, z are unequal.

If a small cube of soil of sides δx, δy, δz is considered it may readily be shown that the condition of continuity requires that

$$\frac{\partial v_x}{\partial x} + \frac{\partial v_y}{\partial y} + \frac{\partial v_z}{\partial z} = 0. \tag{7.15}$$

Use may now be made of equations (7.7) and (7.14) to show that the differential equation describing three-dimensional seepage is given by

$$k_x \frac{\partial^2 h}{\partial x^2} + k_y \frac{\partial^2 h}{\partial y^2} + k_z \frac{\partial^2 h}{\partial z^2} = 0. \tag{7.16}$$

Similar differential equations govern steady flow problems in Chapters 2 (equation 2.4), 6 (equation 6.80), 8 (equation 8.2), and so on.

For isotropic soils, equation (7.16) reduces to the *Laplacian form* [see IV, §8.2]. Stratified soils exhibit significant anisotropic characteristics. However, if required, a scale transformation is possible for two-dimensional, plane problems. Equation (7.16) may be written as follows for plane problems (assuming no flow in the *y* direction):

$$\frac{\partial^2 h}{(k_z/k_x)\,\partial x^2} + \frac{\partial^2 h}{\partial z^2} = 0. \tag{7.17}$$

If, now, the *x* dimension is transformed as follows:

$$x_t = \sqrt{\frac{k_z}{k_x}}\,x, \tag{7.18}$$

the governing equation becomes laplacian with coordinate variables x_t and z.

Special cases

(*a*) *Radial flow.* The equation for radial, symmetrical flow is

$$\frac{1}{r}\frac{\partial}{\partial r}\left(r\frac{\partial h}{\partial r}\right) = 0 \tag{7.19}$$

[see IV, Example 8.2.2]. This equation may be integrated twice to give the general solution

$$h = a \log_e\left(\frac{r}{b}\right) \tag{7.20}$$

[see IV, §7.5], where a and b are constants. A case of practical interest is that of flow in a confined aquifer. If the aquifer of thickness B is perforated by a well, provided the head of water in the well is not drawn below the level of the upper surface of the aquifer, the discharge q is given by

$$q = \frac{2\pi BHk}{\log_e(r_2/r_1)}, \tag{7.21}$$

where H is the head at radius r_2 relative to that at radius r_1 (well radius) and k is the permeability. If the water surface is free or unconfined the condition of one-dimensional radial flow no longer holds and the problem becomes axisymmetric.

(*b*) *Spherical flow.* In spherical coordinates Laplace's equation is

$$\frac{1}{r^2}\frac{\partial}{\partial r}\left(r^2\frac{\partial h}{\partial r}\right) = 0. \tag{7.22}$$

On integrating twice the result is

$$h = \frac{a}{r} + b \tag{7.23}$$

where again a and b are constants [see IV, Example 8.2.5]. The field application of this equation would be flow in an infinite aquifer to a spherical sink.

Methods of solving the Laplacian seepage equation

In geotechnical practice the most popular method of solving two-dimensional seepage problems is by a trial and error graphical method. The method is based upon the mathematical properties of Laplace's equation. For isotropic soils the equipotential lines and the flowlines exhibit the property of *orthogonality* [see IV, §20.4]. In essence the method involves sketching these sets of lines, making use of, and at the same time satisfying, the boundary conditions. The flownet so obtained consists of 'curvilinear rectangles'. The reader is referred to Scott (1963) for a discussion of the various refinements in the method.

Mathematical methods of solution include the *finite difference* [see III, §9.3] and the *finite element* (see Chapter 15) methods, the method of *conformal transformation* [see IV, §9.12] and the *random walk* method (Scott, 1963) [see II, §18.3]. Before the electronic computer became widely available the solution to the finite difference equations was often obtained by *relaxation methods* (Allen and Southwell, 1955; Southwell, 1940).

Today, solutions are most often obtained by discretizing the flow domain into finite elements, whereby the *variational* or the *weighted residual* (e.g. Galerkin) approach may be used to find a numerical solution to the problem. By both finite difference and finite element methods a set of simultaneous equation results, there being an equation for each grid or nodal point in the domain where the head is unknown. The finer the net the larger is the system of equations to be solved. With modern computers the memory capacity is usually such that little restriction is placed on the size of a problem, except perhaps in the solution of three-dimensional seepage.

A mathematical approach to the solution of the seepage equation which is suitable for simple confined flow is provided by the method of conformal transformation (Harr, 1962; Verruijt, 1970). The problem is to find the conjugate potential and stream functions, $\phi(x, z)$ and $\psi(x, z)$, satisfying the boundary conditions of the given problem (see also Chapter 2, Section 2.4). For a finite flow domain it is necessary to make one transformation for the physical geometry to represent it on an infinite half plane. The infinite half plane is then re-mapped onto the ω plane, where ω is the *complex variable* $(\phi + i\psi)$ [see IV, §9.12]. Thus the original flownet is transformed into a simple rectilinear grid. The method is best appreciated with reference to the simple example of flow through homogeneous ground beneath an impervious dam, assuming that the extent of the flow domain corresponds to an infinite semicircle. In this case the first transformation is not needed. The infinite half plane (Figure 7.4a) has coordinates x and z, and is referred to as the w plane, where $w = x + iz$.

Methods are available for arriving at the relationship for transforming the flownet, viz. ω as a function of w. Here the relation is simply given and the boundary conditions are investigated. We consider, therefore, the relation

$$\omega = \frac{2a}{\pi} \cos^{-1}\left(\frac{w}{d}\right). \tag{7.24}$$

By using trigonometric relationships and equating real and imaginary parts [see IV, §§2.12 and 2.13]

$$x = d \cos\left(\frac{\pi\phi}{2a}\right) \cosh\left(\frac{\pi\psi}{2a}\right),$$
$$z = -d \sin\left(\frac{\pi\phi}{2a}\right) \sinh\left(\frac{\pi\psi}{2a}\right). \tag{7.25}$$

Eliminating ϕ we obtain

$$\frac{x^2}{d^2 \cosh^2(\pi\psi/2a)} + \frac{z^2}{d^2 \sinh^2(\pi\psi/2a)} = 1. \tag{7.26}$$

On any streamline the function ψ is constant and thus equation (7.26) defines a family of *confocal ellipses* [see V, §1.3.2] with focus at $x = \pm d, z = 0$. If, now, ψ is eliminated from equation (7.25) a family of *confocal hyperbolae* [see V, §1.3.4] is obtained, representing the equipotential lines, viz.

$$\frac{x^2}{d^2 \cos^2(\pi\phi/2a)} - \frac{z^2}{d^2 \sin^2(\pi\phi/2a)} = 1. \tag{7.27}$$

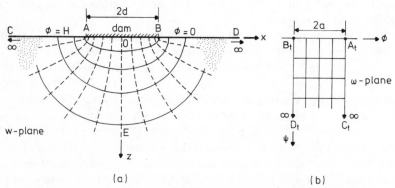

(a)   (b)

Figure 7.4: Flow beneath impervious dam with transformed flownet.

The boundary conditions will now be investigated. With reference to Figure 7.4(a) the bottom of the dam AOB is an impermeable boundary and constitutes a streamline. The potential along the upstream ground surface CA is constant with $\phi = H$, and, on the downstream surface BD, $\phi = 0$. Due to symmetry the potential along the z axis equals $H/2$. We consider first this axis, i.e. $x = 0$. From equation (7.25), since cosh ( ) > 1, $\phi = a$. The variation of the stream function along OE is given by

$$z = -d \sinh\left(\frac{\pi\psi}{2a}\right) \tag{7.28}$$

and at O the stream function ψ is zero. On the other axis ($z = 0$), either $\psi = 0$ or $\phi = 2a$ or 0. For the case $\psi = 0$, the potential function (in this case, the distribution of head along the dam base) is obtained from the relation

$$x = d \cos\left(\frac{\pi\phi}{2a}\right). \tag{7.29}$$

In equations (7.28) and (7.29) the constants d and a have not yet been determined. By substituting for the known conditions of head at A and B, it follows that $d = b$ (the half dam width) and $a = H/2$, where H is the total head difference. If $\psi \neq 0$ the second condition involving ϕ applies, i.e. $\phi = 2a = H$ and $\phi = 0$. It is shown in Figure 7.4(b) how the finite zone bounded by the flowline CED is mapped onto the rectangular region $A_tC_tD_tB_t$.

We now give the solution for a sheet pile wall embedded in a layer of infinite depth (Figure 7.5a). In this case the following relationship gives the correct transformation:

$$\omega = \phi + i\psi = \frac{2a}{\pi} \cos^{-1}\left(\frac{z + ix}{d}\right), \tag{7.30}$$

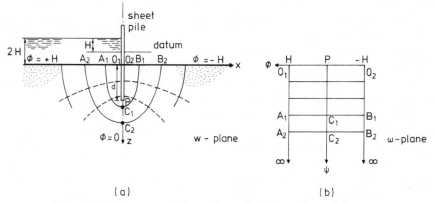

(a)   (b)

Figure 7.5: Flow around sheet pile wall with transformed flownet.

where the constant $a = H$ (the driving head being $2H$) and d equals the depth of the pile. The transformed flownet is a semi-infinite rectangle, a finite portion of which is shown in Figure 7.5(b).

Using the analytical solution for the given problem (e.g. equation 7.25) quantities of interest, such as the hydraulic gradient ($= -\partial\phi/\partial s$, where ∂s is an incremental length measured along a streamline) or seepage quantities, may be derived. For example, on the base of the dam $s = x$ and $z = \psi = 0$. The hydraulic gradient may be found directly by differentiating equation (7.29). The result shows that theoretically at the entrance and exit points (A and B in Figure 7.4a) infinite gradients occur, whereas at the centre point O,

$$i = \frac{H}{\pi d}.$$

Although the method of conformal transformation provides a valuable background to groundwater flow studies it is not as general as the finite element solution techniques and the range of problems where it is applicable is limited.

7.3 Non-steady groundwater flow

If external load is applied to a porous medium a part of it is carried by the solid structure and the rest by the pore fluids. If the pores are completely saturated with water the compressibility of the pore fluid is usually much smaller than that of the solid structure, at least in the case of soils. Thus, initially, the greater part of the load is carried by the water phase. Unless the porous medium is completely enclosed by impermeable boundaries drainage takes place. As water is expelled there is a transfer of load to the grain structure. The phenomenon is a time-dependent transient process called *consolidation*. The time to attain a given degree of consolidation—known as the hydrodynamic time lag—depends upon the soil's permeability and compression characteristics, the drainage boundary conditions and the size of the soil mass. The latter two conditions determine the maximum length of drainage path. For coarse-grained soils consolidation is almost instantaneous, but for fine soils, e.g. fat clays, the hydrodynamic time lag may be several years.

An analogy which is helpful in understanding the process of consolidation is that of a simple spring–piston system as shown in Figure 7.6. It is assumed that the water is incompressible, so that at the first instant of loading all the load is carried by the water and there is no load sharing. In an actual soil system the pressure gradients induced in the water phase cause a flow of water to take place, the water exiting at the pervious boundaries. The flow obeys Darcy's law. In the analogy a valve constriction simulates the permeability effects. The analogy is carried further in that the force applied to the piston represents the *external* stress σ and that in the spring the *effective* stress σ'. At time $t = 0$, the pore water pressure u equals σ, and $\sigma' = 0$. When the consolidation is complete

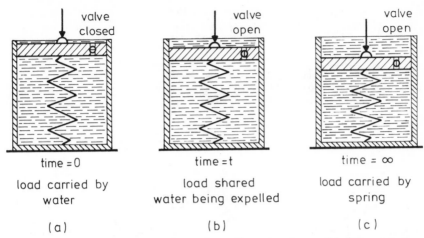

Figure 7.6: Spring-piston analogy of one-dimensional consolidation process.

(theoretically at $t = \infty$) the soil has settled an amount u_z, $u = 0$ and $\sigma' = \sigma$. If the external load is then suddenly removed, i.e. $\sigma = 0$, in the first instant of unloading $u = -\sigma'$. Water is sucked into the system and with increasing time the linear spring element relaxes, i.e. $\sigma' \to 0$, $u \to 0$, and the settlement u_z is reversed. The physical analogy corresponds to a simple Kelvin rheological model. In real soils the deformation characteristics of the grain structure are highly inelastic and the settlement is only partially reversible.

One-dimensional consolidation

In practice, consolidation of a compressible layer may often be assumed to be essentially one dimensional, for example, if a uniformly distributed load due to, say, fill material is applied at the ground surface, if the water table is lowered or if a thin soil layer is at a sufficient depth below the finite loaded area. In such situations one-dimensional theory has been used with a large measure of success.

A laboratory apparatus to test the behaviour of soils under conditions of zero lateral strain is called an *oedometer* (Figure 7.7). Typical test results obtained with this apparatus are shown in Figure 7.8. The points E_0, E_1, etc., correspond to the successive equilibrium points E, denoting final consolidation under the given increment of loading.

Generally the consolidation curve does not approach the asymptotic value—the primary consolidation—as predicted by the simple theory based upon the spring–dashpot analogy. There is a certain creep behaviour called *secondary consolidation*. To investigate secondary consolidation more complicated rheological models have been used, as described in Section 7.5. It will be noted in

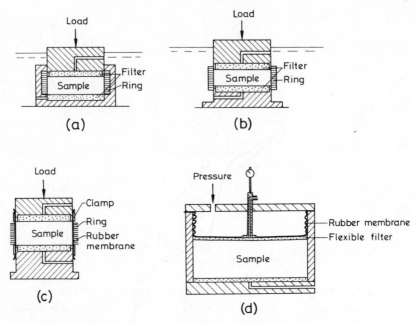

Figure 7.7: Different types of oedometer: (a) fixed ring, (b) floating ring, (c) compressiometer, (d) Rowe consolidation cell.

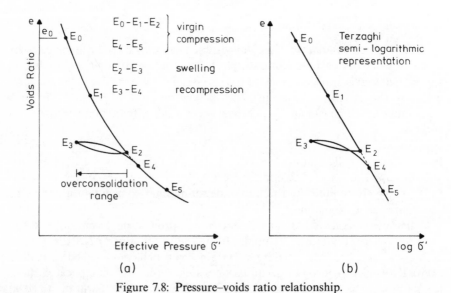

Figure 7.8: Pressure–voids ratio relationship.

Figure 7.8 that the settlement of a specimen is conveniently represented by the voids ratio. Many investigations have shown that the relationship between effective equilibrium pressure and voids ratio is linear in a semilogarithmic plot, provided one is concerned with the virgin part of the compression curve, i.e.

$$e = e_o - C_c \log_{10} \frac{\sigma'}{\sigma'_0}, \tag{7.31}$$

where C_c is called the compression index.

In the overconsolidation pressure range (Figure 7.8b) the compression behaviour departs from linearity, but the curve may be approximated by linearizing using an expansion or swelling index C_e. During reloading the small hysteresis effect is often neglected and the same index C_e is used. Overconsolidation denotes that the soil has experienced a reduction of effective stress. Most naturally occurring soils have undergone some erosion of overburden in their previous loading history, and exhibit overconsolidation effects.

Terzaghi consolidation theory

The theory developed by Terzaghi (1943) leads to an equation of transient flow which is analogous to the diffusion equation describing the flow of heat in a solid (see Chapter 8, equation 8.41). To arrive at this equation certain assumptions are necessary including:

(i) Complete saturation.
(ii) Negligible compressibility of mineral grains and water.
(iii) Validity of Darcy's law.
(iv) The values of certain material parameters remain constant during consolidation. In particular, the effective pressure–void ratio curve is linearized over the range of compression.

The most limiting of the assumptions is the last one, which greatly idealizes the compression relationship (Figure 7.9a). The resulting equation is

$$\frac{de}{d\sigma'} = -(1 + e_o)m_v, \tag{7.32}$$

where m_v is the coefficient of volume compressibility.

The equation of transient flow is obtained from continuity considerations. The change in volume of an element of soil of area $\delta x\, \delta y$ is given by the settlement δu_z, and may be expressed as

$$\delta V = \frac{\delta e}{1 + e_o} \delta x\, \delta y\, \delta z, \tag{7.33}$$

since the volume change is due only to a change in the voids ratio, e. The time

EFFECTIVE VERTICAL PRESSURE (LOG SCALE)

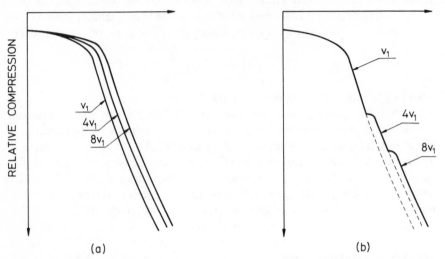

(a)   (b)

Figure 7.9: Influence of strain rate on pressure–voids ratio in constant-rate-of-strain
oedometer test (showing the effect of jump in the strain rate during test).

rate of volume change may be shown to be

$$\frac{\partial V}{\partial t} = k_z \frac{\partial^2 h}{\partial z^2}\, \delta x\, \delta y\, \delta z. \tag{7.34}$$

Consolidation is caused only by hydraulic heads due to the excess pressure u, so
that h may be replaced by u/γ_w in equation (7.34). Substituting equation (7.33)
into equation (7.34) leads to

$$\frac{k_z}{\gamma_w} \frac{\partial^2 u}{\partial z^2} = \frac{1}{1 + e_o} \frac{\partial e}{\partial t}. \tag{7.35}$$

Equation (7.35) is general with respect to the pressure–voids ratio relationship. If
the logarithmic form of the latter is adopted equation (7.35) becomes

$$\frac{(1 + e_o)k_z}{\gamma_w} \frac{\partial^2 u}{\partial z^2} = \frac{C_c}{2.303\,(\sigma - u)}\left(\frac{\partial u}{\partial t} - \frac{\partial \sigma}{\partial t}\right). \tag{7.36}$$

This equation is no longer linear and superposition of solutions is not possible.
If we assume that only small strains occur, as Terzaghi did, we can use equation
(7.32) and the result becomes

$$\frac{k_z}{m_v \gamma_w} \frac{\partial^2 u}{\partial z^2} = \frac{\partial u}{\partial t} - \frac{\partial \sigma}{\partial t}. \tag{7.37}$$

If the applied stress is constant the last term in equation (7.37) is zero and the equation reduces to the well-known *diffusion* equation [see IV, §8.5]. It is a *parabolic-type partial* differential equation [see III, §9.4]. The group of constants $k_z/m_v\gamma_w$ is designated by c_v, the coefficient of consolidation.

If the Terzaghi theory were extended to three dimensions the left-hand side of equation (7.37) would be replaced by

$$\frac{1}{m_v\gamma_w}\left(k_x\frac{\partial^2 u}{\partial x^2} + k_y\frac{\partial^2 u}{\partial y^2} + k_z\frac{\partial^2 u}{\partial z^2}\right).$$

Review of testing procedures

As shown in Figure 7.7, there are different types of oedometer equipment. The main defect of the oedometer test is the presence of friction on the inside of the ring enclosing the sample. This may be reduced by having a smooth surface with a non-corroding ring material and the use of silicon grease helps too. Another method of overcoming the frictional effect is to employ a floating ring set-up (Figure 7.7b) or a compressiometer (Figure 7.7c) in which the sample is enclosed in a rubber membrane reinforced by a number of thin rings offering no resistance to vertical load but acting rigidly in the radial direction.

In the Rowe consolidation cell, which is generally used for larger size samples, the load is applied by means of a flexible rubber diaphragm under the action of pressure, whilst the settlement is measured at the axis of the sample. The apparatus may not work so well for soft, highly compressible soils.

For coarse-grained soils lateral friction effects may be substantial. In this case there may be no alternative to accepting that the stress distribution is non-uniform within the sample and to allow for it by measuring the load transferred to the base of the sample and calculating an average value of stress.

Further modifications to the testing equipment and procedures include measurement of pore water pressure at the base of the sample, drainage being permitted at the top only, and for highly stratified soils, e.g. varved clays, by providing radial drainage at the periphery and in a central core of the sample. The Rowe consolidometer may be easily modified for such purposes.

Apart from the above developments in apparatus and technique, a notable advance in recent years has been to move away from the incremental testing procedure, whereby load increments are applied sequentially, usually in 24-hour intervals. Instead, various forms of continuous load application have been introduced. The advantages of this are that the tests may be run automatically, they may be speeded up and the preconsolidation pressure may be identified due to the continuous increase of effective pressure. The preconsolidation pressure itself, however, may be affected by the loading procedure and, in a constant rate of strain tests, by the strain rate, as shown in Figure 7.9 (Larsson, 1981). The tests may be run by controlling either the rate of strain (CRS test) or the pore water pressure at the base of the sample. In either case a pore pressure transducer is required at the bottom end and drainage is one way only. The most sophisticated test (Janbu, Tokheim and Senneset, 1981) maintains a constant ratio of induced pore pressure to applied stress $\Delta u_b/\Delta\sigma$, equal to a value close to 0.1 by automatically adjusting the strain rate. With the CRS test the strain rate must be set to a value which will give a maximum value of $\Delta u_b/\Delta\sigma < 0.15$. The standard rate will be of the order of 0.002 mm/min for medium to high plasticity clays. According to Smith and Wahls (1969) the average effective stress in the sample is then approximately

$$\sigma' = \sigma - \tfrac{2}{3}u_b \tag{7.38}$$

and the coefficient of consolidation for a specimen of height H is

$$c_v = \frac{d\sigma'}{dt}\frac{H^2}{2u_b}. \tag{7.39}$$

It is assumed here that the pore pressure distribution is parabolic, which may not be a good assumption for an interval of load in which the preconsolidation pressure is passed, there being a higher value of oedometer modulus E_{oed} in the lower part of the sample which has not yet reached the preconsolidation pressure. The oedometer modulus is also called the constrained modulus and is simply the inverse of the coefficient of volume change m_v.

It is possible to determine the coefficient of permeability k_z from the oedometer test. It is related to the coefficients c_v and m_v and using equation (7.39) and noting that equation (7.32) may be written in the form $m_v = d\varepsilon/d\sigma'$, where ε is the axial strain, it follows that

$$k_z = \gamma_w c_v m_v = \gamma_w \frac{H^2}{2u_b}\frac{d\varepsilon}{dt} \tag{7.40}$$

for the constant rate of strain test.

7.4 Mathematical solutions to the Terzaghi consolidation equation

Layer of finite thickness, 2H

A solution may be found assuming that the excess pressure u is a product of some function of z and some function of t using the *separation of variables* approach [see IV, §8.2], i.e.

$$u = F(z)\Phi(t).$$

The solution is found to be of the form

$$u = (C_1 \cos Az + C_2 \sin Az) \exp(-A^2 c_v t),$$

where the constants A, C_1, C_2 must be determined from the given boundary conditions for u and $\partial u/\partial z$ at $z = 0, 2H$, and for the initial distribution u_i of u at $t = 0$. The thickness of the layer is $2H$. The final solution takes the form of an infinite series. For double-sided drainage, $u = 0$, $z = 0, 2H$, we obtain

$$u = \sum_{n=1}^{\infty} \left[\frac{1}{H} \int_0^{2H} u_i \sin\left(\frac{n\pi z}{2H}\right) dz \right] \sin\left(\frac{n\pi z}{2H}\right) \exp\left(-\frac{n^2\pi^2 T_v}{4}\right), \quad (7.41)$$

where T_v is defined as the time factor

$$T_v = \frac{c_v t}{H^2}.$$

Due to the exponential term in equation (7.41) only a few terms of the infinite series are required to achieve sufficient accuracy.

For the special case of constant initial excess pressure, $u_i = u_o$, the distribution of pressure after lapse of time $t = t_1$ and $t = t_2$, as given by equation (7.41), is sketched in Figure 7.10. These pressure distribution lines are called *isochrones*. Their shape is approximately parabolic. The settlement of the surface of the layer at time $t = t_1$ is proportional to the hatched area in Figure 7.10. A quantity of interest is the average degree of consolidation, U, defined as the settlement at time t divided by the final settlement. The quantity U may be expressed in terms of the average excess pressure at time t, u_{av}, i.e.

$$U = 1 - \frac{u_{av}}{u_o}. \quad (7.42)$$

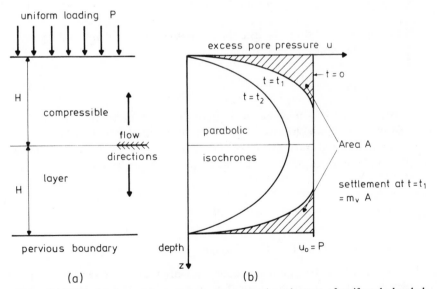

uniform loading P

H   compressible

flow
directions

H   layer

pervious boundary

(a)

excess pore pressure u

$t = 0$

$t = t_1$

$t = t_2$

parabolic

isochrones

Area A

settlement at $t = t_1$
$= m_v$ A

depth

$u_o = P$

z

(b)

Figure 7.10: Dissipation of pore water pressure and settlement of uniformly loaded compressible layer.

Integration of equation (7.41) with respect to the depth z leads to the following expression for U:

$$U = 1 - \sum_{m=0}^{\infty} \frac{2}{M^2} \exp(-M^2 T_v),$$ (7.43)

where $M = (\pi/2)(2m + 1)$.

Layer of infinite thickness

A simple solution exists for the case of a stress u_0 applied uniformly at the surface of a semi-infinite bed of compressible soil, in which drainage is permitted at the upper surface, viz.

$$u = u_0 \operatorname{erf}\left(\frac{z}{2\sqrt{c_v t}}\right),$$ (7.44)

where erf is the *error function* defined as

$$\operatorname{erf}(y) = \frac{2}{\sqrt{\pi}} \int_0^y \exp(-t^2) \, dt$$ (7.45)

and which is tabulated (Jahnke, Emde and Lösch, 1960).

The discharge of water q at the ground surface may be easily calculated using Darcy's law, i.e.

$$q_{z=0,t} = \frac{k_z}{\gamma_w}\left(\frac{\partial u}{\partial z}\right)_{z=0}.$$ (7.46)

Substitution of equation (7.44) and differentiating under the integral sign leads to the following result:

$$q_{z=0,t} = \frac{k_z u_0}{\gamma_w} \frac{1}{\sqrt{\pi c_v t}}.$$ (7.47)

This expression for q may be integrated with respect to time to yield the cumulative flow Q_t, which is also equal to the total settlement of the layer. It may be shown that the average degree of consolidation U is given by the following simple parabolic relation with the time factor T_v (Scott, 1963):

$$U = \sqrt{\frac{4T_v}{\pi}}.$$ (7.48)

Finite difference solution [see III, Chapter 9]

In many cases the best and sometimes the only way to arrive at a solution is by means of approximate numerical procedures, e.g. if the external loading varies with time (cf. equation 7.37) or if the non-linear form of the consolidation equation (equation 7.36) is used. The *finite difference* procedure described in this section is applicable for one or more space dimensions and for simple inhomogeneous situations (e.g. multilayered soil systems), but for generality with respect to geometry and inhomogeneity the finite element method discussed in the next section is better.

The finite difference representation is given here of the equation for one-dimensional consolidation (equation 7.37), viz.

$$c_v \frac{(u_a + u_b - 2u_o)^t}{(\Delta z)^2} = \frac{1}{\Delta t}(u_o^{t+1} - u_o^t - \sigma_o^{t+1} + \sigma_o^t), \qquad (7.49)$$

where Δz and Δt are the space and time increments respectively and the suffices o, a and b signify the space positions (refer to the mesh in Figure 7.11). The above equation is of the *forward time-marching explicit* type (Mitchell and Griffiths, 1980), i.e. the unknown excess pore pressure at time $t + \Delta t$ is found directly from the known values at time t; thus

$$u_o^{t+1} = u_o^t + \beta(u_a + u_b - 2u_o)^t + \sigma^{t+1} - \sigma^t, \qquad (7.50)$$

where

$$\beta = \frac{c_v \Delta t}{(\Delta z)^2}.$$

The solution is convergent and stable only if the value of β is less than some limiting value, which places a condition upon the size of the time step for a given mesh increment. The limiting value of β lies around $\frac{1}{2}$, depending not only upon

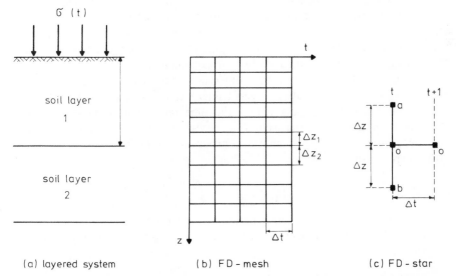

(a) layered system   (b) FD-mesh   (c) FD-star

Figure 7.11: Solution of one-dimensional consolidation problem by the finite difference method: (a) layered system, (b) finite difference mesh, (c) finite difference star.

the diffusion equation but also on the boundary conditions. Superior accuracy has been shown to be provided by $\beta = \frac{1}{6}$. The restriction placed upon the time interval may be somewhat severe, in which case an *implicit* finite difference scheme might be preferred, e.g. the *Crank–Nicolson method* [see III, §9.4]. The advantage of the implicit methods is that they are unconditionally stable, but require the solution of a linear system of simultaneous equations at each time step (Schiffmann and Arya, 1977).

In the case of a multilayered soil system the difference equation (equation 7.50) will apply in each layer, but the factor β changes from layer to layer. It is advantageous to maintain a constant value of time interval Δt in all layers, with a convenient space interval Δz in each layer. At the interface between two layers 1 and 2 at $z = L$, the following continuity conditions hold:

(i) The pore pressures are equal

$$u(L, t)_1 = u(L, t)_2 \tag{7.51}$$

and (ii) using Darcy's law the vertical flow condition is

$$k_1 \left(\frac{\partial u}{\partial z}\right)_1 = k_2 \left(\frac{\partial u}{\partial z}\right)_2. \tag{7.52}$$

The second of these conditions may be written in difference form:

$$\frac{k_1}{(\Delta z)_1} (u_a - u'_b) = \frac{k_2}{(\Delta z)_2} (u'_a - u_b), \tag{7.53}$$

where u'_a and u'_b are imaginary values. If the finite difference star (Figure 7.11) is placed at the interface boundary and equation (7.50) is applied to both layers the imaginary values that must be introduced into the equation can be eliminated and a modified difference expression results at the boundary. The other space boundary conditions are handled in a similar manner.

Higher order, e.g. nine-point, finite difference schemes may be used to improve accuracy [see III, §9.5] or alternatively a finer mesh may be adopted. In the case of unconditionally stable algorithms the time interval may be substantially increased, but a check on the accuracy is necessary. An increase in the time step at later stages in the time domain or a programmed logarithmic change in the time step is also possible.

Finite element solution (see also Chapter 15)

The transient flow equation generalized for variable permeability $k(x, y, z)$ and extended to three dimensions with anisotropy—principal permeabilities k_x, k_y and k_z—is given by

$$\frac{\partial}{\partial x}\left(k_x \frac{\partial h}{\partial x}\right) + \frac{\partial}{\partial y}\left(k_y \frac{\partial h}{\partial y}\right) + \frac{\partial}{\partial z}\left(k_z \frac{\partial h}{\partial z}\right) + \left(\bar{Q} - m_v \gamma_w \frac{\partial h}{\partial t} + m_v \frac{\partial \sigma}{\partial t}\right) = 0, \tag{7.54}$$

where $\bar{Q}$ is the rate at which the fluid is generated or discharged into a unit volume of space.

A *variational functional* [see IV, Chapter 12] χ, corresponding to the governing equation (7.54), is (cf. Desai and Abel, 1972)

$$\chi = \int_V \frac{1}{2}\left[k_x \left(\frac{\partial h}{\partial x}\right)^2 + k_y \left(\frac{\partial h}{\partial y}\right)^2 + k_z \left(\frac{\partial h}{\partial z}\right)^2 - 2Qh \right] dV + \int_{S_2} qh \, dS \quad (7.55)$$

where Q represents the last bracketed term in equation (7.54) and the surface integral on S_2 satisfies the boundary condition

$$k_x \frac{\partial h}{\partial x} l_x + k_y \frac{\partial h}{\partial y} l_y + k_z \frac{\partial h}{\partial z} l_z + q = 0. \quad (7.56)$$

Here l_x, l_y, l_z are the *direction cosines* of the outward normal n to the boundary surface [see V, §13.3.7] and q is the prescribed intensity of fluid flow across the boundary. For an impermeable boundary $q = \partial h/\partial n = 0$. The other boundary condition is h known on S_1, and for unconfined flow on the free surface $h = H$, the elevation head.

The finite element discretization of the problem with h described in terms of the nodal values $\mathbf{h}$ leads to a matrix differential equation of the form

$$\mathbf{Hh} + \mathbf{C}\frac{\partial}{\partial t}\mathbf{h} + \bar{\mathbf{F}} = \mathbf{0}, \quad (7.57)$$

which may be solved using a selected form of time-stepping scheme. Many *recurrence* relations for such *initial value* problems have been proposed [see I, §§14.12 and 14.13].

A wide range of problems have been handled by the finite element method including various types of consolidation problem and confined and unconfined transient flow problems.

7.5 Secondary consolidation

It has long been recognized that Terzaghi's model of consolidation does not always describe satisfactorily the time-settlement behaviour observed in the oedometer test. The deviation from the predicted primary consolidation curve, however, is not very substantial (~ 10 per cent), except in the case of certain organic soils, e.g. peaty soils and organic clays.

Secondary consolidation is best identified using a semilogarithmic plot of the settlement versus time. Instead of approaching a horizontal asymptote as predicted by Terzaghi's theory, the curve tends towards an inclined asymptote (Figure 7.12). Various causes have been adduced to explain this creep effect. The condition of zero lateral strain in the oedometer test creates a stress state in which shear stresses are present. The axial compression is accompanied by

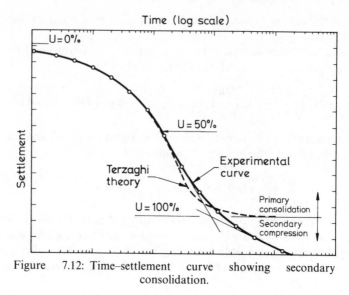

Figure   7.12: Time–settlement   curve   showing   secondary
consolidation.

relative slip and rearrangement of grains. Due to the effects of chemical bonding
at grain contacts and the adsorbed film of water surrounding the clay mineral
particles, secondary time effects are introduced, which are not accounted for in
the simple Terzaghi theory, in which it is assumed that the grain skeleton may be
represented by an elastic body. The viscosity of the bound water is greater than
that of the free pore water and it is squeezed out under the action of the
confining pressure at a much reduced rate. Secondary time effects are present
during the whole of the consolidation process, though they are overshadowed by
the primary effects during the initial stages. This fact simplifies the incorporation
of secondary time effects in a general consolidation model.

Of the various attempts to describe the process using mathematical or
physical models the one which has appealed most to researchers is the extension
of the Terzaghi spring–dashpot model. Rheological models consisting of a
spring element coupled in series to a Kelvin element (see also Chapter 11) have
been proposed by some authors. Here the more general method proposed by
Schiffmann, Ladd and Chen (1964) is described. The authors write the basic
consolidation equation (equation 7.35) in terms of the effective stress σ' and the
strain in the z direction ε_z, viz.

$$\frac{k_z}{\gamma_w} \frac{\partial^2 \sigma'}{\partial z^2} = \frac{\partial \varepsilon_z}{\partial t}. \qquad (7.58)$$

In Terzaghi's approach the pore pressure u and the voids ratio are related by
experimental data (equation 7.32), and elastic parameters are not directly

invoked. It may be shown, however, that the one-dimensional coefficient of volume compressibility is given by

$$m_v = \frac{(1 + v')(1 - 2v')}{E'(1 - v')},$$

(7.59)

where v' and E' are the Poisson ratio and elastic modulus of the grain skeleton based on effective stresses. Other values of m_v apply for two- and three-dimensional strain conditions.

It can be shown that the effective stress–strain relationship for zero lateral strain conditions is

$$3RP(\sigma') = (PS + 2QR)(\varepsilon),$$

(7.60)

where R, P, S, Q are linear differential operators, R applying to the volumetric stress, P to the deviatoric stress, S to the dilatation and Q to the deviatoric strain. The operators depend upon the viscoelastic models chosen to describe the volumetric and the deviatoric behaviour respectively (see also Chapter 11, Section 11.3). For instance, for the deviatoric stresses

$$P(\sigma_{ij}) = Q(\varepsilon_{ij}),$$

(7.61)

where $i \neq j$, and stresses and strains are written in *tensor notation* [see V, Chapter 7], and

$$P = p_m D^m + p_{m-1} D^{m-1} + \cdots + p_0,$$

$$Q = q_n D^n + q_{n-1} D^{n-1} + \cdots + q_0,$$

and

$$D^m = \frac{\partial^m}{\partial t^m}, \text{ etc.}$$

Here p_m, q_n, etc., are functions of the model parameters.

For simple elastic theory $P/Q = 2G$ and $R/S = 3K$, where G is the shear modulus and K the bulk modulus of the skeleton. To account for secondary consolidation a possible choice would be a three-parameter model for volume change and a Maxwell model (a spring in series with a dashpot), which gives effectively a five-parameter model for the soil layer, consisting of two Kelvin elements and an elastic element in series (Figure 7.13).

With a complex arrangement of elements in the models the combined model for the layer can always be reduced to a series of Kelvin elements connected in series, and a general expression for the effective stress–strain relationship would be of the form

$$\varepsilon(z, t) = a\sigma'(z, t) + \sum_{i=2}^{N} \lambda_1 \int_0^t \sigma'(z, \tau) \exp\left[-\left(\frac{\lambda_i}{b_i}\right)(t - \tau)\right] d\tau,$$

(7.62)

where a represents the immediate elastic compressibility, b_i is the retarded

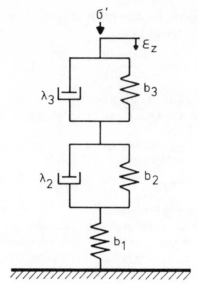

Figure 7.13: Rheological model for
secondary consolidation.

elastic compressibility and λ_i the viscosity of the ith Kelvin element. The time-dependent coefficient of compressibility, $m_v(t)$, may be obtained from equation (7.62), viz.

$$m_v(t) = \frac{\partial \varepsilon}{\partial \sigma'} = a + \sum_{i=2}^{N} b_i \left[1 - \exp\left(-\frac{\lambda_i}{b_i} t \right) \right]. \tag{7.63}$$

The equation governing secondary consolidation is then given by employing the Boltzmann superposition principle. An equation of the *Volterra* type [see III, §10.1.1] will result, viz.

$$\frac{k_z}{\gamma_w} \frac{\partial^2 \sigma'}{\partial z^2} = m_v(0) \frac{\partial \sigma'}{\partial t} + \int_0^t \frac{\partial \sigma'}{\partial \tau} \frac{\partial m_v(t - \tau)}{\partial t} \, d\tau. \tag{7.64}$$

In effect, the use of rheological models is basically a curve-fitting technique and models with as many parametes $b_1, \ldots, b_n, \lambda_1, \ldots, \lambda_n$ as necessary may be found to describe a given set of experimental data. Oedometer test results may be compared directly with results predicted using the relationship for m_v given by equation (7.63). Alternatively, if tests were carried out to investigate the basic rheological properties of the soil, i.e. by isolating volumetric and deviatoric components of stress, the linear operators R, P, S, Q would be obtained. The governing differential equation (equation 7.58) can then be written in the following form by substituting for σ' from equation (7.60), i.e.

$$\frac{k_z}{\gamma_w}(PS + 2QR)\left(\frac{\partial^2 \varepsilon_z}{\partial z^2}\right) = 3RP\left(\frac{\partial \varepsilon_z}{\partial t}\right),\qquad(7.65)$$

which together with the appropriate boundary and initial conditions constitutes a complete mathematical statement of the one-dimensional consolidation of a homogeneous compressible layer exhibiting secondary time effects.

7.6 Biot's theory of three-dimensional consolidation

Rendulic (see Christian, 1977; Lee, Lawson and Donald, 1968) extended Terzaghi's theory to include three-dimensional flow effects. He proposed the following equation:

$$c_3 \nabla^2 u + \tfrac{1}{3}\frac{\partial}{\partial t}(\sigma_1 + \sigma_2 + \sigma_3) = \frac{\partial u}{\partial t},\qquad(7.66)$$

where ∇^2 is the laplacian operator $(\partial^2/\partial x^2 + \partial^2/\partial y^2 + \partial^2/\partial z^2)$,

$$c_3 = \frac{k}{\gamma_w}\frac{E'}{3(1 - 2v')}$$

and σ_1, σ_2, σ_3 are the time-dependent externally applied stresses. The initial excess pore pressure is equated to the total volumetric stress determined from elastic theory. Biot (1941) realized that this treatment of the consolidation problem is too simplified, as the problem of water flow is handled separately from that of stress distribution. In his theory the coupling between stress and flow in the consolidation process is given due consideration. The basic equations are as set out below. As previously, compression is positive.

The equations of equilibrium in terms of effective stresses, omitting body forces due to gravity, are (cf. equation 10.1)

$$\frac{\partial \sigma'_{xx}}{\partial x} + \frac{\partial \sigma_{xy}}{\partial y} + \frac{\partial \sigma_{xz}}{\partial z} + \frac{\partial u}{\partial x} = 0,$$

$$\frac{\partial \sigma_{yx}}{\partial x} + \frac{\partial \sigma'_{yy}}{\partial y} + \frac{\partial \sigma_{yz}}{\partial z} + \frac{\partial u}{\partial y} = 0,\qquad(7.67)$$

$$\frac{\partial \sigma_{zx}}{\partial x} + \frac{\partial \sigma_{zy}}{\partial y} + \frac{\partial \sigma'_{zz}}{\partial z} + \frac{\partial u}{\partial z} = 0.$$

As in the theory of elastic solids, the geometric compatibility relations (equations 10.2) between strain and displacement can be combined with the stress–strain relations; whence equations (7.67) may be written in terms of the displacement field variables u_x, u_y, u_z, viz.

$$\mu' \nabla^2 u_x - (\lambda' + \mu') \frac{\partial \varepsilon_v}{\partial x} + \frac{\partial u}{\partial x} = 0,$$

$$\mu' \nabla^2 u_y - (\lambda' + \mu') \frac{\partial \varepsilon_v}{\partial y} + \frac{\partial u}{\partial y} = 0, \qquad (7.68)$$

$$\mu' \nabla^2 u_z - (\lambda' + \mu') \frac{\partial \varepsilon_v}{\partial z} + \frac{\partial u}{\partial z} = 0,$$

where λ' and μ' are Lamé's elastic constants for *effective stresses*, and ε_v is the volumetric strain

$$\varepsilon_v = \varepsilon_x + \varepsilon_y + \varepsilon_z = \left(\frac{\partial u_x}{\partial x} + \frac{\partial u_y}{\partial y} + \frac{\partial u_z}{\partial z} \right).$$

It should be noted that in Terzaghi's theory the elastic parameters μ', λ' are replaced by a single parameter m_v obtained from the oedometer test. In general three-dimensional stress situations it is necessary to use the elastic parameters. Equation (7.68) contains four unknowns, the three displacement components and the pore water pressure. Using Darcy's law and the continuity requirement the following equation is obtained:

$$\frac{k}{\gamma_w} \nabla^2 u = -\frac{\partial \varepsilon_v}{\partial t} = \frac{1 - 2v'}{E'} \frac{\partial}{\partial t} (\theta - 3u), \qquad (7.69)$$

where θ is the first invariant of the total stress tensor, i.e.

$$\theta = \sigma_{xx} + \sigma_{yy} + \sigma_{zz} = 3\sigma_{oct},$$

and σ_{oct} is the octahedral or volumetric stress again. This equation is similar to Rendulic's, but differs in that the octahedral stress is in terms of the stresses at a point and not of the externally applied stresses, i.e. θ varies not only as the external stresses vary in time but also as the consolidation process proceeds. Terzaghi's theory does not take any account of the coupling between the dissipation and stress distribution effects through the term θ. Equations (7.68) and (7.69), together with the boundary conditions, comprise a complete mathematical statement of the problem. McNamee and Gibson (1960) have found a method of handling the equations for the case of axial symmetry and plane strain. They introduced two displacement functions, by which means the four equations reduced to a *biharmonic* equation and a *Laplacian* equation [see IV, §8.2, and III, §9.5]. Using *integral transform techniques* [see IV, Chapter 13] analytical solutions were obtained for certain boundary conditions. The first finite element solution is due to Sandhu and Wilson (1969). Christian and

Boehmer (1970) have developed an alternative finite element method specifically for plane strain conditions.

The first of these approaches, with the variational formulation due to Sandhu, is more geenral (Christian, 1977; Sandhu, 1976).

Flow of a compressible fluid in a porous elastic medium with mass inertia effects

The basic theory of the mechanics of deformation and stress wave propagation in porous media is described in a paper by Biot (1962). The theory has been implemented in a finite element code by Ghaboussi and Wilson (1973), which has been used for seismic analysis and dynamic consolidation problems. Only a brief summary of the field equations is possible in the context of this article. The formulation is in terms of lagrangian material coordinates.

The equations of motion excluding body forces are as follows:

$$\sigma_{ij,j} = \rho \ddot{u}_i + \rho_f \ddot{w}_i,$$

$$p_{,i} = \rho_f \ddot{u}_i + \frac{1}{n} \rho_f \ddot{w}_i + \frac{\rho_f g}{k} \dot{w}_i, \tag{7.70}$$

in which σ_{ij} are the components of the total stress tensor for the saturated porous solid, p is the fluid pressure, ρ and ρ_f are the mass densities of the bulk solid and fluid respectively, k is the coefficient of permeability and n is the porosity. The quantity u_i is the independent displacement variable for the solid skeleton and w_i that for the relative displacement between the fluid and the solid. The conventional dot notation is used for time derivatives. The first equation is a statement of the equilibrium of the bulk solid (cf. equation 7.67) with added inertia terms. The second equation is the statement of the generalized Darcy flow law. The apparent mass effect incorporated in the origianl theory has been omitted here.

The strain displacement compatibility relations for the solid and fluid are

$$e_{ij} = \tfrac{1}{2}(u_{i,j} + u_{j,i}),$$

$$\zeta = w_{i,i}, \tag{7.71}$$

where e_{ij} are the components of the strain tensor for the skeleton and ζ is the volumetric strain in the fluid.

The constitutive relations for the isotropic case are

$$\sigma_{ij} = \lambda' \delta_{ij} \delta_{kl} e_{kl} + 2\mu' e_{ij} + \alpha M \delta_{ij}(\alpha \delta_{kl} e_{kl} + \zeta),$$

$$p = M(\alpha \delta_{ij} e_{ij} + \zeta), \tag{7.72}$$

where δ_{ij} is the Kronecker delta [see I, (6.2.8)], λ' and μ' are the Lamé elastic constants for the solid skeleton for fully drained conditions, M is the bulk modulus of the fluid and α is a measure of the compressibility of the solid particles (α has a value between n and 1).

7.7 Behaviour of partially saturated soils and swelling soils

In developing the various theories of consolidation the assumption of complete saturation has been invoked. In certain regions of the Earth the climatic conditions give rise to high water tables and this assumption is of practical relevance. In countries in which semiarid climatic conditions prevail, soils are usually encountered in an unsaturated condition. Under these circumstances special attention must be given to the characteristics of partially saturated soils, especially in the construction of highway and airfield pavements in swelling ground conditions.

Mechanics of unsaturated moisture diffusion

Here, only the movement of moisture through the soil will be considered. In a simplified manner the problem of an externally loaded soil mass could be solved by the theory presented in the last section by assuming the water and air phases to be equivalent to a single compressible pore fluid. However, this approach would seriously neglect certain aspects of the problem, such as the concept of *soil suction* with the associated phenomenon of swelling. Soil suction is the ability of an unsaturated soil to absorb water if it comes into contact with a pool of free water.

In partially saturated soils the presence of air in the pores influences the flow of water, since the pores are effectively diminished in size. We may still apply Darcy's law (equation 7.12), but the permeability k is no longer a constant (for a particular soil of a given voids ratio) but depends upon the degree of saturation which in turn is a function of the soil suction h, i.e. $k = k(h)$.

Lytton (1977), for example, gives the following relationship for k:

$$k = \frac{k_0}{1 + A|h|^n} \tag{7.73}$$

where k_0 is the saturated permeability and A and n are constants for a given clay material, e.g. for a particular clay shale Lytton gives the following values: $k_0 = 2.7 \times 10^{-6}$ cm/s, $A = 10^{-9}$ and $n = 3$, with units of h in cm.

The mass conservation or continuity equation combined with Darcy's law results in the following *diffusion equation* [see III, §9.1, and IV, §8.3] which is valid for non-swelling soils:

$$\frac{\partial h}{\partial t} = \frac{\partial h}{\partial \theta} \nabla \cdot (K \nabla \phi), \tag{7.74}$$

where θ is defined as the volumetric water content, $\theta = \theta(h)$ (see Appendix A), ϕ is the soil water potential, K is the permeability tensor and ∇ is the vector differential operator (grad). It should be noted that equation (7.74) is non-linear due to the dependence of K upon h. In most cases $\phi \simeq -h$.

In the case of *swelling* soils the situation is more complicated, since the soil skeleton itself is in motion and the effects of external loading and gravity may be of significance (cf. the consolidation behaviour of soils).

A generalized form of the diffusion equation for a swelling soil has been given by Yong and Warkentin (1975):

$$-\frac{\partial \theta}{\partial t} = \text{div } \bar{\mathbf{v}} + \frac{\theta}{1 + e}\left(\frac{\partial e}{\partial t} + \bar{\mathbf{v}}_\text{s}\cdot\text{grad } e\right) + \bar{\mathbf{v}}_\text{s}\cdot\text{grad } e, \qquad (7.75)$$

where $e = e(h)$, the voids ratio, is an independent function of the soil suction h. Both θ and e are required to describe volume change behaviour for swelling soils, and their relations must be obtained experimentally. Here $\bar{\mathbf{v}}$ is the velocity vector of water relative to solids and $\bar{\mathbf{v}}_\text{s}$ is the velocity vector of the solids. Darcy's equation,

$$\bar{\mathbf{v}} = -K \text{ grad } \phi, \qquad (7.76)$$

may be substituted into equation (7.75). The engineer is more conversant with the water content

$$w = \frac{\theta(1 + e)}{G_\text{s}}, \qquad (7.77)$$

where G_s is the specific gravity of the solids, than with the volumetric water content θ. This value may also be substituted in equation (7.75). In addition, it has been found convenient to work with material coordinates m instead of x, the physical coordinate, whereby the following relation holds between m and x:

$$dm = \frac{dx}{1 + e}. \qquad (7.78)$$

This substitution extends the validity of the equation to large deformations. Neglecting second-order terms (e.g. gravitational effects) the diffusion equation then becomes

$$\frac{\partial w}{\partial t} = \text{div}_m\left[\frac{K}{(1 + e)G_\text{s}}\frac{\partial \phi}{\partial w}\text{ grad}_m\, w\right], \qquad (7.79)$$

the suffix m indicating that the divergence and gradient operations are with respect to material coordinates.

Coupled elasticity and moisture diffusion

Analogous to Terzaghi's consolidation equation the aforementioned diffusion equation may be applied to one-dimensional situations, e.g. to moisture flow in an infinite sheet of finite thickness. In many cases, however, the problem is two or three dimensional, e.g. a footing resting on a thick layer of expansive clay, and it is necessary to couple the equations of elasticity and moisture diffusion in the manner of Biot (1941) for consolidation (see Section 7.6). Richards (1974) modified Biot's equations to obtain a general theory for the consolidation and swelling of partially saturated soils. He writes the equation of transient flow in the form:

$$\frac{\partial \theta}{\partial t} = \frac{1 + e}{[1 + e + \theta(de/d\theta)]} \, \mathbf{V} \cdot (K \mathbf{V} \phi). \tag{7.80}$$

The soil water potential ϕ is made up of suction, gravitational and overburden components. The stress–volume change relation is given as follows:

$$\frac{\partial \sigma'}{\partial t} = \frac{E'}{3(1 - v')} \frac{de}{d\theta} \frac{\partial \theta}{\partial t}. \tag{7.81}$$

The consolidation swelling equation is then obtained by substituting for $\partial \theta / \partial t$ from equation (7.80). This equation must be satisfied simultaneously with the equilibrium equations (cf. equation 7.67).

Based on the approach of Christian and Boehmer (1970), Richards (1974) formulated the problem for an incremental non-linear finite element solution. A similar approach is presented by Lytton (1977).

7.8 Failure of soils

Much of the science of soil mechanics is concerned with the strength and deformation of soils rather than the flow of water through them. Soil must be capable of withstanding, without failure or excessive deformation, the loads imposed upon it, such as buildings, self-weight, water, etc. Unfortunately the nature of the constitutive stress–strain law, even before failure, is more complex than the conventional structural materials of Chapter 10, and may be dependent (Chapter 11) on time also. It is only for low (10^{-5}) strains, for example, that linear laws and constant material properties are strictly valid.

There has been a great deal of effort in recent years to describe the stress–strain behaviour of soils by several alternative means. Three basic approaches have been used:

  (i) non-linear elasticity (Christian and Desai, 1977; Duncan and Chang, 1970; Gudehus and Kolymbas, 1979; Hardin, 1978);
 (ii) elasto-plastic behaviour (Drucker, Gibson and Henkel, 1957; Lade, 1977; Naylor *et al.*, 1981; Prévost, 1978; Schofield and Wroth, 1968);
(iii) endochronic models (Cuellar, Bazant and Krizek, 1977; Dungar and Nuh, 1980).

In addition to a non-linear stress–strain history and a possible complex failure mode, there is the added complication of time-dependent behaviour (creep). This *rheological* behaviour of soils may be simplified as *viscoelastic* (Chapter 11) or elastoviscoplastic (Fritz, 1982; Suklje, 1969; Zienkiewicz and Humpheson, 1977). All realistic models effectively inhibit any form of classical analysis and therefore do not in themselves make any demands on classical mathematical techniques.

If, then, a numerical method is used, neither the geometry nor the stress–strain law present insuperable problems (e.g. see Clough and Tsui, 1977). Because no special mathematical tools are needed, we will exclude here a detailed description of the many numerical and empirical models of the constitutive equations for soils, and instead turn to a more tractable problem—the stability of soil masses (e.g. the slope of a cutting, the collapse of a sheet-pile wall, the bearing capacity of a footing). This problem can be solved in a simplified limit–equilibrium fashion (Chowdhury, 1978), but we turn to the application of upper and lower bound solutions based on plasticity theory (Chen, 1975).

7.9 Plasticity analysis in soil mechanics

In the application of plasticity theory of the rigid-plastic type to soils it is necessary to divide the body into rigid and plastic zones, since, in general, only some parts of the body will be undergoing deformation. As a result, at the boundaries between zones there is a discontinuity of rate of strain (or of velocities, since the one may be derived from the other). To avoid the formation of a gap between the two zones the velocity normal to the interface must be continuous. Similar considerations lead to the continuity of stress components normal to the interface. To obtain the correct solution of a boundary-value problem it may be shown that the following requirements hold (Booker and Davis, 1977; Salençon, 1977):

(i) An equilibrium stress field exists satisfying the yield criterion and the stress boundary conditions in the plastic region.
(ii) A velocity field exists in the plastic regions satisfying the material's flow rule and the kinematic boundary conditions, which involves no negative plastic work and is compatible with the motion of the rigid regions.
(iii) An equilibrium stress field exists in the rigid regions which does not violate the yield criterion.

Upper and lower bounds to the correct load may be obtained by partial fulfilment of the above requirements, i.e. an upper bound giving a collapse load results from the satisfaction of (i) and (ii), whereas satisfaction of (i) and (iii) provides a lower bound solution. In the first case a *kinematically admissible velocity field* is obtained, whereas the second gives a *statically admissible stress field* (Prager, 1959). The limit theorems have been rigorously proven for a perfectly plastic material with an associated flow rule, while limited proofs are available for soils having zero dilatancy (Chen, 1975), there being some empirical evidence from tests on retaining walls that for soils having reached the critical void state plasticity solutions are applicable (Lee and Herington, 1974).

Illustrative example of limit analysis

The simple example is taken of the critical height of a vertical cut in isotropic, homogeneous soil obeying Coulomb's law with the parameters cohesion (c) and friction angle (ϕ). The disturbing force is the self-weight of the material for which the unit weight is γ. The solutions given below are found in Chen (1975) (cf. also Naylor *et al.*, 1981).

(a) *Upper bound solutions.* The basic failure mechanisms involve plane and log-spiral sliding surfaces. The moving body of soil is assumed to be rigid with thin transition zones of Coulomb material. It can be shown (Chen, 1975) that the application of the associated flow rule or normality concept requires that the tangential velocity change δu must be accompanied by a separation velocity change $\delta v = \delta u \tan \phi$, i.e. the kinematic slip condition implies that the relative velocity change for a straight transition zone forms an angle ϕ with the slip planes. If a body of soil performs a rigid body rotation the familiar circular failure surfaces frequently used in the limit equilibrium analysis are not permissible and the trace of the transition zone can only be described by a log-spiral surface with the property that the radii from the pole intersect the surface at an angle $\frac{1}{2}\pi + \phi$.

The failure surface for the first case (Figure 7.14a) passes through the toe making an angle β with the vertical. At the limit state the rate at which the gravity forces perform work is equal to the rate of energy dissipated in the transition zone. The gravity force, i.e. the weight of the wedge, must be multiplied by the vertical component of velocity to give the rate of work

$$E_1 = \tfrac{1}{2}\gamma H^2 \tan \beta V \cos (\phi + \beta).\tag{7.82}$$

The rate of energy dissipation along the surface of discontinuity is given by the product of the cohesion and the tangential velocity change, i.e.

$$E_2 = c \frac{H}{\cos \beta} V \cos \phi.\tag{7.83}$$

Equating E_1 and E_2 leads to the expression for the height:

$$H = \frac{2c}{\gamma} \frac{\cos \phi}{\sin \beta \cos (\phi + \beta)}.\tag{7.84}$$

If H is minimized with respect to β the result is

$$H_{\text{crit}} = \frac{4c}{\gamma} \tan \left(\frac{\pi}{4} + \frac{\phi}{2} \right),\tag{7.85}$$

which is the value obtained by the conventional Rankine analysis (Terzaghi, 1943), which is thus shown to be an upper bound solution.

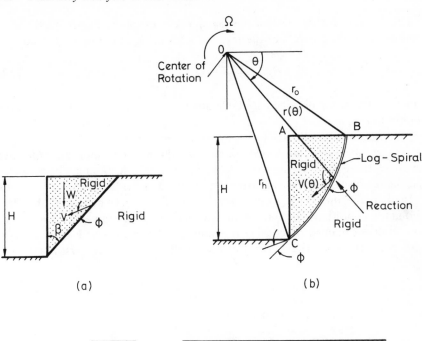

(a)   (b)

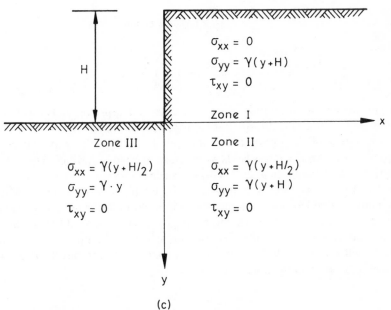

(c)

Figure 7.14: Upper bound mechanisms with (a) straight and (b) log-spiral failure surface. (c) Statically admissible stress field satisfying the plasticity condition for a vertical cut in purely cohesive soil.

The solution for the case of the log-spiral mechanism (Figure 7.14b) is too tedious to produce here, but the result is a lowering of the upper bound to

$$H_{crit} = \frac{3.83c}{\gamma} \tan\left(\frac{\pi}{4} + \frac{\phi}{2}\right), \tag{7.86}$$

which was a result obtained by Fellenius (1927) using the conventional limit equilibrium method.

(b) *Lower bound solution.* A very simple lower bound solution is presented here involving three zones in the stress field (see Figure 7.14c). As may be seen from the figure, discontinuities of stress are allowed provided there is continuity normal to the interfaces. Equilibrium is satisfied everywhere with the stress boundary conditions, and the failure criterion is not violated. The critical condition is obtained by considering failure to occur in zone I at its interface with zone II. Since the horizontal and vertical stresses are assumed to be principal stresses,

$$\sigma_3 = \sigma_{xx} = 0, \qquad \sigma_1 = \sigma_{yy} = \gamma_x.$$

At the interface I/II the Mohr circle at failure gives the condition

$$\frac{\gamma H}{2} = \frac{\gamma H}{2} \sin\phi + c\cos\phi, \tag{7.87}$$

whence

$$H_{crit} = \frac{2c}{\gamma} \tan\left(\frac{\pi}{4} + \frac{\phi}{2}\right). \tag{7.88}$$

Thus the lower bound solution gives a collapse load approximately one half that of the upper bound solution. Obviously, the assumed stress field is very approximate and, to improve the solution and close the gap between the bounds, more complex stress fields would have to be found, which requires some ingenuity. De Josselin de Jong (1978) succeeded in constructing an admissible stress field giving a lower bound $H_{crit} = 3.39\ c/\gamma$ (for $\phi = 0$ material), while an even higher value has been obtained by techniques of *linear programming* [see I, Chapter 11] (see Verruijt, 1980), namely $H_{crit} = 3.64\ c/\gamma$, so that the gap between the bounds is about 5 per cent. In passing, it should be noted that the lower bound solution (equation 7.88) is more appropriate for the case of a cut with a vertical tension crack in the soil mass.

Stress fields and characteristics

In this section the stress equations will be developed for the two-dimensional plane strain case. The assumed material behaviour is the Mohr–Coulomb failure law, which may be written in the form

$$f = R - p \sin \phi - c \cos \phi = 0, \tag{7.89}$$

in which R and p are related to the principal stresses, i.e.

$$R = \tfrac{1}{2}(\sigma_1 - \sigma_3), \qquad p = \tfrac{1}{2}(\sigma_1 + \sigma_3), \tag{7.90}$$

and by means fo the Mohr representation

$$\sigma_{xx} = p + R \cos 2\theta, \qquad \sigma_{yy} = p - R \cos 2\theta, \qquad \tau_{xy} = R \sin 2\theta, \tag{7.91}$$

where θ is the angle between the major principle stress and the direction of the x axis. These values may be substituted into the equilibrium equations

$$\frac{\partial \sigma_{xx}}{\partial x} + \frac{\partial \tau_{xy}}{\partial y} = \gamma \cos v,$$

$$\frac{\partial \tau_{xy}}{\partial x} + \frac{\partial \sigma_{yy}}{\partial y} = -\gamma \sin v, \tag{7.92}$$

in which v is the angle between the direction of gravity and the x axis. The following pair of equations then results (Booker and Davis, 1977):

$$\frac{\partial p}{\partial x}(1 + \sin \phi \cos 2\theta) + \frac{\partial p}{\partial y} \sin \phi \sin 2\theta$$

$$+ 2R\left(-\frac{\partial \theta}{\partial x} \sin 2\theta + \frac{\partial \theta}{\partial y} \cos 2\theta\right) = \gamma \cos v,$$

$$\frac{\partial p}{\partial x} \sin \phi \sin 2\theta + \frac{\partial p}{\partial y}(1 - \sin \phi \cos 2\theta) \tag{7.93}$$

$$+ 2R\left(\frac{\partial \theta}{\partial x} \cos 2\theta + \frac{\partial \theta}{\partial y} \sin 2\theta\right) = -\gamma \sin v.$$

Equations (7.93) and a pair of *quasi-linear hyperbolic equations* [see III, Chapter 9] associated with which are two families of lines known as *characteristics* [see III, §9.1]. These lines have the properties that (i) along the characteristic lines two stress fields which are continuous but analytically different may be joined, (ii) a solution known on one side of a characteristic cannot be extended beyond it without additional information. If the selected field quantities (p, θ) are continuous, discontinuities in their derivatives can only occur across a characteristic line.

The characteristics of equations (7.93) are

$$\frac{dy}{dx} = \tan(\theta - \mu), \qquad \frac{dy}{dx} = \tan(\theta + \mu), \tag{7.94}$$

where $\mu = \dfrac{\pi}{4} - \dfrac{\phi}{2}.$

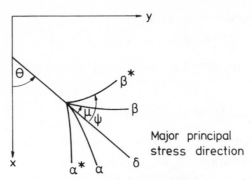

Figure 7.15: Definition of velocity
characteristics.

The characteristics are called the α and β lines (shown schematically in Figure
7.15) and equations (7.93) may be referred to these lines, in which case it is found
that

$$-\sin 2\mu \frac{\partial p}{\partial s_\alpha} + 2R \frac{\partial \theta}{\partial s_\alpha} + \gamma \left[\sin (v + 2\mu) \frac{\partial x}{\partial s_\alpha} + \cos (v + 2\mu) \frac{\partial y}{\partial s_\alpha} \right] = 0$$

and   (7.95)

$$\sin 2\mu \frac{\partial p}{\partial s_\beta} + 2R \frac{\partial \theta}{\partial s_\beta} + \gamma \left[\sin (v - 2\mu) \frac{\partial x}{\partial s_\beta} + \cos (v - 2\mu) \frac{\partial y}{\partial s_\beta} \right] = 0,$$

where $\partial/\partial s_\alpha$ and $\partial/\partial s_\beta$ refer to differentiation with respect to arc length along the
α and β lines respectively. It should be noted that the first of equations (7.95)
involves differentiation along the α line only and the second along the β line
only. This result is useful in developing numerical methods of solution to the
equations (see, for example, Booker and Davis, 1977). There are three
fundamental problems: the *initial value of Cauchy problem* (field quantities
known along a non-characteristic line), the characteristic initial value or *Goursat
problem*, and the *mixed boundary-value* problem [see IV, §8.1]. It is beyond the
limits of this section, however, to discuss the details of the numerical techniques
used to solve these problems (see Booker and Davis, 1977; Harr, 1966; Lee and
Herington, 1974; Scott, 1963; Sokolowski, 1965).

Velocity fields

The solutions based on the stress characteristics above have to be extended
for problems involving kinematic boundary conditions in addition to the static
boundary conditions by determining the velocity characteristics. This will be
necessary in any case for materials obeying a non-associated flow rule since it
needs to be checked that the internal rate of plastic work is everywhere positive.

For an associated flow rule and plane strain conditions it may be shown that the principal plastic strain rates

$$\dot{\varepsilon}_1^p = \lambda; \quad \dot{\varepsilon}_3^p = -\lambda N_\phi, \tag{7.96}$$

where $N_\phi = \tan^2 (\pi/4 + \phi/2)$ and λ is the plastic multiplier or proportionality factor. The factor λ may be eliminated by considering a ratio of the strains, e.g.

$$\frac{\dot{\varepsilon}_v^p}{\dot{\varepsilon}_1^p} = 1 - N_\phi. \tag{7.97}$$

Since equation (7.97) predicts too high a dilatancy, Davis (1968) proposed substitution of the dilatancy angle observed with real frictional materials v in the expression for N_ϕ. If ψ now denotes the angle $\pi/4 - v/2$ it may be shown that for the plane strain condition and a rigid plastic material (zero elastic strain component):

$$\frac{\partial u}{\partial x} = -\lambda(\cos 2\theta - \cos 2\psi),$$

$$\frac{\partial v}{\partial y} = \lambda(\cos 2\theta + \cos 2\psi), \tag{7.98}$$

$$\frac{\partial u}{\partial y} + \frac{\partial v}{\partial x} = -2\lambda \sin 2\theta.$$

If λ is eliminated from equations (7.98) two *quasi-linear hyperbolic* equations in u and v are obtained with the characteristics

$$\frac{dy}{dx} = \tan (\theta - \psi), \quad \frac{dy}{dx} = \tan (\theta + \psi), \tag{7.99}$$

which are called the α^* and β^* lines respectively. For the case of a material obeying an associated flow rule, $v = \phi$ and equations (7.99) are identical to equations (7.94) for the stress characteristics. In general, the stress and velocity characteristics for soils do not coincide as illustrated in Figure 7.15. If the stress field is known then the numerical procedure to determine the velocity characteristics is relatively simple using the property shown in Figure 7.15 (cf. Booker and Davis, 1977).

There may be non-uniqueness of the solution for a non-associated flow rule material. This is illustrated by Verruijt (1980) for a frictional material ($\phi > 0$) for a shear test system. It should be noted, however, as pointed out by Lee and Herington (1974), that the associated flow rule with the same ϕ values provides an upper bound which is greater than that for the non-associated case.

7.10 Further topics on soil mechanics

In the previous sections some fundamental properties of soils have been briefly described, including the flow of water in soil elements together with the associated field equations for seepage, consolidation and deformation analysis. There are various other important properties which are outside the scope of this text, e.g. the mineralogy of clays, the structure and fabric of soils, frost action, and compaction and stabilization (see Ingles, 1974). These topics are covered in several general introductions to soil behaviour to which the reader is referred (e.g. Holtz and Kovacs, 1981; Kezdi, 1970; Lambe and Whitman, 1969; Scott, 1963; Yong and Warkentin, 1975).

From the mathematical viewpoint another aspect of soil mechanics deserving greater consideration is the application of *statistical and probabilistic methods* [see II and VI] (Lumb, 1974). The treatment of soil as a particulate medium employing the theory of probability has made progress in recent years (Harr, 1977), while discontinuum mechanics approaches are also applied in rock mechanics.

Appendix A

Some useful definitions

As an aid to understanding the following quantities a diagram representing the volumetric proportions is shown in Figure 7.16.

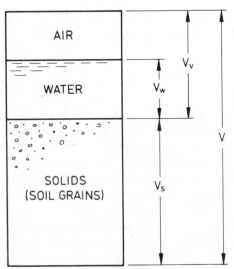

Figure 7.16: Phase diagram showing volumetric relationships.

(i) *Voids ratio.* Ratio of volume of voids to volume of solids:

$$e = \frac{V_v}{V_s}.$$

(ii) *Porosity.* Ratio of volume of voids to total volume:

$$n = \frac{V_v}{V} = \frac{e}{1+e}.$$

(iii) *Volumetric water content.* Ratio of volume of water to total volume:

$$\theta = \frac{V_w}{V} = \frac{wG_s}{1+e}.$$

(iv) *Degree of saturation.* Ratio of volume of water to volume of voids:

$$S = \frac{V_w}{V_v},$$

where $0 < S < 100$ per cent.

(v) *Gravimetric water content.* Ratio of weight of water to weight of solids:

$$w = \frac{W_w - W_d}{W_d},$$

where $\quad W_w =$ wet weight of soil,
$\qquad W_d =$ dry weight of soil.

Note that the following relation holds:

$$eS = wG_s,$$

where G_s is the *specific gravity* of the soil particles, which for common soil minerals lies in the range 2.6 to 2.9. The *unit weight of water* is denoted by γ_w.

(vi) *Bulk* or *wet unit weight of soil.* Ratio of total weight to total volume:

$$\gamma = \frac{W_w}{V} = \frac{(G_s + eS)\gamma_w}{1+e}.$$

(vii) *Dry unit weight of soil.* Ratio of weight of solids to total volume:

$$\gamma_d = \frac{W_d}{V} = \frac{\gamma}{1+w}.$$

(viii) *Submerged unit weight of soil.* Buoyancy effect:

$$\gamma' = \gamma - \gamma_w = \frac{(G_s - 1)\gamma_w}{1+e}, \qquad \text{if } S = 100 \text{ per cent.}$$

Appendix B

Stress invariants and octahedral stresses

The state of stress at a point (x, y, z) of a continuum may be described by a tensor:

$$\sigma_{ij} = \begin{bmatrix} \sigma_{xx} & \tau_{xy} & \tau_{xz} \\ \tau_{yx} & \sigma_{yy} & \tau_{yz} \\ \tau_{zx} & \tau_{zy} & \sigma_{zz} \end{bmatrix}, \tag{B1}$$

in which σ_{xx}, etc., represent the normal stresses on the six faces of an infinitesimal cubical element and τ_{xy}, etc., the shear stresses. By considering the equilibrium of the forces acting on the element it may be shown that $\tau_{xy} = \tau_{yx}$, etc. Further, due to the symmetry of the tensor it possesses three *eigenvalues*, the principal stresses σ_1, σ_2 and σ_3 [see I, §7.8] and three invariants I_1, I_2 and I_3. The principal stresses are obtained as the roots of the equation (cf. Christian and Desai, 1977; Scott, 1963)

$$\sigma^3 - I_1\sigma^2 + I_2\sigma^1 - I_3 = 0 \tag{B2}$$

whereby the stress invariants are defined as

$$I_1 = \sigma_1 + \sigma_2 + \sigma_3 = \sigma_{xx} + \sigma_{yy} + \sigma_{zz} = \sigma_{kk},$$

$$I_2 = \sigma_1\sigma_2 + \sigma_2\sigma_3 + \sigma_3\sigma_1$$

$$= \sigma_{xx}\sigma_{yy} + \sigma_{yy}\sigma_{zz} + \sigma_{zz}\sigma_{xx} - \tau_{xy}^2 - \tau_{yz}^2 - \tau_{zx}^2$$

$$= -\tfrac{1}{2}(\sigma_{ij}\sigma_{ji} - I_1^2),$$

$$I_3 = \sigma_1\sigma_2\sigma_3 \tag{B3}$$

$$= \det \begin{vmatrix} \sigma_{xx} & \tau_{xy} & \tau_{xz} \\ \tau_{yz} & \sigma_{yy} & \tau_{yz} \\ \tau_{zx} & \tau_{zy} & \sigma_{zz} \end{vmatrix}$$

$$= \sigma_{xx}\sigma_{yy}\sigma_{zz} + 2\tau_{xy}\tau_{yz}\tau_{zx} - \sigma_{xx}\tau_{yz}^2 - \sigma_{yy}\tau_{zx}^2 - \sigma_{zz}\tau_{xy}^2.$$

In the above *Einstein's summation convention* is used for suffices i, j and k [see V, §7.1].

The stress tensor may be conveniently split into a spherical part in which non-diagonal terms are zero, with components

$$\tfrac{1}{3}\sigma_{kk}\delta_{ij} = \tfrac{1}{3}I_1\delta_{ij}$$

and a deviatoric part

$$s_{ij} = \sigma_{ij} - \tfrac{1}{3}\sigma_{kk}\delta_{ij}, \tag{B4}$$

in which δ_{ij}, the Kronecker delta, is defined as

$$\delta_{ij} = 1 \quad \text{if } i = j, \qquad \delta_{ij} = 0 \quad \text{if } i \neq j.$$

(See also equations 11.34 to 11.36.)

The deviatoric stress tensor has the invariants

$$J_1 = 0,$$

$$J_2 = \tfrac{1}{6}[(\sigma_1 - \sigma_2)^2 + (\sigma_2 - \sigma_3)^2 + (\sigma_3 - \sigma_1)^2]$$

$$= \tfrac{1}{2}s_{ij}s_{ji}$$

$$= \frac{I_1^2 + 3I_2}{3}, \tag{B5}$$

$$J_3 = \tfrac{1}{3}s_{ij}s_{jk}s_{ki}$$

$$= \frac{2I_1^3 + 9I_1I_2 + 27I_3}{27}.$$

Another common designation is that of octahedral stresses. Nadai (1950) considered planes making equal intercepts with the principal stress axes—of which there are eight forming a regular octahedron. The stresses on these planes are defined in terms of stress invariants only, viz.

$$\sigma_{\text{oct}} = \frac{I_1}{3},$$

$$\tau_{\text{oct}} = \left(\frac{2J_2}{3}\right)^{1/2}. \tag{B6}$$

The terms 'deviator plane' and 'octahedral plane' are synonymous.

E.G.P.

References

Allen, D. N. De G., and Southwell, R. V. (1955). Motion of a viscous fluid past a cylinder, *Quart. J. Mech. Appl. Math.*, **VIII**, 129–145.

Biot, M. A. (1941). General theory of three dimensional consolidation, *J. Appl. Phys.*, **12**, 155–164.

Biot, M. A. (1962). Mechanics of deformation and acoustic propagation in porous media, *J. Appl. Phys.*, **33**, 1482–1498.

Booker, J. R., and Davis, E. H. (1977). Stability analysis by plasticity theory, in *Numerical Methods in Geotechnical Engineering* (Eds C. S. Desai and J. T. Christian), pp. 719–750, McGraw-Hill.

Chen, W. F. (1975). *Limit Analysis and Soil Plasticity*, Elsevier.

Chowdhury, R. N. (1978). *Slope Analysis*, Elsevier.

Christian, J. T. (1977). Two and three dimensional consolidation, in *Numerical Methods in Geotechnical Engineering* (Eds C. S. Desai and J. T. Christian), pp. 399–426, McGraw-Hill.

Christian, J. T., and Boehmer, J. W. (1970). Plane strain consolidation by finite elements, *J. Soil Mech. and Fnd. Eng., ASCE,* **96**, 1435–1457.

Christian, J. T., and Desai, C. S. (1977). Constitutive laws for geologic media, in *Numerical Methods in Geotechnical Engineering* (Eds C. S. Desai and J. T. Christian), pp. 65–115, McGraw-Hill.

Clough, G. W., and Tsui, Y. (1977). Static analysis of earth retaining structures, in *Numerical Methods in Geotechnical Engineering* (Eds C. S. Desai and J. T. Christian), pp. 506–527, McGraw-Hill.

Cuellar, V., Bazant, Z. P., and Krizek, R. J. (1977). Densification and hysterisis of sand under cyclic shear, *J. Geot. Eng. Div. ASCE,* **103**, 399–416.

Davis, E. H. (1968). Theories of plasticity and the failure of soil masses, in *Soil Mechanics—Selected Topics* (Ed. I. K. Lee), Butterworths.

De Josselin de Jong, G. (1978). Improvement of the lower bound solution for the vertical cut-off in a cohesive frictionless soil, *Geotechnique,* **28**, 197–201.

Desai, C. S., and Abel, J. F. (1972). *Introduction to the Finite Element Method,* p. 382, van Nostrand.

Drucker, D. C., Gibson, R. E., and Henkel, D. J. (1957). Soil mechanics and work-hardening theories of plasticity, *Trans. ASCE,* **122**, 338–346.

Duncan, J. M., and Chang, C. Y. (1970). Nonlinear analysis of stress and strain in soils, *J. Soil Mech. Fnd. Div., ASCE,* **96**, 1629–1653.

Dungar, R., and Nuh, S. (1980). Endochronic critical state models for sand, *J. Eng. Mech. Div., ASCE,* **106**, 951–968.

Fellenius, W. (1927). *Erdstatische Berechnungen,* Ernst, Berlin.

Fritz, P. (1982). Modelling rheological behaviour of rock, Paper presented at the Fourth Int. Conf. on Numerical Methods in Geomechanics, Edmonton.

Ghaboussi, J., and Wilson, E. L. (1973). Seismic analysis of earth dam-reservoir systems, *J. Soil Mech. Fnd. Div., ASCE,* **99**, 849–862.

Gudehus, G., and Kolymbas, D. (1979). A constitutive law of the rate type for soils, in *Numerical Methods in Geomechanics* (Ed. W. Wittke), Vol. 1, Balkema.

Hardin, B. O. (1978). The nature of stress–strain behaviour for soils, *ASCE Conf. Earthquake Engineering and Soil Dynamics, Pasadena,* **1**, 3–90.

Harr, M. E. (1966). *Foundations of Theoretical Soil Mechanics,* McGraw-Hill.

Harr, M. E. (1977). *Mechanics of Particulate Media,* McGraw-Hill.

Holtz, R. D., and Kovacs, W. D. (1981). *An Introduction to Geotechnical Engineering,* Prentice-Hall.

Ingles, O. G. (1974). Compaction and Stabilization, in *Soil Mechanics—New Horizons* (Ed. I. K. Lee), pp. 1–43, Newnes–Butterworths.

Jahnke, Emde, and Lösch (1960). *Tables of Higher Functions,* 6th edn, McGraw-Hill.

Janbu, N., Tokheim, O., and Senneset, K. (1981). Consolidation tests with continuous loading, *Proc. Tenth Int. Conf. Soil Mech. Fnd. Eng., Stockholm* **I**, 645–654.

Kezdi, A. (1970). *Handbuch der Bodenmechanik,* VEB Verlag für Bauwesen, Berlin, 4 vols. (In English, *Handbook of Soil Mechanics,* Vol. I, *Soil Physics,* Elsevier, 1974.)

Lade, P. V. (1977). Elasto-plastic stress–strain theory for cohesionless soil with curved yield surfaces, *Int. J. Sol. and Struct.,* **13**, 1019–1035.

Lambe, T. W., and Whitman, R. V. (1969). *Soil Mechanics,* Wiley.

Larsson, R. (1981). Drained behaviour of Swedish clays, Report No. 12, Swed. Geot. Inst., Linköping.

Lee, I. K., and Herington, J. R. (1974). Stability and earth pressures, in *Soil Mechanics—New Horizons* (Ed. I. K. Lee), Newnes–Butterworths.

Lee, I. K., Lawson, J. D., and Donald, I. B. (1968). Flow of water in saturated soil and rockfill, in *Soil Mechanics—Selected Topics* (Ed. I. K. Lee), Butterworths.

Lumb, P. (1974). Application of statistics in soil mechanics, in *Soil Mechanics—New Horizons* (Ed. I. K. Lee), Newnes–Butterworths.

Lytton, R. L. (1977). Foundations in expansive soils, in *Numerical Methods in Geotechnical Engineering* (Eds C. S. Desai and J. T. Christian), Chap. 13, pp. 427–457, McGraw-Hill.

McNamee, J., and Gibson, R. E. (1960). Plane strain and axially symmetric problems of the consolidation of a semi-infinite clay stratum, *Quart. J. Mech. Appl. Math.*, **13**, 210–227.

Mitchell, A. R., and Griffiths, D. F. (1980). *The Finite Difference Method in Partial Differential Equations*, Wiley.

Naylor, D. J., Pande, G. N., Simpson, B., and Tabb, R. (1981). *Finite Elements in Geotechnical Engineering*, Pineridge Press.

Nadai, A. (1950). *Theory of Flow and Fracture of Solids*, McGraw-Hill.

Prager, W. (1959). *An Introduction to Plasticity*, Addison-Wesley.

Prévost, J. H. (1978). Plasticity theory for soil stress–strain behaviour, *J. Eng. Mech. Div. ASCE*, **104**, 1177–1194.

Richards, B. G. (1974). Behaviour of unsaturated soils, in *Soil Mechanics—New Horizons* (Ed. I. K. Lee), Newnes–Butterworths.

Salençon, J. (1977). *Application of the Theory of Plasticity in Soil Mechanics*, Wiley.

Sandhu, R. S. (1976). Variational principles for finite element analysis of consolidation, in *Proc. Second Int. Conf. Num. Meth. in Geomech, ASCE*, (Ed. C. S. Desai), **1**, 20–40.

Sandhu, R. S., and Wilson, E. L. (1969). Finite element analysis of flow in porous saturated media, *J. Eng. Mech. Div., ASCE*, **95**, 641–652.

Schiffmann, R. L., and Arya, S. K. (1977). One-dimensional consolidation, in *Numerical Methods in Geotechnical Engineering* (Eds C. S. Desai and J. T. Christian), Chap. 11, pp. 364–398, McGraw-Hill.

Schiffmann, R. L., Ladd, C. C., and Chen, A. T. F. (1964). The secondary consolidation of clay, in *Rheology and Soil Mechanics, IUTAM Symp.* (Eds J. Kravtchenko and P. M. Sirieys), pp. 273–304, Grenoble (1964), Springer (1966).

Schofield, A., and Wroth, P. (1968). *Critical State Soil Mechanics*, McGraw-Hill.

Scott, R. F. (1963). *Principles of Soil Mechanics*, Addison-Wesley.

Smith, R. E., and Wahls, H. E. (1969). Consolidation under constant rates of strain, *J. Soil Mech. Fnd. Div., ASCE*, **95**, 519–539.

Sokolowski, V. V. (1965). *Static of Granular Media*, Pergamon.

Southwell, R. V. (1940). *Relaxation Methods in Engineering Science: A Treatise on Approximate Computation*, Oxford University Press.

Šuklje, L. (1969). *Rheological Aspects of Soil Mechanics*, Wiley.

Terzaghi, K. (1925). *Erdbaumechanik auf bodenphysikalischer Grundlage*, Franz Deuticke, Leipzig and Wien.

Terzaghi, K. (1943). *Theoretical Soil Mechanics*, Wiley.

Verruijt, A. (1970). *Theory of Groundwater Flow*, Macmillan.

Verruijt, A. (1980). The weak foundation of slipline analysis in soil mechanics, Delft Report XXI No. 2 (Tribute to Prof. De Josselin de Jong).

Yong, R. N. and Warkentin, B. P. (1975). *Soil Properties and Behaviour*, Elsevier.

Zienkiewicz, O. C., and Humpheson, C. (1977). Visco-plasticity: a generalised model for description of soil behaviour, in *Numerical Methods in Geotechnical Engineering* (Eds C. S. Desai and J. T. Christian), Chap. 3, pp. 116–147, McGraw-Hill.

Mathematical Methods in Engineering
Edited by G. A. O. Davies
© 1984, John Wiley & Sons, Ltd.

8

Heat Transfer

8.1 Conduction

Heat transfer by conduction takes place in a medium from a region of high temperature to that of a low temperature by the exchange of the kinetic energy of motion of the molecules by direct communication. The molecules vibrate about their mean positions and energy in the form of heat flows from the higher energy molecules to the lower energy ones. In the case of metals, the drift of large numbers of free or mobile electrons considerably enhances the transfer of energy in the substance.

The macroscopic theory of heat conduction is based on the Fourier conduction law which simply relates a heat flux to a temperature gradient via a proportionality constant k which is called the thermal conductivity of the material, i.e.

$$\frac{q}{A} = -k\frac{\partial T}{\partial x},$$ (8.1)

where q = heat being transferred through the area A,
      T = temperature at some point x in the material.

Equation (8.1) can be used to derive a general heat conduction equation for a three-dimensional body in which the temperature at any point may be varying with time. By making a simple volumetric heat balance on a typical element of the material, the following equation may be derived in differential form [see IV, §5.2]:

$$\left[\frac{\partial}{\partial x}\left(k\frac{\partial T}{\partial x}\right) + \frac{\partial}{\partial y}\left(k\frac{\partial T}{\partial y}\right) + \frac{\partial}{\partial y}\left(k\frac{\partial T}{\partial z}\right)\right] + \dot{q} = \rho C_{\mathrm{p}}\frac{\partial T}{\partial \tau}.$$ (8.2)

The terms in the brackets represent the difference between the energy being conducted in and out of the element and $\dot{q}$ represents the rate at which heat may be generated in the element per unit volume. The expression on the right-hand side of the equation represents the change in internal energy of the element with time τ, where ρ and C_{p} are the density and specific heat of the solid respectively.

Generally, unless the body is subjected to very high temperature gradients, the thermal conductivity may be assumed to be constant, in which case equation (8.2) becomes

$$\nabla^2 T + \frac{\dot{q}}{k} = \frac{1}{\alpha} \frac{\partial T}{\partial \tau}, \tag{8.3}$$

where ∇^2, the *Laplacian operator*, represents $\partial^2/\partial x^2 + \partial^2/\partial y^2 + \partial^2/\partial z^2$ [see IV, §8.2] and the thermal diffusivity $\alpha = k/\rho C_p$. In cylindrical coordinates

$$\nabla^2 = \frac{\partial^2}{\partial r^2} + \frac{1}{r} \frac{\partial}{\partial r} + \frac{1}{r^2} \frac{\partial^2}{\partial \theta^2} + \frac{\partial^2}{\partial z^2}.$$

If there is no dependence on time in a particular problem, *steady-state conditions* prevail and equation (8.3) reduces to *Poisson's equation* [see IV, §8.1]:

$$\nabla^2 T + \frac{\dot{q}}{k} = 0, \tag{8.4}$$

and if internal heat generation is absent, Laplace's equation is obtained:

$$\nabla^2 T = 0. \tag{8.5}$$

Boundary conditions

The above equations may be solved for any particular problem provided conditions are known along the boundaries of the system. These are generally specified in terms of either the dependent variable, i.e. the boundary temperature, or its gradient.

If the heat is transferred either to or from the system by a fluid then quite often it is convenient to express the gradient in terms of a *heat transfer coefficient*. The calculation of this coefficient requires a thorough understanding of thermo-hydrodynamics and in general is extremely difficult to predict. The complexities involved in such analytical approaches are dealt with later in this chapter and it is sufficient here to invoke the so-called Newton law of cooling which relates the heat transfer coefficient h to the heat flux q/A through the system boundary and temperature difference between that of the boundary T_w and the fluid T_∞. Hence

$$\frac{q}{A} = h(T_w - T_\infty). \tag{8.6}$$

Since at the surface, by Fourier's conduction law,

$$\frac{q}{A} = -k \left(\frac{\partial T}{\partial n} \right), \tag{8.7}$$

where n is the normal to the surface, then

$$\frac{\partial T}{\partial n} = -\frac{h}{k} (T_w - T_\infty). \tag{8.8}$$

8.2 One-dimensional steady-state heat conduction

In this section, the simplest form of system is considered, in which the temperature and heat flow are functions of one spatial variable only. The problems that will be examined are: a plane wall, infinite in extent, of finite thickness, and a hollow cylinder of infinite length. If heat generation is present then the equations to be solved are:

Plane wall:
$$\frac{d^2T}{dx^2} + \frac{\dot{q}}{k} = 0, \tag{8.9}$$

Hollow cylinder:
$$\frac{d^2T}{dr^2} + \frac{1}{r}\frac{dT}{dr} + \frac{\dot{q}}{k} = 0$$

or
$$\frac{d}{dr}\left(r\frac{dT}{dr}\right) + \frac{\dot{q}}{k} = 0. \tag{8.10}$$

The solutions of these equations are straightforward [see IV, §7.3] and two fundamental problems are shown in Figures 8.1 and 8.2 for either specified temperatures on the boundaries (T_1 and T_2) or in terms of a fluid temperature (T_A and T_B) and heat transfer coefficients h_1 and h_2 when $\dot{q} = 0$.

In the plane wall problem,

$$T(x) = T_1 + (T_2 - T_1)\frac{x}{L} - \frac{\dot{q}L^2}{2k}\left[\frac{x}{L} - \left(\frac{x}{L}\right)^2\right]$$

on $x = 0$:
$$\frac{q}{A} = \frac{k(T_1 - T_2)}{L} - \frac{\dot{q}L}{2},$$

on $x = L$:
$$\frac{q}{A} = \frac{k(T_1 - T_2)}{L} + \frac{\dot{q}L}{2}.$$

For the hollow cylinder in Figure 8.2 the solution of equations (8.10) is

$$T(r) = T_1 + (T_2 - T_1)\frac{\ln(r/R_1)}{\ln(R_2/R_1)} + \frac{\dot{q}}{4k}\left[(R_2^2 - R_1^2)\frac{\ln(r/R_1)}{\ln(R_2/R_1)} - (r^2 - R_1^2)\right]$$

on $r = R_1$:
$$\frac{q}{L} = \frac{2\pi k(T_1 - T_2)}{\ln(R_2/R_1)} - \pi R_1^2\dot{q}\left\{\frac{[(R_2/R_1)^2 - 1]}{2\ln(R_2/R_1)} - 1\right\},$$

on $r = R_2$:
$$\frac{q}{L} = \frac{a\pi k(T_1 - T_2)}{\ln(R_2/R_1)} + \pi R_1^2\dot{q}\left\{1 - \frac{[1 - (R_1/R_2)^2]}{2\ln(R_2/R_1)}\right\}.$$

Although these results are trivial, the consequences, when there is no heat generation present (i.e. $\dot{q} = 0$), find much practical usage in engineering heat transfer problems for analysing composite walls and cylinders. They may also be used to introduce the concept of the overall heat transfer coefficient (U). This enables the heat transferred from one fluid through a solid body to another fluid

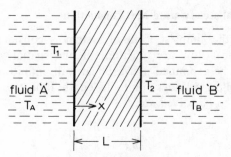

Figure 8.1: Plane wall.

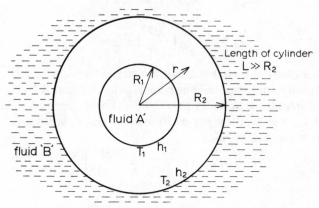

Figure 8.2: Hollow cylinder.

to be expressed in terms of the overall temperature difference of the fluids. Thus putting $\dot{q} = 0$ in the above equations:

For the wall:

$$\frac{q}{A} = \frac{k(T_1 - T_2)}{L} = h_1(T_A - T_1) = h_2(T_2 - T_B)$$

or

$$\frac{q}{A} = \frac{T_A - T_B}{1/h_1 + L/k + 1/h_2} = U(T_A - T_B),$$

where $U = \dfrac{1}{1/h_1 + L/k + 1/h_2}.$

For the cylinder:

$$q = \frac{2\pi k L(T_1 - T_2)}{\ln(R_2/R_1)} = h_1\, 2\pi R_1 L(T_A - T_1) = h_2\, 2\pi R_2 L(T_2 - T_B)$$

or

$$q = \frac{T_A - T_B}{1/h_1 \, 2\pi R_1 L + \ln(R_2/R_1)/2\pi kL + 1/h_2 2\pi R_2 L}$$

$$= U_1 \, 2\pi R_1 L (T_A - T_B)$$

$$= U_2 \, 2\pi R_2 L (T_A - T_B)$$

where $$U_1 = \frac{1}{1/h_1 + R_1/k \ln(R_2/R_1) + R_1/R_2 h_2},$$

$$U_2 = \frac{1}{R_2/R_1 h_1 + R_2/k \ln(R_2/R_1) + 1/h_2}.$$

8.3 Extended surfaces

In some instances it is desirable to enhance the heat transfer from a solid body to a fluid by attaching 'fins' or 'extended surfaces' to the main body, thereby effectively increasing the heat transfer area. Some examples of extended surface cooling are the fins of air-cooled engines, the fin extensions to the tubes of radiators of automobiles, heat sinks for the cooling of electronic components, etc. For such problems it is required to predict, for a given fin configuration, the rate of heat dissipation by the fin and the temperature variation along its length. If the fin is assumed to be surrounded by a fluid of fixed temperature T_∞ and is attached to a body of constant temperature T_0, the problem becomes one dimensional and steady state provided the only variation in fin temperature is in a direction perpendicular to the body being cooled.

Considering first a straight fin of uniform thickness and a fin in the form of a spine of uniform cross-section, then by a simple energy balance on an element of the fin (Figure 8.3) the following equation for the fin temperature is obtained:

$$\frac{d^2 T}{dx^2} - \frac{hP}{kA}(T - T_\infty) = 0, \tag{8.11}$$

where P is the perimeter of the fin and A its cross-sectional area.

By defining a temperature difference variable $\theta = T - T_\infty$ and letting $m = \sqrt{hP/kA}$, equation (8.11) becomes

$$\frac{d^2 \theta}{dx^2} - m^2 \theta = 0, \tag{8.12}$$

for which the general solution is [see IV, Example 7.4.1]

$$\theta = C_1 e^{-mx} + C_2 e^{mx}, \tag{8.13}$$

where C_1 and C_2 are constants.

Generally three physical cases are considered:

(i) The fin is very long and its temperature at its tip is essentially that of the surrounding fluid. Hence,

at $x = 0$: $\theta = T_0 - T_\infty = \dot{\theta}_0,$

at $x = \infty$: $\theta = 0.$

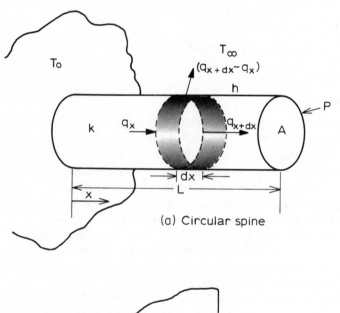

(a) Circular spine

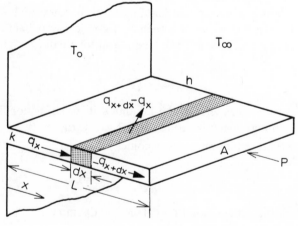

(b) Straight fin

Figure 8.3: Cooling fins.

(ii) The fin is of finite length and loses heat by convection at its tip so that

at $x = 0$: $\qquad\qquad\qquad\qquad\qquad\qquad \theta = \theta_0,$

at $x = L$: $\qquad\qquad\qquad\qquad -kA\left(\dfrac{d\theta}{dx}\right) = hA(T_L - T_\infty)$

$$= hA\theta_L,$$

where T_L is the tip temperature.

(iii) The end of the fin is insulated giving

at $x = 0$: $\qquad\qquad\qquad\qquad\qquad\qquad \theta = \theta_0,$

at $x = L$: $\qquad\qquad\qquad\qquad\qquad\qquad \dfrac{d\theta}{dx} = 0.$

As an example, the solution for case (ii) becomes:
Temperature distribution:

$$\frac{\theta}{\theta_0} = \frac{e^{m(L-x)} + e^{-m(L-x)} + (h/km)(e^{m(L-x)} - e^{-m(L-x)})}{(e^{mL} + e^{-mL}) + (h/km)(e^{mL} - e^{-mL})}, \qquad (8.14)$$

Heat transfer from the fin:

$$Q_0 = -kA\left(\frac{d\theta}{dx}\right)_{x=0}$$

$$= mkA\theta_0 \left| \frac{\tanh mL + h/km}{1 + (h/km)\tanh mL} \right|. \qquad (8.15)$$

In some cases the cross-sectional area of the fin is not uniform as, for example, in a tapered or wedge-shaped fin, as shown in Figure 8.4. The energy balance for the element now becomes

$$\frac{d}{dx}\left(kA_x \frac{dT}{dx}\right) - hP_x(T - T_\infty) = 0, \qquad (8.16)$$

where A_x and P_x are the local cross-sectional area and perimeter of the fin respectively. The area A_x can be directly expressed in terms of a fin–length ratio x/L as $A_x = tl(x/L)$, and for $l \gg t$, $P_x = 2l$, where t is the root thickness of the fin.

These values are substituted in equation (8.16) with the result that in terms of the temperature difference variable:

$$\frac{d^2\theta}{dx^2} + \frac{1}{x}\frac{d\theta}{dx} - n^2 \frac{\theta}{x} = 0, \qquad (8.17)$$

where $\qquad n = \sqrt{\dfrac{2hL}{kt}}.$

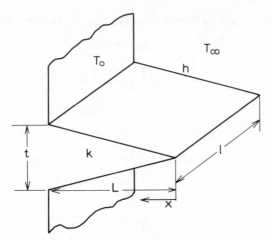

Figure 8.4: Wedge-shaped fin.

Equation (8.17) is a *modified Bessel equation* and has the solution:

$$\theta = C_1 I_0(2nx^{1/2}) + C_2 K_0(2nx^{1/2}), \tag{8.18}$$

where C_1 and C_2 are constants [see IV, §10.4.3].

Since the modified Bessel function of the second kind, K_0, approaches infinity as the argument goes to zero, then C_2 must be zero to achieve a defined temperature. Hence for a fixed body temperature difference θ_0,

$$\frac{\theta}{\theta_0} = \frac{I_0(2nx^{1/2})}{I_0(2nL^{1/2})} \tag{8.19}$$

and

$$Q_0 = l\theta_0 \sqrt{2hkt}\, \frac{I_1(2nL^{1/2})}{I_0(2nL^{1/2})}. \tag{8.20}$$

Another type of extended surface that is of considerable engineering importance is that of the annular fin of constant thickness. In this case the fin temperature is assumed to be a function of the radial coordinate r only, as shown in Figure 8.5.

The appropriate expression for the temperature variation as a function of r now becomes

$$\frac{d^2\theta}{dr^2} + \frac{1}{r}\frac{d\theta}{dr} - n^2\theta = 0, \tag{8.21}$$

where $n = \sqrt{\dfrac{2h}{kt}}.$

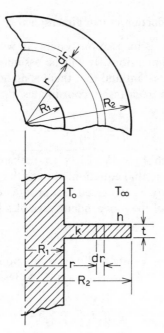

Figure 8.5: Annular fin.

This is Bessel's equation of zero order for which the solution is

$$\theta = C_1 I_0(nr) + C_2 K_0(nr), \tag{8.22}$$

where again C_1 and C_2 are constants [see IV, §§7.6.1 and 10.4.2].

For the case where the main body to which the fin is attached is at constant temperature and assuming no heat loss from the fin tip, the boundary conditions are

at $r = R_1$: $\qquad\qquad T = T_0, \qquad \theta = \theta_0,$

at $R = R_2$: $\qquad\qquad \dfrac{dT}{dr} = \dfrac{d\theta}{dr} = 0.$

This leads to a temperature distribution given by

$$\frac{\theta}{\theta_0} = \frac{I_0(nr)K_1(nR_2) + K_0(nr)I_1(nR_2)}{I_0(nR_1)K_1(nR_2) + K_0(nR_1)I_1(nR_2)} \tag{8.23}$$

and the heat transferred from the fin by

$$Q_0 = 2\pi knt\theta_0 R_1 \frac{K_1(nR_1)I_1(nR_2) - I_1(nR_1)K_1(nR_2)}{K_0(nR_1)I_1(nR_2) + I_0(nR_1)K_0(nR_2)}. \tag{8.24}$$

8.4 Steady-state heat conduction in two dimensions

In this section attention is focused on systems in which the temperature is a function of two spatial dimensions. It will be assumed that the coefficient of thermal conductivity is constant and that there is no heat generation term. The heat conduction equation in cartesian coordinates, therefore, reduces to

$$\frac{\partial^2 T}{dx^2} + \frac{\partial^2 T}{dy^2} = 0. \tag{8.25}$$

One classical approach [see IV, §8.3] to the solution of a linear and homogeneous partial differential equation, such as the above, is to assume that the dependent variable may be expressed as a product of two functions, each of which involves only one of the independent variables [see IV, §8.2]. Hence,

$$T = X(x)Y(y), \tag{8.26}$$

where $X(x)$ is a function of x only and $Y(y)$ a function of y only.

Substituting equation (8.26) into equation (8.25) results in

$$-\frac{1}{X}\frac{d^2 X}{dx^2} = \frac{1}{Y}\frac{d^2 Y}{dy^2}. \tag{8.27}$$

Each side of equation (8.27) is independent of the other because x and y are independent variables and this requires that each side be equal to some constant λ^2. Equation (8.27) therefore reduces to two ordinary equations:

$$\frac{d^2 X}{dx^2} + \lambda^2 X = 0, \tag{8.28}$$

$$\frac{d^2 Y}{dy^2} - \lambda^2 Y = 0. \tag{8.29}$$

The respective solutions to these equations are [see IV, §7.4.1]

$$X = A \cos \lambda x + B \sin \lambda x$$

and

$$Y = C \sinh \lambda y + D \cosh \lambda y$$

and therefore the temperature distribution becomes

$$T = (A \cos \lambda x + B \sin \lambda x)(C \sinh \lambda y + D \cosh \lambda y). \tag{8.30}$$

The constants A, B, C and D and the separation constant λ^2 are determined by application of the boundary conditions. This may be demonstrated by considering the case of a thin rectangular plate that is subjected to constant temperatures along three edges while the fourth is subjected to some specified temperature variation $\Gamma(x)$, as depicted in Figure 8.6.

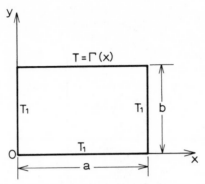

Figure 8.6: Rectangular plate.

The solution for this problem may be written in the form of an infinite series as

$$T = T_1 + \frac{2}{a} \sum_{n=1}^{\infty} \frac{\sinh(n\pi y/a)}{\sinh(n\pi b/a)} \sin\left(\frac{n\pi x}{a}\right) \int_0^a \Gamma(x) \sin\left(\frac{n\pi x}{a}\right) dx, \qquad (8.31)$$

where $\Gamma(x)$ has been expanded as a (sine) Fourier series [see IV, §§10.2 and 20.5].

For the particular case where the edge is maintained at a constant temperature T_2 at $y = b$ the solution reduces to

$$\frac{T - T_1}{T_2 - T_1} = \frac{2}{\pi} \sum_{n=1}^{\infty} \frac{1 - (-1)^n}{n} \frac{\sinh(n\pi y/a)}{\sinh(n\pi b/a)} \sin\left(\frac{n\pi x}{a}\right). \qquad (8.32)$$

If the function is in the form of a sine wave the solution for the problem is

$$T = T_1 + (T_m - T_1) \frac{\sinh(\pi y/a)}{\sinh(\pi b/a)} \sin\left(\frac{\pi x}{a}\right), \qquad (8.33)$$

where T_m is the amplitude of the temperature.

Similar procedures can be followed when using polar coordinates to describe the system shown in Figure 8.7. Assuming that axial symmetry exists, the heat conduction equation now becomes

$$\frac{\partial^2 T}{\partial r^2} + \frac{1}{r} \frac{\partial T}{\partial r} + \frac{\partial^2 T}{\partial z^2} = 0. \qquad (8.34)$$

The solution will be of the form

$$T = R(r)Z(z), \qquad (8.35)$$

which when substituted into equation (8.34) yields

$$\frac{1}{R} \frac{d^2 R}{dr^2} + \frac{1}{r} \frac{1}{R} \frac{dR}{dr} = -\frac{1}{Z} \frac{d^2 Z}{dz^2}. \qquad (8.36)$$

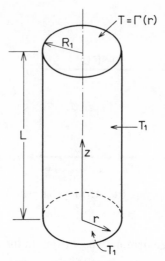

Figure 8.7: Axisymmetric
problem.

Applying the same arguments used earlier, two ordinary differential equations
are obtained in place of the original partial differential equation:

$$\frac{d^2R}{dr^2} + \frac{1}{r}\frac{dR}{dr} + \lambda^2 R = 0, \tag{8.37}$$

$$\frac{d^2Z}{dz^2} - \lambda^2 Z = 0. \tag{8.38}$$

Equation (8.37) is recognized as *Bessel's equation of zero order* [see IV,
§10.4.2] whereas equation (8.38) can be solved as before, using hyperbolic
functions [see IV, §2.13]. The temperature distribution is therefore given by

$$T = [AJ_0(\lambda r) + BY_0(\lambda r)](C \sinh \lambda z + D \cosh \lambda z), \tag{8.39}$$

where the constants A, B, C and D and the separation constant λ are determined
by the boundary conditions for the problem under consideration. For the case
depicted in Figure 8.7, the temperature distribution is

$$T = T_1 + \frac{2}{R_1^2} \sum_{n=1}^{\infty} \frac{\sinh \lambda_n z}{\sinh \lambda_n L} \frac{J_0(\lambda_n r)}{J_1^2(\lambda_n R_1)} \int_0^{R_1} r\Gamma(r)J_0(\lambda_n r)\, dr. \tag{8.40}$$

8.5 Transient conduction in one spatial dimension

The solution of transient conduction in solids in which the temperature
depends on only one spatial variable is very similar to the two-dimensional
conduction problems discussed above, since both problems involve the

determination of the temperature in terms of two independent variables. In this type of problem it is convenient to express the heat conduction equation in terms of the temperature difference θ so that equation (8.3) in cartesian coordinates may be cast in the form

$$\frac{\partial^2 \theta}{\partial x^2} = \frac{1}{\alpha} \frac{\partial \theta}{\partial \tau}. \tag{8.41}$$

A product solution of the form

$$\theta = F(\tau)G(x) \tag{8.42}$$

is assumed, where F is a function of τ only and G is a function of x only. Substitution in equation (8.41) gives

$$\frac{1}{\alpha F} \frac{dF}{d\tau} = \frac{1}{G} \frac{d^2 G}{dx^2} = \pm \lambda^2. \tag{8.43}$$

The solutions of these equations are [see IV, §7.4.1]

$$F(\tau) = C_1' e^{-\alpha \lambda^2 \tau}$$

and

$$G(x) = C_2' \sin \lambda x + C_3' \cos \lambda x, \tag{8.44}$$

which when combined gives

$$\theta = e^{-\lambda^2 \alpha \tau}(C_1 \sin \lambda x + C_2 \cos \lambda x). \tag{8.45}$$

The C's are arbitrary constants and the negative value of λ^2 has been taken since it is known that the temperature difference θ will decrease with time.

Considering the problem of the infinite plane wall that has been initially heated to some prescribed temperature distribution then, if the two surfaces of the wall are suddenly reduced and held at a constant temperature, the initial and boundary conditions are (see Figure 8.8)

$$\theta_i = \Gamma(x) - T_1 \qquad \text{at } \tau = 0,$$

$$\theta = 0 \qquad \text{at } x = 0 \text{ and } x = L \text{ for } \tau > 0.$$

The temperature distribution for these conditions is given by

$$\theta = \frac{2}{L} \sum_{n=1}^{\infty} e^{-(n\pi/L)^2 \alpha \tau} \sin\left(\frac{n\pi x}{L}\right) \int_0^L \Gamma(x) \sin\left(\frac{n\pi x}{L}\right) dx. \tag{8.46}$$

For the particular case of the wall being heated to a constant initial temperature T_i the temperature distribution is given by

$$\frac{\theta}{\theta_i} = \frac{2}{\pi} \sum_{n=1}^{\infty} \frac{1 - (-1)^n}{n} e^{-(n\pi/L)^2 \alpha \tau} \sin \frac{n\pi x}{L}, \tag{8.47}$$

where $\theta_i = T_i - T_1$.

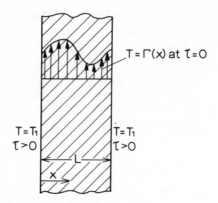

Figure 8.8: Infinite plane wall.

The rate of heat transfer from both surfaces of the wall for this case is given by employing the Fourier conduction equation (i.e. $q = -kA(\partial T/\partial x)$) at $x = 0$ and $x = L$ and integrating it over the required time interval τ. Hence,

$$Q = \int_0^\tau q \, d\tau. \tag{8.48}$$

This yields

$$\frac{Q}{Q_i} = \frac{4}{\pi^2} \sum_{n=1}^\infty \frac{1 - (-1)^n}{n^2} (1 - e^{-(n\pi/L)^2 \alpha \tau}), \tag{8.49}$$

where Q_i is the initial internal energy of the wall. Again a similar procedure can be applied to an infinitely long cylinder which is subjected to a sudden boundary temperature change and where the body temperature is a function of radial coordinate only (Figure 8.9).

Here the general solution is

$$\theta = Ce^{-\lambda^2 \alpha \tau} J_0(\lambda r), \tag{8.50}$$

where C is a constant.

For the case where the cylinder is initially at a uniform temperature, the distribution is given by

$$\frac{\theta}{\theta_i} = 2 \sum_{n=1}^\infty e^{-\lambda^2 \alpha \tau} \frac{1}{\lambda_n R_1} \frac{J_0(\lambda_n r)}{J_1(\lambda_n R_1)} \tag{8.51}$$

and the heat transfer by

$$\frac{Q}{Q_i} = 4 \sum_{n=1}^\infty \left(\frac{1}{\lambda_n R_1} \right)^2 (1 - e^{-\lambda_n^2 \alpha \tau}). \tag{8.52}$$

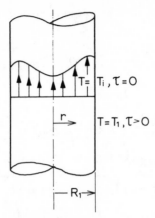

Figure 8.9: Infinite long
cylinder.

Other cases for the above geometries which are of more practical significance can be considered by assuming that the heat transferred from the surfaces to the surroundings is by convection. For the infinite wall of thickness $2L$ then, by invoking the Newton law of cooling (equation 8.8), the boundary conditions for the wall at an initial uniform temperature before being exposed to a fluid of lower temperature become (cf. Figure 8.10)

$$\theta_i = T_i - T_\infty \qquad \text{at } \tau = 0,$$

at $x = 0$:
$$\frac{\partial \theta}{\partial x} = 0,$$

at $x = L$:
$$\frac{\partial \theta}{\partial x} = -\frac{h}{k}\theta.$$

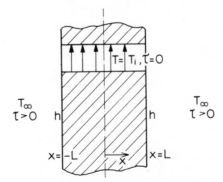

Figure 8.10: Infinite wall, transient
heating.

Equation (8.45) may then be solved to give the temperature distribution in the wall as

$$\frac{\theta}{\theta_i} = 2 \sum_{n=1}^{\infty} e^{-\lambda_n^2 \alpha \tau} \frac{\sin^2 \lambda_n L}{\lambda_n L + \sin \lambda_n L \cos \lambda_n L} \cos \lambda_n x \qquad (8.53)$$

and the heat transfer by

$$\frac{Q}{Q_i} = 2 \sum_{n=1}^{\infty} \frac{1}{\lambda_n L} \frac{\sin^2 \lambda_n L}{\lambda_n L + \sin \lambda_n L \cos \lambda_n L} (1 - e^{-\lambda_n^2 \alpha \tau}). \qquad (8.54)$$

Similar conditions can be imposed on the infinitely long cylinder problem for the solution of equation (8.50) which will yield the following temperature distribution and heat transfer rates respectively:

$$\frac{\theta}{\theta_i} = 2 \sum_{n=1}^{\infty} \frac{1}{\lambda_n R_1} e^{-\lambda_n^2 \alpha \tau} \frac{J_0(\lambda_n r) J_1(\lambda_n R_1)}{J_0^2(\lambda_n R_1) + J_1^2(\lambda_n R_1)}, \qquad (8.55)$$

$$\frac{Q}{Q_i} = 4 \sum_{n=1}^{\infty} \frac{1}{(\lambda_n R_1)^2} \frac{J_1^2(\lambda_n R_1)}{J_0^2(\lambda_n R_1) + J_1^2(\lambda_n R_1)} (1 - e^{-\lambda_n^2 \alpha \tau}). \qquad (8.56)$$

Analytical solutions of equation (8.41) can be obtained for other simple geometric shapes, such as the semi-infinite solid depicted in Figure 8.11. If its surface temperature is suddenly lowered from its initial value of T_i to some value T_0 the initial and boundary conditions are

$$\theta_i = T_i - T_0 \qquad \text{at } \tau = 0,$$

$$\theta = 0 \qquad \text{at } x = 0 \text{ and } \tau > 0.$$

Equation (8.41) can now be solved subject to the above conditions using *Laplace transform* [see IV, §13.4.8] techniques to give a temperature distribution of

$$\frac{\theta}{\theta_0} = \text{erf} \frac{x}{2\sqrt{\alpha \tau}}, \qquad (8.57)$$

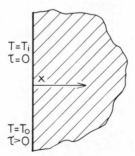

Figure 8.11: Semi-
infinite solid.

where erf $x/2\sqrt{\alpha\tau}$ is the *Gauss error function* [see IV, §13.4.1]. At the surface the
heat transfer rate is given by

$$Q = -kA\left(\frac{\partial\theta}{\partial x}\right)_{x=0} = \frac{kA\theta_i}{\sqrt{\pi\alpha\tau}}. \tag{8.58}$$

Multidimensional systems

It is apparent from the foregoing sections that analytical solutions are only
available for simple geometries with relatively straightforward initial and
boundary conditions. For the transient cases the temperature and heat transfer
histories are usually presented in graphical form in most engineering heat
transfer textbooks. The data are presented in dimensionless form using a
dimensionless distance and two dimensionless numbers which are known as
Biot and *Fourier moduli*. For the case of the infinite cylinder, for example, the
dimensionless distance parameter is r/R, with the two moduli referred to above
being defined as:

Biot modulus: $$B_i = \frac{hR_1}{k},$$

Fourier modulus: $$F_o = \frac{\sigma\tau}{R_1^2}.$$

Under some conditions, however, the solutions obtained for the one space
dimension cases can be used to predict the behaviour of problems which involve
two or three space dimensions by the principle of superposition. For example, a
finite cylinder may be formed by the intersection of an infinite cylinder and the
infinite plane wall. Hence the superposition of the one spatial dimensional
solutions discussed earlier will produce the transient temperature distributions
for a finite cylinder.

Generally, multidimensional heat conduction problems, whether steady state
or transient, are not amenable to analytical solutions. This is brought about by
complex geometrics and by boundary and initial conditions which cannot be
simply specified. A limited number of problems can be solved satisfactorily using
analogue methods. However, by far the most powerful technique used for
calculating temperature distributions and heat fluxes in engineering heat
conduction problems are based on the *finite difference* [see III, Chapter 9] or
finite element methods (see Chapter 15).

8.6 Convection

When heat is transferred from a solid to a fluid the process is defined as
convection. If, for example, the fluid is being pumped past the body the
mechanism by which the heat is transferred is known as *forced convection*. When

the fluid movement past the surface is due entirely to density changes in the fluid brought about by temperature variations, the process is called *free* or *natural convection*. In both cases, if the fluid velocities are low, the motion of the fluid past the heat transfer surface may be streamline or laminar in form. At higher velocities the streamline character of the fluid motion may break down, resulting in turbulent flow. For either type of flow, it is obvious that the prediction of heat transfer rates in any convection heat transfer situation requires the simultaneous solution of the fundamental equations governing the motion of the fluid in conjunction with the equation which is based on the conservation of energy in the fluid (i.e. energy in the form of heat). In all problems, however, the prime objective of any theoretical approach must be the calculation of the heat transfer coefficient *h*, which has already been mentioned in the preceding section. There it was related to the heat being transferred from within a solid body to the surrounding fluid by Newton's law of cooling. On the fluid side, the heat transfer coefficient can be related to the temperature gradient in the fluid itself by the following equation:

$$h = -\frac{k'(\partial T/\partial n)}{\Delta T},\tag{8.59}$$

where k' = thermal conductivity of the fluid,
$\quad\quad\;\; T$ = temperature at some point in the fluid near to the surface,
$\quad\quad\;\; n$ = the normal to the surface through which heat is being transferred,
$\quad\;\, \Delta T$ = some representative temperature difference between the fluid and the surface.

The above equation assumes that the fluid motion immediately adjacent to the wall is laminar in form.

The governing equations for laminar fluid flow can be found in Section 2.1 and their thermal equivalents in Section 6.5 on atmospheric circulation. Simplified forms of these equations will be used in the examination of various problems which will be considered in the following sections. Accordingly, for laminar flow the discussion will be restricted to flat plate and circular tube geometries.

8.7 Laminar flow systems

Free or natural convection on a vertical flat plate

The first case to be considered is that of natural convection from a vertical flat plate which is heated to a constant temperature. If the usual boundary layer assumptions are made for the plate, aligned as in Figure 8.12, we have first the

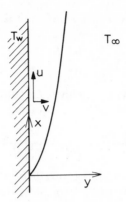

Figure 8.12: Natural
convection on vertical
surface.

incompressibility condition:

$$\frac{\partial u}{\partial x} + \frac{\partial v}{\partial y} = 0 \tag{8.60}$$

Second, the Navier–Stokes equations become

$$u\frac{\partial u}{\partial x} + v\frac{\partial u}{\partial y} = F_x + v\frac{\partial^2 u}{\partial y^2}. \tag{8.61}$$

Third, the thermal equivalent of (8.61) is a simplified version of (6.79):

$$u\frac{\partial T}{\partial x} + v\frac{\partial T}{\partial y} = \alpha\frac{\partial^2 T}{\partial y^2}, \tag{8.62}$$

valid for moderate temperature changes. The constant α is the thermal diffusivity of the fluid and v is again its kinematic viscosity. Here the body force F_x, which sustains the fluid motion, may be written as the product of the coefficient of cubical expansion β, the gravitational constant g and the local temperature difference $(T - T_w)$.

The boundary conditions for the system illustrated in Figure 8.12 are

along $y = 0$: $\qquad\qquad T = T_w, \qquad u = 0,$

at $y = \infty$: $\qquad\qquad T = T_\infty, \qquad u = 0.$

The governing equations can now be reduced to ordinary forms by the introduction of a *similarity parameter*, so that

$$\eta = \left(\frac{g\beta\,\Delta T x^3}{4v^2}\right)^{1/4}\frac{y}{x}, \tag{8.63}$$

where $\Delta T = T_w - T_\infty$.

If the velocity component in the x direction is assumed to be given by the expression

$$u = 4v \left(\frac{g\beta\,\Delta T}{4v^2}\right)^{1/4} x^{1/2} f_1(\eta), \tag{8.64}$$

where $f_1(\eta)$ is some function of η only, then, for continuity to prevail, the velocity component v must satisfy the following equation:

$$v = v \left(\frac{g\beta\,\Delta T}{4v^2}\right)^{1/4} x^{-1/4}[\eta f'_1(\eta) - 3f_1(\eta)]. \tag{8.65}$$

Introducing a dimensionless temperature distribution $\theta = (T - T_\infty)/(T_w - T_\infty)$ and assuming that it is a function of η only, then substituting it and equations (8.64) and (8.65) into equations (8.61) and (8.62) yields

$$f'''_1 + 3ff''_1 - 2(f'_1)^2 + \theta = 0 \tag{8.66}$$

and

$$\theta'' + \frac{3v}{\alpha} f_1\theta' = 0 \tag{8.67}$$

with the boundary conditions becoming

$$f'_1(0) = 0, \qquad \theta(0) = 1,$$

$$f'_1(\infty) = 0, \qquad \theta(\infty) = 0.$$

The above equations are coupled and must be solved simultaneously to give velocity and temperature distributions. It is then necessary to evaluate $\theta'(0)$ as a function of x for substitution into equation (8.59) for the determination of the heat transfer coefficient (Ostrach, 1952). The average heat transfer coefficient h_{av} is obtained by integrating the local value over the whole length of the plate L and presenting the result in dimensionless form, so that

$$N_u = \frac{0.902 G_r^{1/4} P_r^{1/2}}{(0.861 + P_r)^{1/4}}, \tag{8.68}$$

where the dimensionless groups are

Nusselt number:
$$N_u = \frac{h_{av}L}{k'},$$

Grasshof number:
$$G_r = \frac{g\beta\,\Delta T L^3}{v^2},$$

Prandtl number:
$$P_r = \frac{v}{\alpha}.$$

The above theoretical result also applies to the natural convection from the outside of vertical cylinders.

Forced convection on a flat plate

It has already been demonstrated, in Section 3.2, that the isothermal hydrodynamic laminar boundary layer on a flat plate may be analysed using the similarity method proposed by Blasius (1908). If the plate is maintained at a constant temperature T_w the rate of heat transferred to the fluid, passing over it at a constant velocity U_∞ and temperature T_∞, may be derived by employing the energy equation. It is also assumed that there is no viscous dissipation in the boundary layer and that the hydrodynamic and thermal boundary layers are coincident.

Using the Blasius similarity parameter, viz.

$$\eta = y\sqrt{\frac{U_\infty}{vx}}$$

and assuming that the temperature distribution perpendicular to the plate is a function only of η, the energy equation becomes

$$\theta'' + \frac{1}{2}\frac{v}{\alpha} f\theta' = 0, \tag{8.69}$$

where, this time,

$$\theta = \frac{T - T_w}{T_\infty - T_w}.$$

The boundary conditions are

on $\eta = 0$: $\qquad\qquad\qquad\qquad \theta = 0$,
at $\eta = \infty$: $\qquad\qquad\qquad\qquad \theta = 1.0$.

In deriving the above it is further assumed that the velocity distributions determined by the Blasius solution are applicable here, i.e.

$$u = U_\infty f',$$

$$v = \frac{1}{2}\sqrt{\frac{U_\infty v}{x}}\,(\eta f' - f).$$

Using the Pohlhausen (1921) method of solution, the following temperature distribution is obtained:

$$\theta = \theta'(0) \int_0^\eta \exp\left(-\frac{1}{2}\frac{v}{\alpha}\int_0^\eta f\,d\eta\right) d\eta, \tag{8.70}$$

where $\theta'(0)$ gives the required temperature gradient at the surface which can be evaluated by setting $\theta = 1$ at $\eta = \infty$ and using the Blasius values of $f(\eta)$ by *numerically integrating* equation (8.70) [see III, Chapter 7]. In dimensionless form the expression for the heat transfer rates to the plate may be written as

$$N_u = 0.664 P_r^{0.343} R_e^{0.5},$$ (8.71)

where the Reynolds number $R_e = V_\infty L / v$ and L is again the length of the plate.

Forced convection in circular tubes

The classical solutions for forced convection heat transfer in circular pipes are due to Leveque (1928) and Gratz (1885). The Leveque solution strictly applies to the unidirectional fluid flow region in the near vicinity of the wall. It assumes that, for constant fluid properties, the velocity distribution near the wall is linear and that the heat is transferred by conduction only in a direction perpendicular to that of the fluid flow. Hence, the energy equation reduces to

$$u \frac{\partial T}{\partial x} = C_1 y \frac{\partial T}{\partial x} = \frac{\partial^2 T}{\partial y^2}$$ (8.72)

with the boundary conditions (see Figure 8.13)

at $x = 0$: $y > 0$, $T = T_\infty$,
for $x > 0$: $y = 0$, $T = T_w$.

By introducing a new variable X such that

$$X = y \left(\frac{C_1}{9\alpha x} \right)^{1/3}$$ (8.73)

then equation (8.72) can be reduced to the ordinary differential equation

$$\frac{d^2 T}{dx^2} + 3X^2 \frac{dT}{dx} = 0$$ (8.74)

with the new boundary conditions becoming

at $X = 0$: $T = T_w$,
at $X = \infty$: $T = T_\infty$.

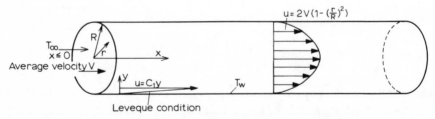

Figure 8.13: Boundary conditions for Leveque and Gratz solutions.

The solution of equation (8.74) for the above conditions yields [see IV, §7.3]

$$\frac{T - T_w}{T_\infty - T_w} = \frac{1}{0.893} \int_0^X e^{-X^3}\, dX. \tag{8.75}$$

The integral in equation (8.75) is a function of x and y and has been evaluated by Abramowitz (1951). Equation (8.75) may be differentiated to obtain the temperature gradient at the wall. Thus

$$\left(\frac{\partial T}{\partial y}\right)_{y=0} = \frac{T_\infty - T_w}{0.893} \left(\frac{C_1}{9\alpha x}\right)^{1/3}. \tag{8.76}$$

C_1 can be evaluated for the circular tube if the velocity distribution is assumed to be the fully developed profile of parabolic form, viz.

$$u = 2V\left[1 - \left(\frac{r}{R}\right)^2\right] = \frac{2V}{R^2}(2Ry - y^2), \tag{8.77}$$

where V is the average velocity of flow in the tube. Hence,

$$C_1 = \left(\frac{\partial u}{\partial y}\right)_{y=0} = \frac{4V}{R}. \tag{8.78}$$

Substituting for C_1 in equation (8.76) and then combining the resulting expression with equation (8.59) yields

$$N_u = 1.077(P_e)\left(\frac{D}{x}\right)^{1/3}. \tag{8.79}$$

For the circular tube of diameter D the dimensionless groups are

Nusselt number: $\qquad\qquad N_u = \dfrac{hD}{k'},$

Peclet number: $\qquad\qquad P_e = \dfrac{VD}{\alpha}.$

Also, $\qquad\qquad P_e = R_e P_r,$

where $\qquad\qquad R_e = \dfrac{VD}{\nu}.$

The original Gratz (1885) solution was one of the first theoretical analyses of laminar fluid flow heat transfer in tubes. He assumed that the fluid properties were constant and that the velocity profile was the fully developed form given by equation (8.77). The energy equation then becomes, in polar coordinates and assuming axisymmetry,

$$2V\left[1 - \left(\frac{r}{R}\right)^2\right]\frac{\partial T}{\partial x} = \alpha\left[\frac{1}{r}\frac{\partial}{\partial r}\left(r\frac{\partial T}{\partial r}\right)\right] \tag{8.80}$$

with the boundary conditions

at $x = 0$: $0 < r < R$, $T = T_\infty$,
for $x > 0$: $r = R$, $T = T_w$.

Equation (8.80) can be solved using the product method similar to those discussed in Sections 8.4 and 8.5. Hence,

$$\frac{T - T_\infty}{T_w - T_\infty} = \sum_{n=0}^{\infty} C_n \phi_n \left(\frac{r}{R}\right) \exp\left[-\frac{\lambda_n^2(x/R)}{P_e}\right], \tag{8.81}$$

where C_n are coefficients and $\phi_n(r/R)$ and λ_n^2 are functions and exponents respectively which are determined by boundary conditions. The local Nusselt number variation along the length of the tube is then given by evaluating the gradient of equation (8.81) at $r = R$ and employing equation (8.59). Hence,

$$N_u = \frac{\displaystyle\sum_{n=0}^{\infty} \frac{C_n \phi_n'(1)}{2} \exp\left[-\frac{\lambda_n^2(x/R)}{P_e}\right]}{2 \displaystyle\sum_{n=0}^{\infty} \frac{C_n \phi_n'(1)}{2\lambda_n^2} \exp\left[-\frac{\lambda_n^2(x/R)}{P_e}\right]}. \tag{8.82}$$

The classical Gratz solution has been extended for other boundary conditions by Sellars, Tribus and Klein (1956). For example, for a linearly varying wall temperature the Nusselt number variation is derived as

$$N_u = \frac{\frac{1}{2} + 4 \displaystyle\sum_{n=0}^{\infty} \frac{C_n \phi_n'(1)}{2\lambda_n^2} \exp\left[\frac{-\lambda_n^2(x/R)}{P_e}\right]}{\frac{88}{768} + 8 \displaystyle\sum_{n=0}^{\infty} \frac{C_n \phi_n'(1)}{2\lambda_n^4} \exp\left[\frac{-\lambda_n^2(x/R)}{P_e}\right]}. \tag{8.83}$$

For the mathematical details of this problem and others of this type the reader should consult Sellars, Tribus and Klein (1956).

8.8 Turbulent flow systems

When the laminar nature of a fluid flow system breaks down and becomes turbulent, it is accompanied by substantial increases to the resistance to the fluid motion and rates of heat transfer. Turbulence is characterized by the random chaotic motion of the fluid particles. At any point in the fluid field the velocities, pressures and temperatures vary with respect to both time and distance. It is convenient to consider that the instantaneous values of these dependent variables should be made up of a time-*averaged* quantity and a time-*dependent* fluctuating value. If the discussion is restricted to steady-state two-dimensional flow only, the substitution of instantaneous velocities, pressures and temperatures into the laminar equations of motion and energy result in the following equations:

Continuity:
$$\frac{\partial u_i}{\partial x_i} = 0 \tag{8.84}$$

Momentum:
$$\rho u_j \frac{\partial u_i}{\partial x_j} = F_i - \frac{\partial P}{\partial x_i} + \frac{\partial}{\partial x_j}(-\rho\overline{u_i'u_j'})$$

$$+ \mu \frac{\partial}{\partial x_j}\left(\frac{\partial u_i}{\partial x_j} + \frac{\partial u_j}{\partial x_i}\right) \qquad \text{for } i, j = 1, 2, \tag{8.85}$$

Energy:
$$\frac{\partial}{\partial x_i}(u_i T + \overline{u_i' T'}) = \frac{\partial}{\partial x_i}\left(\frac{\mu}{\sigma}\frac{\partial T}{\partial x_i}\right), \tag{8.86}$$

where u_i, P and T now denote time-averaged values of velocity, pressure and temperature; the overbar indicates time-averaging and dashes denote fluctuations from the mean. σ is the Prandtl number based on the molecular quantities of the fluid.

The terms $-\rho\overline{u_i'u_j'}$ are referred to as the Reynolds stresses and may be written as

$$-\rho\overline{u_i'u_j'} = \mu_T\left(\frac{\partial u_i}{\partial x_j} + \frac{\partial u_j}{\partial x_i}\right) - \tfrac{1}{3}\rho\overline{u_i'^2}\,\delta_{ij}, \tag{8.87}$$

where μ_T is known as the eddy or turbulent viscosity and δ_{ij} is the usual Kronecker delta [see I, (6.2.8)]. The last term on the right-hand side of equation (8.87) is coupled with the pressure P so that substituting for the Reynolds stresses in equation (8.85) yields

$$u_j \frac{\partial u_i}{\partial x_j} = F_i - \frac{1}{\rho}\frac{\partial p}{\partial x_i} + \frac{\partial}{\partial x_j}\left[\frac{\mu_e}{\rho}\left(\frac{\partial u_i}{\partial x_j} + \frac{\partial u_j}{\partial x_i}\right)\right] \tag{8.88}$$

where $p = P + \tfrac{1}{3}\overline{u'^2}$ and $\mu_e = \mu + \mu_T$, $\qquad$ (8.89)

the so-called effective viscosity.

A similar hypothesis exists for the fluctuating velocity–temperature terms, viz.

$$\overline{u'T'} = -\frac{\mu_T}{\sigma_T}\frac{\partial T}{\partial x_i} \tag{8.90}$$

where σ_T is the turbulent Prandtl or Schmidt number. The energy equation becomes

$$u_i \frac{\partial T}{\partial x_i} = \frac{\partial}{\partial x_i}\left[\left(\frac{\mu}{\sigma} + \frac{\mu_T}{\sigma_T}\right)\frac{\partial T}{\partial x_i}\right]. \tag{8.91}$$

The derivation of the last equation has, of course, neglected any dissipation terms. Normally it is assumed that σ_T is a constant for fully turbulent flows and appropriate values for σ and σ_T for various fluids and flow geometries have been tabulated elsewhere (Lauder and Spalding, 1972). Equations (8.84), (8.88) and (8.90) can only be solved, however, if the distribution of μ_T is specified.

The most popular means of depicting μ_T for many years was the Prandtl mixing length concept. He used his mixing length concept to model the Reynolds stresses in the following way:

$$\mu_T = \rho l^2 \left(\frac{\partial u_i}{\partial x_j} + \frac{\partial u_j}{\partial x_i} \right), \tag{8.92}$$

where *l, the mixing length*, is the distance traversed by a particle of fluid across the flow before losing its momentum. However, to be useful the mixing length variation must be specified as a function of position and this is geometry-dependent. Before the advent of modern-day electronic computers, the mixing length hypothesis found wide usage for simple boundary layer type flows and fully developed duct flows such as the circular pipe. The assumption of the distribution of mixing length had to be such that, on substitution into the appropriate equations, the resulting expressions utilizing it could be integrated simply. If heat transfer was involved, it was usually assumed that the mechanisms for the transport of momentum and energy were analogous and that the turbulent viscosity was substituted directly into the energy equation whilst further assuming that the turbulent Prandtl number was unity.

Clearly, if more complicated flows are to be analysed, a more sophisticated turbulent viscosity model is required which will reflect the local level of turbulence intensity and its structure at the point under consideration. Such a model is that proposed by Prandtl (1945) and Kolgomorov (1942) as

$$\mu_T = C_\mu K^{1/2} l, \tag{8.93}$$

where l is representative of the turbulent length scale, C_μ is a constant and

$$K = \tfrac{1}{2} \overline{u_i'^2} \tag{8.94}$$

is the fluctuating turbulent kinetic energy.

By suitable manipulation of the Navier–Stokes (Taylor and Morgan, 1981) or momentum equations the turbulence kinetic energy term can be shown to satisfy equation (8.9):

$$\rho u_j \frac{\partial K}{\partial x_j} = \frac{\partial}{\partial x_j} \left[\left(\mu + \frac{\mu_T}{\sigma} \right) \frac{\partial K}{\partial x_j} \right] + \mu_T \frac{\partial u_i}{\partial x_j} \left(\frac{\partial u_i}{\partial x_j} + \frac{\partial u_j}{\partial x_i} \right) - \rho \frac{C_D K^{3/2}}{l}, \tag{8.95}$$

where C_D is a constant. The above represents the so-called *one-equation model* of turbulence but still requires specification of the mixing length and a knowledge of the constants appearing in equations (8.93) and (8.95).

The turbulence model can be further refined by introducing a dissipation rate

ε which is linked to the turbulence kinetic energy and mixing length by the expression

$$l = \frac{K^{2/3}}{\varepsilon}. \tag{8.96}$$

The derivation of the transport equation for ε is long and complicated and has been performed by Harlow and Nakayama (1968). Their results may be simplified to

$$\rho u_j \frac{\partial \varepsilon}{\partial x_j} = \frac{\partial}{\partial x_j}\left[\left(\mu + \frac{\mu_T}{\sigma_\varepsilon}\right)\frac{\partial \varepsilon}{\partial x_j}\right] + C_1 \frac{\varepsilon}{K}\frac{\mu_T}{\rho}\frac{\partial u_i}{\partial x_j}\left(\frac{\partial u_i}{\partial x_j} + \frac{\partial u_j}{\partial x_i}\right) - \frac{C_2}{C_\mu}\frac{\varepsilon^2}{K}, \tag{8.97}$$

where σ_ε, C_1 and C_2 are constants.

The derivation of the two-equation model of turbulence necessitates the use of constants, as seen above, in regions very close to solid walls where laminar viscosity dominates. However, these quantities may not be constant but, in these flow regions, the governing equations can be simplified and solved analytically. The solutions obtained can then be compared with experimental evidence to evaluate the constants. Similar arguments apply to the treatment of the energy equation (8.90) if the value of σ_T varies in the region of a solid surface, as it will do when the temperature gradients in these regions are very steep. Each problem, whether it is of a free or forced convection type, must be treated on its own merits. Once the constants are determined equations (8.84), (8.88), (8.90), (8.93), (8.95), (8.96) and (8.97) may be solved using numerical techniques to yield, ultimately, the velocity, pressure and temperature fields. Much work is being conducted on this subject at the moment using both finite difference [see III, Chapter 9] and the finite element methods of Chapter 15, and for further information the reader should consult specialist research literature and journals.

However, many occasions arise where it is far more convenient to resort to empirical formulae for the prediction of heat transfer rates. These are generally presented in dimensionless groups which are based on specified temperatures or temperature analyses for the evaluation of fluid properties. For example, in the case of external flows, the fluid properties are usually determined at a film temperature which is the arithmetic mean of the surface temperature and the free stream or undisturbed fluid temperature. The fluid properties for forced convection inside ducts are usually evaluated at a bulk temperature which is defined as

$$T_B = \frac{\displaystyle\int_0^R Tur\,dr}{\displaystyle\int_0^R ur\,dr}. \tag{8.98}$$

This temperature may also be used when calculating the heat transfer coefficient using equation (8.59). If there are large temperature variations in the flow, appreciable differences can arise in the fluid properties. These are generally compensated for by further including a ratio of fluid viscosities, evaluated at the bulk temperature conditions and the wall temperature. Empirical and practical relations for free convection and forced convection heat transfer for a wide variety of fluids and geometries may be found in the many heat transfer textbooks which are currently on the market (see, for example, Chapman, 1974; Holman, 1976; and Knudsen and Katz, 1958).

J.O.M.

References

Abramowitz, J. (1951). Table of the integral $\int_0^x e^{-u^3}\, du$, *J. Math. and Phys*, **30**, 162.

Blasius, H. (1908). Grenzschichten in Flüssigkeiten mit Kleiner Reibung, *Z. Math. u. Phys.*, **56**.

Chapman, A. J. (1974). *Heat Transfer*, 3rd ed., Macmillan.

Gratz, L. (1885). Uber die Wärmeleitungs fährigkeiten der Flussigkeiten, *Ann. Phys. v. Chem.*, **25**, 337.

Harlow, F. H., and Nakayama, P. I. (1968). Transport of turbulence energy decay rate, Los Alamos Sci. Lab. Rep. No. L.A. 3854.

Holman, J. P. (1976). *Heat Transfer*, 4th ed., McGraw-Hill.

Knudsen, J. G., and Katz, D. L. (1958). *Fluid Dynamics and Heat Transfer*, McGraw-Hill.

Kolmogorov, A. H. (1942). Equations of turbulent motions of an incompressible fluid, *Izv. Acad. Nauk. SSSR Ser. Phys.*, **6**, No. 1–2.

Launder, B. E., and Spalding, D. B. (1972). *Mathematical Models of Turbulence*, Academic Press.

Leveque, J. (1928). Les Lois de la Transmission de al Chaleur, *Ann. Mines*, **12**, 13, 201, 305, 381.

Ostrach, S. (1952). An analysis of laminar convection flow and heat transfer about a flat plate parallel to the direction of the generating body forces, NACA Tech. Note No. 2635.

Pohlhausen, E. (1921). Der Wärmeaustausch zwischen festen Körpern und Flüssigkeiten mit kleiner Reibung und kleiner Warmeleitung, *Z. Math. u. Phys*, **1**, 115.

Prandtl, L. (1945). *Uber ein neues Farmelsystem für die ausgebildete Turbulenz*, Nachr. Akad. der Wissenschaft in Gottingen.

Sellars, J. R., Tribus, M., and Klein, J. S. (1956). Heat transfer to laminar flow in a round tube or flat conduit—Gratz problem extended, *Trans ASME*, **78**, 441.

Taylor, C., and Morgan, K. (1981). *Computational Techniques in Transient and Turbulent Flow*, Pineridge Press.

Mathematical Methods in Engineering
Edited by G. A. O. Davies
© 1984, John Wiley & Sons, Ltd.

9

Chemical Engineering

9.1 Introduction

This chapter is a very concise account of chemical engineering. The interested reader will find a much fuller treatment in any of the six references quoted at the end of the chapter.

The subject of chemical engineering is primarily concerned with the large-scale production of chemicals by the conversion of other chemical raw materials. This invariably involves the transfer of heat and mass to and from flowing fluids, both to enhance and sustain the chemical reactions and also to separate and purify the products. In a chemical laboratory, the complete process may be carried out in one vessel: mixing the raw materials, heating to initiate the reaction, evaporating to distill off the product, etc. Some valuable chemicals, which are only required in relatively small quantities, are produced commercially in this way by taking a convenient quantity of the raw material and converting it to the desired product as a single batch. In this type of batchwise operation, the conditions in the vessel change with time and *unsteady-state mathematical equations* have to be used to describe such a process.

When large quantities of a product are produced commercially, it is important to obtain consistency of product quality and maximum utilization of the equipment. Such large-scale processes are normally performed continuously with the process materials flowing through a series of vessels of various shapes, each vessel being specifically designed for one part of the process. Each unit, mixer, reactor, heat exchanger, absorber, distillation column, condenser, extractor, drier, etc., can thus operate efficiently. The aim is to run the complete plant continuously, at steady state, with raw materials fed into it and products leaving it at a constant rate. The behaviour of the units in such a process will be described by *steady-state mathematical equations*. However, time-dependent equations sometimes have to be used to consider the *stability* of continuous steady-state systems and also the control of them. This brief article will be devoted to some of the individual units which comprise the continuous type of process.

The fundamental ideas upon which the mathematical equations are based are the laws of conservation of energy and mass, including conservation of specific materials, and the rate equations governing the rate of chemical reaction, the rate of heat transfer and the rate of mass transfer. Most scientists and engineers are familiar with Fourier's law of heat conduction in solids (equation 8.1), written again as

$$q_x = -\kappa A_x \frac{\partial T}{\partial x},$$ (9.1)

where q_x = rate of flow of heat in the direction parallel to the x axis,
κ = thermal conductivity,
A_x = area of the surface normal to the x axis through which heat is conducted, and
$\partial T/\partial x$ = temperature gradient in the x direction.

In a flowing fluid, heat can be conducted across the streamlines to adjacent layers of fluid and to solid walls. However, the detailed temperature distribution is probably unknown and instead of the temperature gradient, the known finite temperature difference ΔT between two well-defined points or regions is divided by a hypothetical distance L to obtain a representative gradient. Thus equation (9.1) becomes

$$q_x = -\left(\frac{\kappa}{L}\right) A_x \, \Delta T.$$ (9.2)

The thermal conductivity, which is strictly defined for solids or stagnant fluids, is of doubtful applicability, particularly in turbulent flow when eddies also transport heat between fluids and boundaries. The ratio (κ/L) between two dubious quantities is replaced by the film heat transfer coefficient h which is a more useful property both for experimental determination and for theoretical prediction. Equation (9.2) thus becomes

$$q_x = hA_x(-\Delta T).$$ (9.3)

Chemical engineering texts pay little attention to the negative sign and it will also be ignored here. Heat transfer between two bulk fluids separated by a flat wall of thickness d and conductivity κ is governed by

$$q = UA \, \Delta T,$$ (9.4)

where U is the overall heat transfer coefficient given by

$$\frac{1}{U} = \frac{1}{h_1} + \frac{d}{\kappa} + \frac{1}{h_2},$$ (9.5)

which is just an equation for the total resistance to heat transfer expressed as the sum of the three resistances in series along the heat transfer path and ΔT is now the overall temperature difference between the two bulk fluids. In the mathematical models which follow, U must be used for the overall heat transfer coefficient when ΔT is the overall temperature difference, but when a detailed temperature distribution is being modelled on one side of a solid boundary, the heat transfer coefficient h must be used to represent the remaining heat transfer resistance between the wall and the other bulk fluid.

For mass transfer, the equation analogous to equation (9.1) is *Fick's law* of mass diffusion

$$N_x = -DA_x \frac{\partial c}{\partial x},\qquad(9.6)$$

where
N_x = molar flow rate of the diffusing species in the x direction,
D = molecular diffusivity, and
$\partial c/\partial x$ = molar concentration gradient.

In a manner analogous to heat transfer, $(\partial c/\partial x)$ is often replaced by $(\Delta c/L)$, the negative sign is dropped and equation (9.6) is written

$$N_x = kA_x \Delta c,\qquad(9.7)$$

where k is the film mass transfer coefficient. It is unfortunate that the symbol k is used conventionally for both thermal conductivity and the mass transfer coefficient. Furthermore, reaction rate constants are also signified by k. In this chapter, κ will represent thermal conductivity, k_2 with a numerical suffix will denote a reaction rate constant and k_G with a letter suffix will denote a mass transfer coefficient.

9.2 Reactors

Chemical reactions are classified as either homogeneous or heterogeneous. In a homogeneous reaction, each molecule of reactant is equally likely to react at any point in space whereas in a heterogeneous reaction, some points in space are reaction sites and others are not. For example, a reaction between a solid and a gas can only occur at the solid surface or, more frequently, a catalysed reaction can only occur at the catalyst surface.

Homogeneous reactions

If the rate of reaction R is proportional to the concentration of one reactant and independent of any other reactant (e.g. because the other reactant is in significant excess) the equation can be written

$$R = k_1 c_A.\qquad(9.8)$$

Such a reaction is of *first order* and its reaction rate constant (k_1) has dimensions of units per second. If the rate of reaction is proportional to the concentrations of each of two reactants, it is said to be of *second order*. Thus

$$R = k_2 c_A c_B,\qquad(9.9)$$

where k_2 has dimensions of cubic metres per kilogram-mole-second (m³/kg mol s). In general,

$$R = k_3 c_A^m c_B^n \tag{9.10}$$

if the reaction is mth order in A and nth order in B. Although m and n are usually integers there is no necessity that they should be, but in a particular case they will be determined experimentally and such experimental results are usually fitted adequately with integer values. In the general case of equation (9.10) the units of k_3 are not simple.

Heterogeneous reactions

It is assumed that there are active sites on the surface of a catalyst and that reactions occur by attaching the reactants to these active sites, the captured molecules then react and the products are detached. This frees the active sites to capture further reactant molecules. For example, the reaction

$$A + B \rightarrow C$$

may occur in three stages involving active sites S:

$$A + S \rightarrow AS,$$

$$B + S \rightarrow BS,$$

$$AS + BS \rightarrow C + 2S.$$

If one of these steps is significantly slower than the other two, it will control the overal rate of reaction. The rate of the reaction

$$B + S \rightarrow BS$$

could well be proportional to the concentration of B and the concentration of S, just as for a homogeneous reaction. However, the concentration of the reactant B in the neighbourhood of the active surface may well differ from the concentration of B in the bulk fluid phase, particularly since B may have to diffuse through the pores to the interior of a porous catalyst particle. The product C must also diffuse out of the pores in the opposite direction and both diffusion processes will require a concentration gradient. The concentration of active sites S can be expressed in volume terms by multiplying the surface concentration of vacant sites by the surface area per unit volume of the catalyst bed. The reaction rate can then be expressed in terms similar to equation (9.10) above, but the vacant sites will not be consumed by the reaction as a normal reactant would be. The rate of reaction equation therefore looks different and in terms of real reactants (vacant site concentrations having been eliminated algebraically) may take a form such as

$$R = \frac{k_1 c_A c_B}{(1 + K_1 c_A + K_2 c_C)(1 + K_3 c_B)}. \tag{9.11}$$

The constants K_1, K_2 and K_3 are related to the equilibrium between molecules in space and molecules of the same species on active sites. Assuming that the catalyst particles are small and the fluid velocities are large enough, then the mass transfer resistances will also be small. If the chemical reactions are also slow (as is likely if a catalyst is required) then the bulk fluid concentrations can be used in equation (9.11) without introducing too much error provided that the values of k_1 and the K_i's are suitably adjusted. When all of these approximations are valid, a heterogeneous reaction can be treated as a homogeneous reaction with reaction sites uniformly distributed in space.

Most reaction rates are temperature dependent and these effects are included in the rate constants (k_i) in the form [see IV, §2.11]:

$$k_1 = k_0 \exp\left(\frac{-E}{RT}\right), \tag{9.12}$$

where E = activation energy,
           R = gas constant,
and T = absolute temperature.

Equations introduced so far require concentrations to be expressed in moles per unit volume and rates of reaction in moles per unit volume per unit time [see I, §17.2.1]. For consistency, it is as well to have flow rates expressed either in molar or volume terms, and when the law of conservation of mass is applied it is only necessary to introduce either molecular weights or densities to convert the units.

Tubular reactors

The design of a typical tubular reactor is illustrated in Figure 9.1. The outer vessel is divided into three compartments with the upper and lower chambers connected together by a bundle of similar tubes which pass through the central chamber. These cylindrical tubes contain the catalyst pellets if a catalyst is required. Reactants are fed into the lower chamber which distributes them amongst the tubes. Products and unused reactants emerge into the top chamber and are led off to the next piece of equipment in the process. The coolant for an exothermic reaction or hot fluid for an endothermic reaction are fed into the shell or central chamber to surround each tube and induce heat transfer with the reactants through the tube walls. Fairly general equations will be derived for this type of reactor so that they can be simplified subsequently to forms which are appropriate to some special cases.

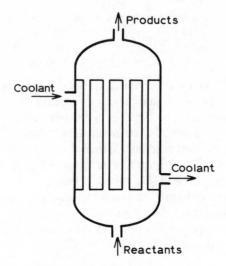

Figure 9.1: Tubular reactor.

A single reactor tube (turned horizontally for convenience) is illustrated in Figure 9.2 to show how *cylindrical polar coordinates* [see V, §2.1.5] are used to select a typical annular ring element of space. Material and energy balances are applied to this element in order to obtain differential equations describing the behaviour inside the tube. The linear flow velocity in the z direction is denoted by u (m/s) which can be a function of both r and z. The function of r will describe the velocity profile, and volume changes due to reaction will cause variation with z. The other two velocity components in the radial and angular directions are assumed to be zero. Random motions in the radial direction are assumed to contribute to the radial effective diffusivity. Other symbols used [cf. I, Table 17.7.1] are:

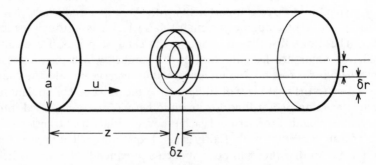

Figure 9.2: Reactor tube.

c concentration of limiting reactant A (kg mol/m³),

C_p specific heat of reaction mixture (J/kg K),

D_L axial effective diffusivity (m²/s),

D_R radial effective diffusivity (m²/s),

h heat transfer coefficient outside the tube wall (J/m² s K),

ΔH heat of reaction (J/kg mol),

κ effective thermal conductivity of tube contents (W/m K),

R_c rate of chemical reaction (kg mol/m³ s),

T temperature of reaction mixture (K),

T_s temperature of fluid in surrounding jacket (K),

ρ_F fluid density (kg/m³).

(Note that the diffusivities in the axial and radial directions are allowed to be unequal because the random fluid movements in the two directions may be unequal and diffusion occurs in the fluid phase. In contrast, there is only one thermal conductivity because the heat is primarily conducted within the solid catalyst particles which are stationary.)

At a plane of constant z, the annular ring element shown in Figure 9.2 has a surface area of $2\pi r\,\delta r$ through which reactant A enters the element by two mechanisms, bulk transport and axial diffusion, at a combined rate of

$$N_z = 2\pi r\,\delta r u c - D_L\,2\pi r\,\delta r\,\frac{\partial c}{\partial z}.$$

The rate at which the reactant leaves the annular element through the adjacent plane at $z + \delta z$ will be

$$N_{z+\delta z} = N_z + \frac{\partial N_z}{\partial z}\,\delta z,$$

provided that δz is arbitrarily small. Reactant A diffuses radially into the element through the cylindrical surface of radius r at a rate of

$$N_r = -D_R\,2\pi r\,\delta z\,\frac{\partial c}{\partial r}$$

and out at the cylindrical surface at $r + \delta r$ at a rate of

$$N_{r+\delta r} = N_r + \frac{\partial N_r}{\partial r}\,\delta r,$$

provided that δr is arbitrarily small. The volume of the annular element is $2\pi r\,\delta r\,\delta z$ and the reaction consumes reactant A at a rate

$$N_c = 2\pi r\,\delta r\,\delta z R_c.$$

Equating these terms to account for reactant A gives

$$N_z - N_{z+\delta z} + N_r - N_{r+\delta r} = N_c,$$

which simplifies to

$$-\frac{\partial N_z}{\partial z}\,\delta z - \frac{\partial N_r}{\partial r}\,\delta r = N_c$$

or substituting for N_z, N_r and N_c,

$$-2\pi r\,\delta r\,\delta z\,\frac{\partial}{\partial z}\left(uc - D_L\frac{\partial c}{\partial z}\right) + 2\pi\,\delta z D_R\,\delta r\,\frac{\partial}{\partial r}\left(r\frac{\partial c}{\partial r}\right) = 2\pi r\,\delta r\,\delta z R_c. \qquad (9.13)$$

Dividing throughout by $-2\pi r\,\delta r\,\delta z$ gives

$$\frac{\partial}{\partial z}(uc) - D_L\frac{\partial^2 c}{\partial z^2} - D_R\left(\frac{\partial^2 c}{\partial r^2} + \frac{1}{r}\frac{\partial c}{\partial r}\right) + R_c = 0. \qquad (9.14)$$

Similar considerations for the heat balance on the element give

$$q_z = 2\pi r\,\delta r u\rho_F C_p T - \kappa\,2\pi r\,\delta r\,\frac{\partial T}{\partial z},$$

$$q_r = -\kappa\,2\pi r\,\delta z\,\frac{\partial T}{\partial r},$$

$$q_c = 2\pi r\,\delta r\,\delta z R_c\,\Delta H,$$

and hence

$$-2\pi r\,\delta r\,\delta z\,\frac{\partial}{\partial z}\left(u\rho_F C_p T - \kappa\frac{\partial T}{\partial z}\right) + 2\pi\,\delta z\kappa\,\delta r\,\frac{\partial}{\partial r}\left(r\frac{\partial T}{\partial r}\right) = 2\pi r\,\delta r\,\delta z R_c\,\Delta H,$$

which simplifies to

$$\rho_F u C_p\frac{\partial T}{\partial z} - \kappa\frac{\partial^2 T}{\partial z^2} - \kappa\left(\frac{\partial^2 T}{\partial r^2} + \frac{1}{r}\frac{\partial T}{\partial r}\right) + R_c\,\Delta H = 0. \qquad (9.15)$$

Although ρ_F and u are both functions of z, their product, which is the total mass velocity, must remain constant and can be taken out of the derivative with respect to z. It has been assumed that κ, C_p, D_L and D_R are independent of temperature and concentration. The heat of reaction ΔH will be positive for an endothermic reaction and negative for an exothermic reaction. The rate of reaction R_c is positive for a reactant such as A.

Equations (9.14) and (9.15) together with a further equation relating ρ_F (and hence u) to temperature and extent of reaction are the equations governing the behaviour of the reactant A within the tube. The equation determining density depends upon whether the fluid is a liquid or a gas. If it is a liquid and the density of the products is similar to the density of the reactants, ρ_F and hence u can be considered to be constant. For a gas, which is the more usual situation, the density will be inversely proportional to the absolute temperature and, for a narrow range, an average constant value can be used again. However, if the reaction changes the number of moles in the system there may be a considerable density change in a gaseous reacting mixture.

It has been tacitly assumed in the above derivation that only one reaction takes place in the tube. In many processes there are competing or consecutive reactions when the term R_c in equation (9.14) must be composed of all reactions consuming A offset by any reactions which produce A. Similarly, the term $R_c \Delta H$ in equation (9.15) must be replaced by a series of terms containing contributions from all reactions, not only those concerning A.

Boundary conditions can be derived from first principles by moving the annular ring element to each boundary in turn and taking material and energy balances including the special conditions at the boundaries. Conditions at radial boundaries are

at $r = 0$:
$$\frac{\partial c}{\partial r} = 0, \qquad \frac{\partial T}{\partial r} = 0,$$

at $r = a$:
$$\frac{\partial c}{\partial r} = 0, \qquad -\kappa \frac{\partial T}{\partial r} = h(T - T_s).$$

If T_s is a function of z, a further equation is required to relate $\partial T_s / \partial z$ to the coolant flow rate and the number of tubes in the bundle.

In the axial direction, the boundary conditions at $z = 0$ require a little care because there will be step changes in both concentration and temperature if the linear velocity is small compared with the conductivity or axial diffusivity. Thus if c_0 is the concentration in the lower chamber of the reactor, the input to the annular ring element will be

$$2\pi r \, \delta r u_0 c_0,$$

but the axial output will still be given by

$$2\pi r \, \delta r u c - D_L \, 2\pi r \, \delta r \frac{\partial c}{\partial z} + \frac{\partial}{\partial z}\left(2\pi r \, \delta r u c - D_L \, 2\pi r \, \delta r \frac{\partial c}{\partial z}\right) \delta z$$

and the other terms in equation (9.13) will also be the same. Thus equation (9.13) becomes

$$2\pi r \, \delta r u_0 c_0 - 2\pi r \, \delta r u c + D_L \, 2\pi r \, \delta r \frac{\partial c}{\partial z} - 2\pi r \, \delta r \, \delta z \frac{\partial}{\partial z}\left(u c - D_L \frac{\partial c}{\partial z}\right)$$
$$+ 2\pi \, \delta z D_R \, \delta r \frac{\partial}{\partial r}\left(r \frac{\partial c}{\partial r}\right) = 2\pi r \, \delta r \, \delta z R_c.$$

Dividing throughout by $2\pi r \, \delta r$ (but not $-\delta z$) gives

$$u_0 c_0 - u c + D_L \frac{\partial c}{\partial z} - \delta z \frac{\partial}{\partial z}\left(u c - D_L \frac{\partial c}{\partial z}\right) + \delta z D_R\left(\frac{\partial^2 c}{\partial r^2} + \frac{1}{r}\frac{\partial c}{\partial r}\right) = \delta z R_c.$$

In the limit as $\delta z \to 0$, all terms except the first three become negligibly small. Therefore,

at $z = 0$:
$$u_0 c_0 - uc + D_L \frac{\partial c}{\partial z} = 0 \qquad (9.16)$$

and the concentration c just inside the tube will be lower than the concentration c_0 in the lower chamber due to diffusion along the tube. Similarly, the temperature just inside the tube will be different from that in the lower chamber according to the equation

$$u_0 \rho_{F0} C_p T_0 - u \rho_F C_p T + \kappa \frac{\partial T}{\partial z} = 0. \qquad (9.17)$$

Equations similar to (9.16) and (9.17) can be derived for the outlet of the reactor tube but the interpretation is different. It is *physically* unreasonable to expect a stepwise concentration increase and equation (9.16) becomes

at $z = L$:
$$\frac{\partial c}{\partial z} = 0. \qquad (9.18)$$

Similarly, equation (9.17) becomes

at $z = L$:
$$\frac{\partial T}{\partial z} = 0. \qquad (9.19)$$

A final point on this general case concerns the application of *numerical methods* to solve equations (9.14) and (9.15) [see III, Chapter 9]. At the axis where $r = 0$, $\partial c/\partial r = 0$ and $\partial T/\partial r = 0$, one term in each equation becomes indeterminate. This is resolved by application of *L'Hôpital's rule* [see IV, §3.4] which gives

$$\lim_{r \to 0} \frac{1}{r} \frac{\partial c}{\partial r} = \frac{\partial^2 c}{\partial r^2} \quad \text{and} \quad \lim_{r \to 0} \frac{1}{r} \frac{\partial T}{\partial r} = \frac{\partial^2 T}{\partial r^2}$$

and these minor modifications must be made to equations (9.14) and (9.15) before numerical calculations are performed along the axis.

The following three special cases of the above mathematical model of a tubular reactor are worth considering.

(a) *Homogeneous reactions in tubular reactors.* If the reactor tubes contain no catalyst pellets and *the flow is laminar*, the velocity profiles will be parabolic (see Chapter 3). Thus

$$u = u_m \left(1 - \frac{r^2}{a^2} \right),$$

where u_m may be a function of z only. The serious disadvantage of using such an empty reactor tube is that some of the reactants will flow rapidly up the axis of the tube whilst the rest travels slowly along the walls. This leads to a great spread of residence times within the reactor tube. Such behaviour can only be tolerated if the desired product does not deteriorate or further react to form undesired products due to spending too long in the reaction zone. In any event, such operation will be inefficient and is therefore seldom used.

(b) *Adiabatic plug-flow reactors.* In this mode, there is no heat transfer through the walls of the tubes and the coolant jacket is not used during normal operation. The resistance to flow in all parts of the tube is the same due to the presence of catalyst pellets so that the velocity profile will be uniform except for a sharp drop to zero at the tube walls. However, this is offset to some extent by the presence of natural channels at one pellet diameter from the walls due to the way in which the pellets tend to become ordered against the solid wall. Provided that the ratio of pellet diameter to tube diameter is small enough, there will be little error in assuming that the fluid velocity is uniform, i.e. that u is independent of r. A little thought will reveal that neither c nor T will vary with r and the standard equations (9.14) and (9.15) degenerate to the pair of *ordinary differential equations* [see IV, §7.1, and III, §8.1.2]

$$\frac{d}{dz}(uc) - D_L \frac{d^2c}{dz^2} + R_c = 0, \tag{9.20}$$

$$\rho_F u C_p \frac{dT}{dz} - \kappa \frac{d^2T}{dz^2} + R_c \Delta H = 0, \tag{9.21}$$

with the boundary conditions (9.16), (9.17), (9.18) and (9.19). If u is independent of z, equations (9.20) and (9.21) would be linear were they not coupled together by the non-linear reaction rate terms.

Equations (9.8), (9.9) and (9.10) show how the rate of reaction varies with concentration and equation (9.12) shows how it varies with temperature. Since the rate of reaction can only be linear with concentration for a first-order reaction and cannot be linear in temperature, equations (9.20) and (9.21) need to be solved *numerically as boundary-value problems* [see III, §8.2.2] in all but the most trivial case of temperature-independent reaction rate, when the equations are not even coupled.

(c) *Negligible axial mixing.* Because diffusion is usually a slow process compared with bulk transport, the second term in each of equations (9.14) and (9.15) is usually ignored in comparison with the first term. When this is done, the boundary conditions at $z = 0$ have no step changes and the boundary conditions at $z = L$ are no longer required. The problem changes its mathematical form from a *boundary-value problem* to an *initial-value problem*, which can be solved by the *Crank–Nicolson* or *Runge–Kutta* methods [see III, §§8.2.2 and 9.4].

Continuous stirred tank reactors

These are often used for homogeneous liquid phase reactions if it is important to conduct the reactions at a closely controlled temperature. The apparatus consists of a tank equipped with a stirrer, turbine, or recycling pump which ensures that the contents are always thoroughly mixed. Reactants are fed into the tank at a constant rate and material is withdrawn from the tank at the same mass flow rate to maintain the same quantity of material within the tank. If operated in this mode for a sufficiently long time, the system will reach a steady state. The composition and temperature of the stream leaving the reactor will be identical to that of the reacting mixture in the tank so that the rate of reaction is determined by the composition of this product stream. This feature makes it easy to calculate the size of reactor required for a given duty, but it also limits the conversion which can be achieved efficiently.

If the volumetric flow rate and molar concentration of the limiting reactant A are L_0 and c_0 into the tank and L_1 and c_1 out of the tank, then $L_0 c_0 - L_1 c_1$ moles of A must be reacted per unit time in the tank of volume V. The rate of reaction per unit volume R_c will be a known function of c_1, depending upon the kinetics of the reaction. Thus

$$L_0 c_0 - L_1 c_1 = R_c V, \qquad (9.22)$$

from which the volume of the reactor can be determined.

To illustrate further features of this kind of reactor, assume that there is no volume change during the reaction ($L_1 = L_0$) and that it is of first order ($R_c = k_1 c_1$). Equation (9.22) becomes

$$V = \frac{L}{k_1}\left(\frac{c_0}{c_1} - 1\right). \qquad (9.23)$$

It is clear that the size of reactor required for 90 per cent conversion will be nine times the volume required for 50 per cent conversion, and 90 per cent is not particularly good. For a second-order reaction the increase in size with conversion is much more severe. The problem is that the reaction rate is low because the concentration in the reactor is low. One way of overcoming this is to put two reactors in series so that the first one can operate at a higher concentration and the second one can achieve the required conversion. Retaining the same assumptions that led to equation (9.23) gives

$$V_1 = \frac{L}{k_1}\left(\frac{c_0}{c_1} - 1\right), \qquad (9.24)$$

$$V_2 = \frac{L}{k_1}\left(\frac{c_1}{c_2} - 1\right). \qquad (9.25)$$

If the reactors are to be of equal volume, equations (9.24) and (9.25) can be solved for c_1 to give

$$c_1 = \sqrt{c_0 c_2}$$

and the volume of each reactor is

$$V_1 = V_2 = \frac{L}{k_1}\left(\sqrt{\frac{c_0}{c_2}} - 1\right). \qquad (9.26)$$

The size of reactors needed for 90 per cent conversion is now only five times the size required for 50 per cent conversion. Furthermore, for 90 per cent conversion equation (9.23) shows that the volume of the single reactor is $9L/k_1$ but for two reactors in series equation (9.26) shows that each is of volume $2.16L/k_1$ (i.e. one quarter of the size). Further economy can be made by putting more reactors in series, but the law of diminishing returns shows that the extra cost of multiple reactors soon outweighs the savings of each being smaller.

It can be proved by differentiation that, for the above system of a first-order reaction with no volume change during reaction, equal-sized reactors give a minimum total volume for a given overall conversion. It is by no means clear that equal-volume reactors would be optimum for a second-order reaction scheme, and the method of *dynamic programming* [see IV, Chapter 16] would have to be used to determine the size ratios in any individual problem. However, the convenience of having all reactors of the same size is probably more important in many cases than a slight improvement in efficiency which could be obtained with reactors of different sizes.

Once the conversion to be achieved in each reactor is known, the heat load can be calculated from the heat of reaction. If it is small it can be provided by putting heat exchangers between the reactors so that the heat content of each feed can be used to control the temperature in each reactor. With higher heat loads or temperature-sensitive reactants, some form of heating or cooling must be fitted in the reactor itself.

9.3 Vapour liquid equilibria

The saturated vapour pressure (p_i^0) exerted by a pure liquid on its surroundings at equilibrium can be expressed as a function of temperature T by the *Antoine equation*

$$\ln p_i^0 = A_i - \frac{B_i}{C_i + T}, \qquad (9.27)$$

where A_i, B_i, C_i are constants for a given pure liquid (i). This correlation is valid over a wide temperature range and there is sufficient flexibility in the three constants to allow any units for p_i^0 and T.

For an ideal mixture of volatile liquids, each component exerts a partial pressure p_i on its surroundings in proportion to its mole fraction x_i in the liquid. Thus

$$p_i = x_i p_i^0. \tag{9.28}$$

The total vapour pressure exerted by all components in the liquid is obtained as a sum of the partial pressures. If this sum is equal to the pressure P of the surroundings, the liquid will be at its *bubble point*. Hence the bubble point of an ideal mixture of N components is given by the solution of the equation

$$\sum_{i=1}^{N} x_i p_i^0 = P, \tag{9.29}$$

with the subsidiary condition, of course, that

$$\sum_{i=1}^{N} x_i = 1. \tag{9.30}$$

In an ideal gas mixture, the mole fraction of component I in the vapour phase (y_i) is equal to the ratio of its partial pressure to the total pressure. Thus

$$y_i = \frac{p_i}{P} = \frac{x_i p_i^0}{P}. \tag{9.31}$$

When equation (9.29) is satisfied at the bubble point, it can be rearranged using equation (9.31) to give

$$\sum_{i=1}^{N} y_i = 1.$$

Unless all values of p_i^0 are equal, the composition of the vapour in equilibrium with a liquid mixture will always be different from the liquid composition, thus leading to distillation as a method for separating liquid mixtures.

The converse of the above calculation of the bubble point of a liquid mixture is the calculation of the dew point of a vapour mixture. Rearranging equation (9.31),

$$x_i = \frac{P y_i}{p_i^0}$$

and the dew point is the temperature which satisfies the equation

$$\sum_{i=1}^{N} x_i = P \sum_{i=1}^{N} \frac{y_i}{p_i^0} = 1. \tag{9.32}$$

Again, the liquid composition will differ from the vapour composition and partial condensation can be used to achieve a measure of separation of the components.

If the liquid mixture is not ideal, equations (9.28) and (9.29) have to be modified by introducing *activity coefficients* (γ_i) to allow for the volatility of each component being influenced by the nature of the other components which are present in the liquid. Thus

$$p_i = \gamma_i x_i p_i^0 \qquad (9.33)$$

and the bubble point will be given by

$$\sum_{i=1}^{N} \gamma_i x_i p_i^0 = P, \qquad (9.34)$$

where γ_i is a function of the mole fractions of all other components. The mole fraction of component i in the vapour phase is now given by

$$y_i = \frac{p_i}{P} = \left(\frac{\gamma_i p_i^0}{P} \right) x_i.$$

This is also known as *Henry's law*: $y_i = m_i x_i$ where $m_i = \gamma_i p_i^0 / P$.

The above equations are of the form $f(T) = 0$ and can be solved by *Newton–Raphson* or other root-finding methods [see III, §5.4].

Flash vaporization

This situation frequently arises when a hot liquid mixture at high pressure is fed into a vessel in which the pressure is much lower. The liquid may be well above its boiling point at the lower pressure so it partially vaporizes by taking latent heat of vaporization at the expense of sensible heat to lower the temperature of the remaining liquid phase. It is assumed that the vapour and liquid leaving the flash tank are in equilibrium with one another. Some of the symbols used to describe this process are illustrated in Figure 9.3. An overall molar balance gives

$$F = G + L. \qquad (9.35)$$

A component molar balance gives

$$Ff_i = Gy_i + Lx_i \qquad (9.36)$$

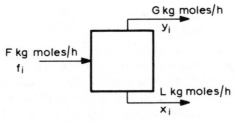

Figure 9.3: Flash tank.

for each of the N components where f_i, x_i, y_i are the mole fractions of component i. Equation (9.35) is the sum of the N equations (9.36). If $m_i = \gamma_i p_i^0 / P$ is the Henry law constant for component i, the N equilibrium equations are

$$y_i = m_i x_i. \tag{9.37}$$

Substituting equations (9.37) into equations (9.36) and eliminating F using equation (9.35) gives

$$(G + L)f_i = (m_i G + L)x_i,$$

which can be rearranged to

$$x_i = \frac{f_i(1 + \beta)}{m_i + \beta}, \tag{9.38}$$

where $\beta = L/G$.

Although equation (9.38) is valid for any mixture, unless it is ideal m_i will be a function of the unknown x_i values. It is therefore assumed from this point that the mixture is ideal.

If the temperature and pressure are both known for the flash tank, m_i will be calculable and the remaining problem is to find the value of β in the set of equations (9.38) which makes

$$\sum x_i = 1.$$

A more difficult problem arises when the flash tank pressure is known but the temperature is not known and has to be calculated from a heat balance. If the feed temperature is T_F and the unknown tank temperature is T, a simplified heat balance can be constructed as follows.

Heat available by cooling the feed from T_F to T is

$$FMC_p(T_F - T),$$

where M is the mean molecular weight of the feed and C_p is its heat capacity (in joules per kilogram). Since G kilogram-moles per hour of this cooled liquid evaporate with a latent heat of Q joules per kilogram-mole,

$$FMC_p(T_F - T) = GQ$$

or

$$(1 + \beta)(T_F - T) = \frac{Q}{MC_p}. \tag{9.39}$$

It has been assumed here that Q has the same value for all components, but if it does not, an average value based on the expected vapour composition can be used. The problem is to find β and T which simultaneously satisfy equations (9.30), (9.38) and (9.39). A value is usually estimated for T; equation (9.39) is solved for β and is substituted into equation (9.27) for p_i^0 and equation (9.38) to find x_i. If $\sum x_i \neq 1$, a fresh estimate is chosen for T.

An example of flash vaporization of this type occurs at the well-head on an oilfield where the crude oil emerges from the well at a high pressure but is stored and transported at atmospheric pressure. The more volatile components evaporate when the pressure is reduced and these are either burnt off or purified for use as a natural gas fuel supply. A larger fraction of the feed can be recovered as a liquid product if the reduction in pressure takes place in a series of three or four stages instead of in a single tank. This phenomenon is due to two effects: first, the vapour leaves the first stage at a higher temperature, thus removing more heat per mole evaporated, and second, there is an element of fractionation whereby the most volatile components are preferentially evaporated, leaving those of intermediate volatility in the liquid phase, thus raising the boiling point. The choice of intermediate stage pressures to yield maximum liquid product is also an interesting problem which can be solved by *dynamic programming* [see IV, Chapter 16].

9.4 Distillation

This is the most common method for separating liquid mixtures into their constituent components. The distillation column itself is designed to keep a rising vapour stream in intimate contact with a falling liquid stream so that the more volatile components are preferentially transferred to the vapour stream and the less volatile components into the liquid stream. Vapour is supplied by boiling part of the liquid which leaves the base of the column and liquid is supplied by condensing the vapour which leaves the top of the column and returning part of the condensate to the column. The mixture to be separated is fed into the column at a point where the composition and state of the feed most closely matches that of the corresponding stream within the column. More than one feed stream can enter a column, each at its appropriate point, and material which is usually liquid can be withdrawn at intermediate points of the column. In its simplest form, however, a distillation column has a single feed and products are only removed from the two ends. It is clear in principle that a binary mixture can be separated in such a column, but a mixture of three or more components can only be split into two mixtures, one of which may consist of a nearly pure component. If intermediate, or side stream, products are removed from the column they cannot consist of only one component. Thus, a mixture of, say, four components is usually split into four products by using three distillation units. Depending upon the materials involved and their non-ideality, the first column would either split the mixture into two binary mixtures, each to be separated in a further column, or split one component from the other three.

Distillation columns are of two main types: plate or packed. In a plate column, the liquid is delayed as pools on a series of plates and the vapour from one plate is dispersed as bubbles through the liquid on the plate above. If the plate is ideal, the vapour leaving a plate will be in equilibrium with the liquid leaving that plate. However, if the vapour bubbles do not mix the liquid sufficiently, or do not remain in contact with the liquid long enough to reach equilibrium, the plate will operate at less than 100 per cent efficiency. A plate column is thus described mathematically by *finite difference equations* [see I, §14.12, and III, §1.4] because the liquid and vapour compositions change by discrete amounts on each plate. In a packed column, specially shaped inert solids are put into an otherwise empty column to provide an extended surface for the liquid to flow over with the vapour passing through the narrow spaces between the packing. This brings the liquid and vapour into close contact, but because they are moving in opposite directions they cannot come to equilibrium. Such a packed column is described mathematically by *differential equations* [see IV, Chapters 7 and 8].

Plate columns

Allowing for multiple feeds and side stream off-takes, the equations describing a plate column can be derived in terms of the following symbols:

$f_{i,n}$ mole fraction of component i in feed to plate n,
$x_{i,n}$ mole fraction of component i in liquid leaving plate n,
$y_{i,n}$ mole fraction of component i in vapour leaving plate n,
L_n liquid flow rate from plate n (kg mol/h),
G_n vapour flow rate from plate n (kg mol/h),
F_n feed rate to plate n (kg mol/h),
S_n off-take rate from plate n (kg mol/h),
H_n enthalpy of vapour leaving plate n (J/kg mol),
h_n enthalpy of liquid leaving plate n (J/kg mol),
h_{Fn} enthalpy of feed to plate n (J/kg mol).

Plates are counted from the top of the column so that the condenser is considered to be plate zero and the reboiler is plate $N + 1$, where N is the number of ideal plates in the column. A balance for component i on plate n gives

$$G_{n+1}y_{i,n+1} + L_{n-1}x_{i,n-1} + F_nf_{i,n} = G_ny_{i,n} + (L_n + S_n)x_{i,n}. \quad (9.40)$$

Using the equilibrium equation (9.37) in the form

$$y_{i,n} = m_{i,n}x_{i,n} \quad (9.41)$$

to allow for non-ideality gives

$$G_{n+1}m_{i,n+1}x_{i,n+1} - (G_nm_{i,n} + L_n + S_n)x_{i,n} + L_{n-1}x_{i,n-1} + F_nf_{i,n} = 0. \quad (9.42)$$

The overall material balance can be obtained by summing equation (9.40) for all components i; thus

$$G_{n+1} + L_{n-1} + F_n = G_n + L_n + S_n. \tag{9.43}$$

There is no external feed to the condenser, nor liquid flow from any higher plate; hence

$$F_0 = L_{-1} = 0. \tag{9.44}$$

If uncondensed vapour leaves the condenser at a rate G_0 and liquid top product is withdrawn at a rate S_0, equation (9.43) for the condenser becomes

$$G_1 = G_0 + L_0 + S_0. \tag{9.45}$$

Adding together equation (9.45) and equation (9.43) for $n = 1$ to $n = n$ gives

$$G_{n+1} + \sum_{k=1}^{n} F_k = G_0 + L_n + \sum_{k=0}^{n} S_k$$

which can be rearranged to

$$L_n = G_{n+1} + \sum_{j=1}^{n} (F_j - S_j) - S_0 - G_0, \tag{9.46}$$

which relates L_n to G_{n+1} in terms of known values. If liquid product is taken from the reboiler at a rate S_{N+1}, there is no other liquid stream leaving the reboiler. There is no plate below the reboiler to provide vapour and no external feed to the reboiler. Therefore

$$L_{N+1} = G_{N+2} = F_{N+1} = 0 \tag{9.47}$$

and equation (9.40) written for the reboiler becomes

$$L_N x_{i,N} = G_{N+1} y_{i,N+1} + S_{N+1} x_{i,N+1}. \tag{9.48}$$

A heat balance on plate n gives

$$G_{n+1} H_{n+1} + L_{n-1} h_{n-1} + F_n h_n = G_n H_n + (L_n + S_n) h_n. \tag{9.49}$$

If Q_c joules per hour is the heat removed by the condenser, equation (9.49) for $n = 0$ becomes

$$G_1 H_1 = G_0 H_0 + (L_0 + S_0) h_0 + Q_c. \tag{9.50}$$

There is little point in summing equations (9.49) and (9.50) because the compositions of the side streams and hence their enthalpies are unknown. If Q_r joules per hour is the heat supplied to the reboiler, equation (9.49) for $n = N + 1$ becomes

$$L_N h_N + Q_r = G_{N+1} H_{N+1} + S_{N+1} h_{N+1}. \tag{9.51}$$

This completes the description of a distillation column in mathematical terms. If there are j components, there will be j equations of type (9.42), equation

(9.49) and the two equations

$$\sum_{i=1}^{j} x_{i,n} = 1 = \sum_{i=1}^{j} m_{i,n} x_{i,n},$$

making $j + 3$ equations for each plate from which the j components, G_n, L_n and T_n, can be determined.

Because this system of equations is so complicated, there are three simplifications, some or all of which are often introduced.

(a) *Ideal mixtures.* Some simplification of equation (9.42) is possible if the mixture is ideal because Henry's law constant $m_{i,n}$ in equation (9.41) will be given by

$$m_{i,n} = \frac{p_{i,n}^0}{P_n},$$

where P_n is the total pressure on plate n and $p_{i,n}^0$ is the saturated vapour pressure of component i at the temperature on plate n. If any component (r) is chosen as a reference component, its Henry law constant will be

$$m_{r,n} = \frac{p_{r,n}^0}{P_n}.$$

Considering $m_{i,n}$ as a measure of the volatility of component i, the relative volatility $(\alpha_{i,n})$ of component i will be the ratio of its volatility to that of the reference component. Thus

$$\alpha_{i,n} = \frac{m_{i,n}}{m_{r,n}} = \frac{p_{i,n}^0}{p_{r,n}^0}. \tag{9.52}$$

Equation (9.27) gives $p_{i,n}^0$ as a function of temperature only; hence $\alpha_{i,n}$ will be a function of temperature only. Furthermore, if equation (9.27) is substituted into equation (9.52),

$$\alpha_{i,n} = \exp\left[A_i - A_r - \frac{B_i}{C_i + T_n} + \frac{B_r}{C_r + T_n} \right] \tag{9.53}$$

If C_i and C_r are approximately the same for all components and B_i and B_r are also of similar value, equation (9.53) shows that $\alpha_{i,n}$ will be independent of temperature. For the few mixtures which are close to ideal, these further assumptions regarding B_i and C_i are also valid so that α_i can be considered constant for each component throughout the column. Because $\Sigma m_{i,n} x_{i,n} = 1$, equation (9.52) can be used to show that

$$m_{r,n} \sum \alpha_{i,n} x_{i,n} = 1. \tag{9.54}$$

Hence, the equilibrium constant $m_{i,n}$ becomes

$$m_{i,n} = \alpha_{i,n} m_{r,n} = \frac{\alpha_i}{\sum \alpha_i x_{i,n}}. \tag{9.55}$$

At the start of the calculation, equation (9.53) can be used to determine α_i for each component, then equation (9.55) will determine $m_{i,n}$ for any liquid composition independent of temperature.

(b) *Constant molar overflow.* If the molar latent heats of evaporation of all components are equal, the heat balance equation (9.49) and overall material balance equation (9.43) can be replaced by two simpler equations which conserve the molar flow rates of the liquid and the vapour independently at each plate. If the feeds are all liquids at their bubble points, the vapour molar flow rate will be constant throughout the column and the pair of equations (9.43) and (9.49) take the forms

$$G_{n+1} = G_n = G \qquad (1 \leqslant n \leqslant N), \tag{9.56}$$

$$L_{n-1} = L_n + S_n - F_n \qquad (1 \leqslant n \leqslant N), \tag{9.57}$$

which enable the liquid and vapour rates at all plates in the column to be determined in terms of the flow rates at the ends of the column once the feed and off-take rates have been decided upon. The value for L_0 is still arbitrary at this stage and is usually expressed in terms of the external reflux ratio (R) given by

$$R = \frac{L_0}{G_0 + S_0}, \tag{9.58}$$

i.e. the ratio of the molar flow returned to the column to the molar flow removed from the top of the column.

(c) *Single feed with no side streams.* With this simplification, $S_n = 0$ $(1 \leqslant n \leqslant N)$, but S_0 is the liquid top product rate and S_{N+1} is the bottom product rate. The column can also be divided into two sections: the rectifying section above the feed plate and the stripping section below the feed plate. In each section separately, $F_n = 0$ and a calculation with the full equations needs only to be done at the feed plate itself. There is also some point now in summing the heat balance equation (9.40) over the top n plates. Thus

$$G_{n+1} y_{i,n+1} - L_n x_{i,n} = G_n y_{i,n} - L_{n-1} x_{i,n-1}$$

$$= \ldots\ldots\ldots\ldots$$

$$= G_0 y_{i,0} + S_0 x_{i,0}$$

and therefore

$$G_{n+1} m_{i,n+1} x_{i,n+1} - L_n x_{i,n} = (G_0 m_{i,0} + S_0) x_{i,0} \tag{9.59}$$

and the heat balance

$$G_{n+1}H_{n+1} - L_n h_n = G_n H_n - L_{n-1}h_{n-1}$$

$$= \dots\dots\dots$$

$$= G_0 H_0 + S_0 h_0 + Q_c. \tag{9.60}$$

These are first-order difference equations instead of the previous second-order difference equations.

(d) *Single feed, ideal mixtures with constant molar overflow and a total condenser.* With all three of the above simplifications plus $G_0 = 0$, equation (9.56) gives constant vapour flow rates and equation (9.57) gives constant liquid flow rates in each section of the column above and below the feed plate. Equation (9.45) with equation (9.58) shows that

$$\frac{L}{G} = \frac{R}{R+1}$$

above the feed plate, and equation (9.59) becomes

$$\frac{\alpha_i x_{i,n+1}}{\sum \alpha_i x_{i,n+1}} - \frac{R x_{i,n}}{R+1} = \frac{x_{i,0}}{R+1}. \tag{9.61}$$

Therefore, for a chosen reflux ratio (R), if the calculation is started from the feed plate, where the composition is assumed to be that of the feed, and $x_{i,0}$ is the assumed top product required, equation (9.61) can be used to calculate $x_{i,n}$ from the known $x_{i,n+1}$ until values close to the assumed $x_{i,0}$ are reached. It is clear that the number of plates (n) required will be determined by this calculation. Also, the values calculated will never match $x_{i,0}$ for all components simultaneously. Therefore the nearest calculated top product composition $x_{i,0}$ will have to be used when the calculation is repeated stepwise from the feed plate. The easiest calculation is to assume that $R \to \infty$, with zero feed rate and no product removal at either end of the column. Equation (9.61) reduces to

$$x_{i,n} = \frac{\alpha_i x_{i,n+1}}{\sum \alpha_i x_{i,n+1}} \tag{9.62}$$

or equation (9.59) with $R \to \infty$ becomes

$$x_{i,n} = y_{i,n+1},$$

which can also be obtained from equation (9.62). Using the equilibrium equation (9.37) to eliminate x instead of y eventually gives

$$y_{i,n+1} = \frac{y_{i,n}/\alpha_i}{\sum y_{i,n}/\alpha_i}, \tag{9.63}$$

which can be used from the feed plate downwards until the bottom product composition is reached.

Hence it is clear that a solution can be found as $R \rightarrow \infty$. It is equally clear that $R = 0$ will not yield a useful solution; indeed in any problem there will be a minimum value of the reflux ratio R which will still yield a solution but with an infinite number of plates n. The most economical value to use for R is usually one which lies between $1.2R_{min}$ and $1.5R_{min}$.

Defining the problem

So far, the equations required have been derived but the specific problem to be solved has not yet been considered. Clearly, complete descriptions of all feeds will be given, and if the reflux ratio, number of plates and total flow rates in all off-take streams are specified, everything else including the compositions of all products is calculable. If the designer wishes to exercise control over any part of the top product composition, some of these values will have to be allowed to float. It is difficult to see how the calculation can proceed if the side stream flow rates are not specified, and in many cases there are no side streams leaving the column. So the parameter which is usually allowed to float is the number of plates. This enables the designer to specify the mole fraction of one component in the one-product stream. If he also wishes to control the mole fraction or distribution of another component the combined top product flow rate must be allowed to vary also. The reflux ratio is usually fixed by economic considerations. Great care must be exercised in laying down the specifications to make sure that they are consistent and practicable. A preliminary calculation at total reflux usually provides an invaluable check that the specifications are reasonable.

In all of these problems, the ordering of the equations is the vital step which determines the efficiency of the numerical calculation. The general procedure is to assume the temperature on all plates, determine (or assume) all flow rates, calculate the distribution of components on all plates and then close the iterative loop by revising the temperatures and flow rates.

Packed columns

The same symbols, with a few exceptions, can be used to describe packed columns as have just been used for plate columns, but symbols which previously had a plate suffix n now become functions of the coordinate (z) defining distance up the column from its base, as illustrated in Figure 9.4. A packed column rarely has any side streams because of practical difficulties and it is assumed for simplicity that the column has one feed, a total condenser and thus two product streams. With no side streams the heat balance over the top section of the column is analogous to equation (9.60) for the plate column without side

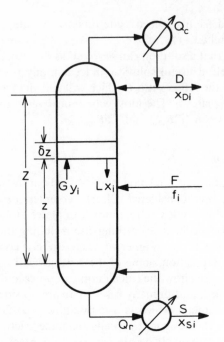

Figure 9.4: Packed distillation column.

streams and no vapour product ($G_0 = 0$). Thus

$$GH - Lh = Dh_D + Q_c, \tag{9.64}$$

where the top product is now signified by D kilogram-moles per hour (instead of S_0) and h_D is its enthalpy. G, L, H and h are all functions of z. Similarly, the overall material balance for any section above the feed plate, including the condenser, is analogous to equation (9.46):

$$L = G - D \tag{9.65}$$

A similar overall component balance takes the form

$$Lx_i = Gy_i - Dx_{Di}, \tag{9.66}$$

but x_i and y_i are not in equilibrium so a balance also has to be taken over the differential element of height δz. The following additional symbols are needed:

A cross-sectional area of column (m^2),
a surface area of packing per unit column volume (m^{-1}),
K_{Gi} overall mass transfer coefficient based on the vapour phase (kg mol/m^2h),

K_{Li} overall mass transfer coefficient based on the liquid phase (kg mol/m^{2}h),
k_{Ci} vapour phase mass transfer coefficient (kg mol/m^{2}h),
k_{Li} liquid phase mass transfer coefficient (kg mol/m^{2}h),
N_i moles transferred to vapour phase (kg mol/h),
$x_i^* = y_i/m_i$ liquid mole fraction in equilibrium with bulk vapour,
$y_i^* = m_i x_i$ vapour mole fraction in equilibrium with bulk liquid.

The volume of the element is $A\,\delta z$ and the interfacial area available for mass transfer is $aA\,\delta z$. Therefore

$$N_i = K_{Gi}aA\,\delta z(y_i^* - y_i) = K_{Li}aA\,\delta z(x_i - x_i^*).$$

These transfer rates must be accounted for in the bulk flow; thus

$$N_i = \frac{d}{dz}(Lx_i)\,\delta z = \frac{d}{dz}(Gy_i)\,\delta z.$$

There is only one new equation here for each component since the two mass transfer coefficient terms are alternative descriptions of the same process, and the equation

$$\frac{d}{dz}(Lx_i) = \frac{d}{dz}(Gy_i)$$

on integration regenerates the component mass balance, equation (9.66). The new equation is usually written as either

$$\frac{d}{dz}(Gy_i) = K_{Gi}aA(y_i^* - y_i) \qquad (9.67)$$

or

$$\frac{d}{dz}(Lx_i) = K_{Li}aA(x_i - x_i^*). \qquad (9.68)$$

For calculations below the feed plate, equation (9.64) becomes

$$GH - Lh = Dh_D + Q_c - FH_F = Q_r - Sh_S \qquad (9.69)$$

and equations (9.65) and (9.66) become

$$L = G - D + F = G + S \qquad (9.70)$$

and

$$Lx_i = Gy_i - Dx_{Di} + Ff_i = Gy_i + Sx_{Si}. \qquad (9.71)$$

Therefore equations (9.64), (9.65), (9.66) and (9.67) or (9.68) have to be solved above the feed plate, and equations (9.67) or (9.68) with equations (9.69), (9.70) and (9.71) need solving below the feed plate. In each set there are $(j + 2)$ independent equations for the j components and the two flow rates L and G. There is, however, a nasty problem hidden in the determination of K_{Gi} and K_{Li}.

Overall mass transfer coefficients

Omitting the suffix i for the purposes of this section, and using the suffix I to denote interface concentrations, the mass transfer rate per unit area between two phases where the bulk mole fractions of a component are denoted by x and y can be written as

$$k_G(y_1 - y) = k_L(x - x_1) = K_G(mx - y) = K_L\left(x - \frac{y}{m}\right) \qquad (9.72)$$

and the equilibrium achieved at the interface is

$$y_1 = mx_1.$$

Therefore

$$\frac{y_1 - y}{m(x - x_1)} = \frac{k_L}{mk_G}$$

and hence

$$\frac{y_1 - y}{y_1 - y + mx - mx_1} = \frac{k_L}{k_L + mk_G} = \frac{y_1 - y}{mx - y}.$$

However, from equation (9.72)

$$\frac{y_1 - y}{mx - y} = \frac{K_G}{k_G}.$$

Hence

$$\frac{1}{K_G} = \frac{1}{k_G} + \frac{m}{k_L} \qquad (9.73)$$

and similarly

$$\frac{1}{K_L} = \frac{1}{mk_G} + \frac{1}{k_L}. \qquad (9.74)$$

Referring back to the origins of k_L and k_G in equation (9.7) it is seen that they are dependent upon diffusivity and hence will vary from one component to another. Also, the presence of m in equations (9.73) and (9.74) makes K_G and K_L dependent upon the components. Since the separation process relies upon the component values for m (the volatility) being different, it is unreasonable in general to assume that K_L and K_G are constant.

However, from a practical point of view, if the values of m vary widely amongst the components, their separation is relatively easy and the column will not be critical in size or cost and it will not be so important to be able to design it carefully. Conversely, when the m values are close to one another, the components will be difficult to separate and it may well be that their other physical properties such as diffusivity are also similar. Therefore, in the latter situation, where accurate design is essential, the approximation that K_L and K_G are independent of concentration involves least error. They may still be functions of z due to temperature, pressure and flow rates varying with height.

Constant molar overflow

Just as for plate columns, if the molar latent heats are equal, equation (9.64) is not needed, L and G become constants and can be taken out of the derivatives in equations (9.67) and (9.68) to give

$$\frac{G}{A}\frac{dy_i}{dz} = K_{Gi}a(m_i x_i - y_i), \tag{9.75}$$

$$\frac{L}{A}\frac{dx_i}{dz} = K_{Li}a\left(x_i - \frac{y_i}{m}\right). \tag{9.76}$$

The other simplification of considering ideal mixtures leads to equation (9.55) as before, so that calculations may be possible in terms of constant relative volatilities; but this does not overcome the problems associated with variable K_{Gi} and K_{Li}.

Equations (9.75) and (9.76) comprise a *boundary-value problem* [see III, Chapter 8] because conditions are fixed by the condenser at one end of the column and the reboiler at the other. In a design problem, the overall column length will be unknown and the pair of equations are usually solved numerically by moving in both directions from the feed point until the end conditions are reached approximately. This technique uses methods such as *Runge–Kutta for initial-value problems* [see III, §8.2.2].

9.5 Gas absorption

Columns for gas absorption are similar to distillation columns in that they are designed to bring a rising gas stream into intimate contact with a falling liquid stream so that material can be transferred between the phases. The objective is to remove one or more components from the vapour phase by selective absorption into a liquid solvent. Since the vapour mixture is fed directly into the bottom of the column and the solvent is fed to the top of the column, there is no need for a condenser, reboiler or intermediate feed plate. It is inevitable that there will be at least three components present in such a column: the solvent and at least two gases in the mixture. Either packed or plate columns can be used to bring the phases into contact and the same sort of equilibrium relationships apply as for distillation. However, an absorption column usually operates at a low ambient temperature to increase the solubility of most gases and to lower the vapour pressure of the solvent. Heat balances are needed less often because the only sources of heat are the latent heats and possible heats of solution.

Absorbers are usually used either for extracting a gaseous product from a process stream so that it can be purified further in the liquid phase or for removing toxic or noxious vapours from a waste gas before it is vented to the atmosphere. In the latter case, especially, the concentration of the transferred component will be low, and simplifying linear approximations can be made. The calculations to be performed can be similar to those for distillation but usually advantage is taken of the special features of gas absorption. First, the solvent is chosen to be of very low volatility at the operating temperature to avoid solvent losses. Second, the other gases in the mixture which are not to be absorbed (such as air) will pass up the column as inerts. Thus in most cases each phase contains at least one component which is present to a negligible extent in the other phase. The mathematical model is based on this feature by assuming that the inert gases rise at a constant flow rate and carry with them the transferable components. Similarly, the solvent falls down the column at a constant flow rate and picks up the transferable components. In such a model, concentrations expressed as mole fractions are not convenient and mole ratios are used instead. Whereas the mole fraction is the number of moles of the component expressed as a fraction of the total moles in the phase, the mole ratio is the number of moles of the component expressed as a ratio to the number of moles of the carrier (solvent or inerts) in that phase. If only one component is being absorbed, the vapour mole ratio (Y) is related to its mole fraction (y) by

$$Y = \frac{y}{1 - y}.$$

If more than one component is being absorbed, the relationship between y_i and Y_i for each component involves all absorbed components.

After absorbing the desired components, the solvent leaves the column and often goes to a desorption or stripping column in which the absorbed components are driven off, usually by heating the solution. This regenerates the solvent so that it can be cooled and recycled to the top of the absorber. An allowance must be made in the model for a low concentration of absorbed components in the solvent feed due to the stripping being less than 100 per cent efficient.

Plate absorbers

The derivation of the governing equations follows the same pattern as for plate distillation columns, but by defining flow rates on a solute-free basis they become constants. The symbols to be used are as follows:

$H_{i,n}$ enthalpy of component i in vapour leaving plate n (J/kg mol),

$h_{i,n}$ enthalpy of component i in liquid on plate n (J/kg mol),

H_{1n} enthalpy of inert vapour leaving plate n (J/kg mol),

h_{Sn} enthalpy of inert solvent leaving plate n (J/kg mol),
G inert gas flow rate (kg mol/h),
L solvent flow rate (kg mol/h),
$m_{i,n}$ Henry's law constant for component i on plate n,
N number of plates in the column,
Q_n heat removed (or lost) from plate n,
$X_{i,n}$ mole ratio of component i in liquid on plate n,
$Y_{i,n}$ mole ratio of component i in vapour leaving plate n.

A component balance over the top n plates gives

$$GY_{i,n+1} + LX_{i,0} = GY_{i,1} + LX_{i,n}, \tag{9.77}$$

where $X_{i,0}$ is the composition of the solvent feed and $Y_{i,1}$ is the composition of the gas stream leaving the top of the absorber. The multicomponent equilibrium equation (9.37) in terms of mole ratios is

$$\frac{Y_{i,n}}{1 + \sum Y_{i,n}} = \frac{m_{i,n}X_{i,n}}{1 + \sum X_{i,n}}, \tag{9.78}$$

where the summations are taken over all transferable components (i). Since the heat balance cannot affect the flow rates, the only effect of temperature is to alter the values of $m_{i,n}$ from plate to plate. If there is no heat of mixing the components, the heat content of each stream in the column can be obtained by adding together the heat contents of the individual components. A heat balance over the top n plates gives

$$G\left(H_{In+1} + \sum_i H_{i,n+1}Y_{i,n+1}\right) + L\left(h_{S0} + \sum_i h_{i,0}X_{i,0}\right)$$
$$= \sum_1^n Q_n + G\left(H_{I1} + \sum_i H_{i,1}Y_{i,1}\right) + L\left(h_{Sn} + \sum_i h_{i,n}X_{i,n}\right). \tag{9.79}$$

In principle, this set of equations (9.77), (9.78) and (9.79) can be solved numerically for any given pair of feed streams as follows:

(i) Guess the composition and temperature of the outlet gas stream.
(ii) Calculate $X_{i,n}$ from equation (9.77) with $n = N$ first time.
(iii) Solve equation (9.79) for T_n.
(iv) Calculate $m_{i,n}$ for all components.
(v) Solve equation (9.78) for $Y_{i,n}$.
(vi) Repeat from step (ii) with n reduced by unity until the top plate is reached.
(vii) Check calculated values against the assumed composition and temperature of the outlet gas stream and repeat with revised values if necessary.

The first estimates for step (i) can be made by assuming temporarily that the column has an infinite number of plates. Thermal equilibrium can be assumed therefore at the top of the column so that the outlet gas temperature is equal to the solvent feed temperature. With approximate values for $m_{i,N}$ assume that the solvent stream leaving the base of the column is in equilibrium with the gas feed and calculate from equation (9.78) the maximum amount of each component which can be absorbed. For some components this will be more than the total quantity of that component fed to the column in both feed streams and for others it will be less. For each component which can be completely absorbed, put its concentration in the gas outlet stream in equilibrium with the incoming solvent feed, and for those which cannot be completely absorbed, put the excess into the gas outlet stream. This procedure will usually provide realistic estimates for use in step (i).

The above description illustrates approximately that if $m_{i,N} < L/G$ for a particular component, then it can be absorbed completely in the solvent in a column of infinite height, whereas if $m_{i,N} > L/G$ the component cannot be completely absorbed. In a practical situation, the value of G will be known and it will be desired to use as small a value of L as possible for the duty required. If the designer wishes to absorb as much as possible of one particular key component, L should be chosen such that it is greater than the appropriate $m_{i,N}$ multiplied by G. All components having values of $m_{i,N}$ less than this key component will be more completely absorbed, whilst those with larger values for $m_{i,N}$ will remain in the gas phase to a larger extent.

Since $m_{i,N}$ increases with temperature for most materials, it is essential that the solvent temperature should be kept low for efficient operation. The temperature of both feed streams will be known and the temperature of the outlet gas will be close to that of the solvent feed, so the solvent stream will collect all of the latent heat released by the absorption unless it has been removed in the column and accounted for in the terms Q_n.

If the latent heats are small and concentrations are low, natural heat losses may be enough to ensure that the temperature does not vary very much within the column so that isothermal conditions can be assumed. It follows that equation (9.79) will not be needed and the values of $m_{i,n}$ will be constant throughout the column provided the components absorbed are not too far from ideal.

Finally, if only one component is being absorbed, equation (9.78) takes the much simpler form

$$\frac{Y_n}{1 + Y_n} = \frac{m_n X_n}{1 + X_n},$$

which can be rearranged and solved for Y_n. Thus

$$Y_n = \frac{m_n X_n}{1 + X_n - m_n X_n}. \tag{9.80}$$

Equation (9.80) substituted into the single-component version of equation (9.77) gives a non-linear first-order finite difference equation which can usually be put into *Riccati form* [cf. IV, (7.10.8)].

Packed absorbers

In general, there is no advantage to be gained by calculating in terms of mole *ratios*, with flow rates expressed on a solute-free basis, because the expression for non-equilibrium mass transfer is defined in terms of mole *fractions*; thus

$$N_i = K_{Gi}aA \ \delta z(y_i^* - y_i).$$

Consequently, the equations to be solved are usually written in terms of mole fractions and total molar flow rates as for distillation in Section 9.4. Using the same symbols again, equations (9.67) and (9.68) are unchanged but the heat balance and component balance over the top section of the column have to be replaced by

$$Lx_i = Gy_i - [Gy_i - Lx_i]_{z=Z} \tag{9.81}$$

and

$$GH - Lh = [GH - Lh]_{z=Z}. \tag{9.82}$$

The same difficulties arise when evaluating the overall mass transfer coefficients. If all concentrations are low, as they will be when cleaning up an effluent gas, there will not be enough absorption to cause significant temperature changes or to alter flow rates. Under these conditions L, G and m_i become independent of z and the pair of equations (9.67) or (9.68) with (9.81) become linear with constant coefficients and can be integrated in terms of logarithmic or exponential functions [see IV, §7.4].

Reacting absorbers

The capacity of an absorber can be greatly increased by using a solvent which reacts with one or more of the components being absorbed. The effect is to keep the concentration in the solvent very small so that the driving force for absorption is kept high. In general, there will still be a liquid phase mass transfer resistance because the reaction does not take place at the liquid surface but in the bulk liquid. There is likely to be a significant heat of reaction as well as the latent heat release so that temperature control becomes important. One disadvantage of reacting absorbers is the increased difficulty of regenerating the solvent. If the product of reaction is saleable this may not matter too much but solvent costs can be prohibitive otherwise unless the reaction can be readily reversed in a cheap regeneration unit.

If a plate column is used, there is some difficulty in defining a meaningful perfect plate, because the gas leaving the plate can hardly be in equilibrium with the reacting liquid solution. If the phases were kept in contact, the only equilibrium they could reach in an irreversible reaction would be when all of the reactant had been consumed. Hence, in terms of perfect plates, the absorber can only contain one plate. A reacting absorber with plates needs to be designed like a series of stirred tank reactors with the residence time of both the gas and the liquid on each plate taken into account.

The design of a packed reacting absorber can be carried out using the modified equations for an ordinary packed absorber, but the liquid hold-up, which provides the reaction volume, has to be included in the model. However, if the reaction is fast and irreversible, the concentration of the absorbed gas in the bulk liquid phase will be essentially zero. Assuming that only one component is being absorbed and reacted in this way, equation (9.67) becomes

$$\frac{d}{dz}(Gy) = -K_G aAy.$$

In this idealized situation, an advantage can be gained by defining G on a solute-free basis (G_s) and working in terms of mole ratios again. Thus

$$G_s \frac{dY}{dz} = -K_G aA \frac{Y}{1+Y}. \tag{9.83}$$

Due to the assumed high rate of reaction, the liquid phase mass transfer resistance will be low and the process should be gas phase controlled. In these circumstances, K_G will be constant and equation (9.83) can be integrated [see IV, §7.2] to give

$$Y + \ln Y = C - \beta z, \tag{9.84}$$

where C is an arbitrary constant and $\beta = K_G aA/G_s$. Since $Y = Y_F$ at $z = 0$ and $Y = Y_Z$ at $z = Z$, the height of the column (Z) will be given by

$$Z = \left(\frac{G_s}{K_G aA}\right) Y_F - Y_Z + \ln\left(\frac{Y_F}{Y_Z}\right). \tag{9.85}$$

V.G.J.

References

Coulson, J. M., and Richardson, J. F. (1968). *Chemical Engineering*, Vols I and II, Pergamon.

Denbigh, K. G. (1981). *Principles of Chemical Equilibrium*, Cambridge University Press.

Jenson, V. G., and Jeffreys, G. V. (1977). *Mathematical Methods in Chemical Engineering*, Academic Press.

Levenspiel, O. (1972). *Chemical Reaction Engineering*, Wiley.

Smith, J. M. (1981). *Chemical Engineering Kinetics*, McGraw-Hill.

Treybal, R. E. (1968). *Mass Transfer Operations*, McGraw-Hill.

Mathematical Methods in Engineering
Edited by G. A. O. Davies
© 1984, John Wiley & Sons, Ltd.

10

Structural Mechanics

10.1 Introduction

The above title describes the art of idealizing and being able to predict the 'behaviour' of those structures which occur in the fields of civil, mechanical, industrial, aeronautical, nuclear and marine engineering. This brief list gives some idea of the vast range of structural sizes, shapes, materials and environments. 'Behaviour' here means the way in which stresses, strains and displacements vary throughout the structure and with time, and consequently how failure to stand up to the environment may take place. The exact nature of the *failure* at the material's atomic level is conventionally understood to be the province of the physicist, material scientist or metallurgist. We will simply introduce 'failure' and other material properties as predetermined laws involving stress, strain, temperature, time, etc., with experimentally deduced parameters.

The analysis of all structures, whether simple or complex, consists of just four statements, three of which always lead to equations of some sort, and it is important to establish this at the onset of any summary of structural analysis.

First, the structure and its environment have to be modelled before we can make any further statements. The modelling of the applied loads is not discussed here since in many cases it is based on experimental observation or on other branches of engineering analysis such as fluid dynamics or thermodynamics. However, even the description of an experimentally observed environment may need analytical techniques in such areas as *statistics* and *stochastic processes* [see II, Chapters 18 to 20], *curve fitting* [see III, Chapter 6], etc. The modelling of the structure itself is simply an assumption about the nature of the stress, strain or displacement fields within the structure. This may be very simple for structures which are one dimensional such as rods, wires, 'pinjointed' frameworks, beams, and so on. It may be a little more complicated for two-dimensional tubes, plates and shells, and it may be almost impossible analytically for three-dimensional structures, such as power plants, reactors, dams, etc., without using numerical descriptions.

Second, having assumed the nature of the internal fields the *equations of equilibrium* must be found—including inertia and damping terms for dynamic problems. These equations may be for overall equilibrium (or motion) of the complete body or for equilibrium between applied forces and internal stresses either on the surface or in the interior. It is frequently possible—and certainly convenient—to analyse the motion of some stiff bodies by *rigid body mechanics* and then to incorporate the solved inertia terms into a separate 'static' analysis of the stresses, strains and deformations. Clearly this is only possible if these deformations do not significantly modify the rigid body solution. The equations of equilibrium may not be simple if the idealized structure has a complex shape; they will certainly not be simple if the stresses themselves modify that shape. At the other extreme, however, they may be so simple that we are able to solve the equilibrium equations immediately and produce stresses in terms of the applied loads—the structure is *statically determinate*.

Third, the strains may be found in terms of the displacements using entirely geometric arguments—the *equations of compatibility*. The definition of *strain* is arbitrary but the generally accepted measure is the fractional change in extensional deformation (direct strain) or change in a right angle (shear strain). The derivation of compatible strains from displacements is as simple—or as difficult—as the derivation of the equations of equilibrium. In fact, either of the equations of equilibrium or compatibility can be deduced from the other using energy or work arguments described later.

Fourth, the link between stress–equilibrium equations and the strain–displacement compatibility equations has to be forged—i.e. the *stress–strain* law. There is no other way to link displacements to stresses and forces, but it should be remembered always that this law has no part at all in the separate equations of equilibrium and compatibility. If the material's constitutive equations are simple—say a linear stress–strain law—it may be possible for the combined equations to be readily solved, but if the constitutive laws are complex then so inevitably will be the structural analysis. The relationship between stress and strain may be non-linear, for example, or may vary with time (creep). The time-dependent creep problem is examined with other viscoelastic and flow problems, important in polymers, in Chapter 11. Ductile structural materials like alloys of steel and aluminium cease to be linear above values of a yield or proof stress, and the analytical treatments of plasticity is briefly mentioned later. Brittle materials will fail by unstable crack growth (and so may notched ductile materials), but this science of fracture mechanics is not discussed here (see, for example, Knott, 1973) although many of the mathematical techniques used in elasticity can be applied in the presence of cracks (Sneddon and Lowengrub, 1969). Briefly, then, non-linear stress–strain behaviour inhibits analytical solutions of all structures but the geometrically simple (such as axisymmetrical problems, frameworks and thin-walled tubes and beams); otherwise numerical solutions like finite-element methods have to be used.

Fortunately most engineering materials are obligingly linear over their useful range, i.e. strain is proportional to all the stress components and to temperature.

Moreover, when the material ceases to be linear, and yields or fractures, the strain is unlikely to exceed 1 per cent. This fact is crucial to structural analysis since it means that the equations of equilibrium and compatibility can be deduced from the geometry of the *undeformed structure*. This is so important that it is worth dwelling a little on these equations.

Consider, then, any *continuum* (to avoid having to describe the sort of structure) with cartesian stress and strain components at a point written as *column matrices* **σ** and **ε**. To save space we write them in *transpose* form [see I, §6.2(iv)]:

$$\boldsymbol{\sigma}^T = [\sigma_{xx}\ \sigma_{yy}\ \sigma_{zz}\ \sigma_{xy}\ \sigma_{yz}\ \sigma_{zx}]$$

and

$$\boldsymbol{\varepsilon}^T = [\varepsilon_{xx}\ \varepsilon_{yy}\ \varepsilon_{zz}\ \varepsilon_{xy}\ \varepsilon_{yz}\ \varepsilon_{zx}].$$

The double suffix refers first to the direction of the stress component and second to the normal of the plane on which it acts: thus σ_{xx} is a direct stress and σ_{xy} a shear stress. To be pedantic 'stress' is a *tensor* [see V, §7.1] and not a vector since it possesses two directions in addition to its magnitude. In this book *both* forms are used (Chapters 6 and 11, for example) the 3×3 array and the 6×1 column matrix. The latter is used here since it is so convenient to express the *scalar* quantity *work* or *energy* as a product $\boldsymbol{\sigma}^T\boldsymbol{\varepsilon}$. Fortunately we find that equilibrium requires that $\sigma_{xy} = \sigma_{yx}$, etc., and so we have only the above six components.

With the proviso that strains are small we obtain two linear sets of equations:

Equilibrium equations: $\quad \mathbf{D}\boldsymbol{\sigma} + \boldsymbol{\omega} = \mathbf{0},$ (10.1)

Compatibility equations: $\quad \boldsymbol{\varepsilon} = \mathbf{D}^T\mathbf{u}.$ (10.2)

The vector

$$\boldsymbol{\omega}^T = [\omega_x\ \omega_y\ \omega_z]$$

contains the body force components per unit volume, and the vector

$$\mathbf{u}^T = [u\ v\ w]$$

contains the displacement components. The matrix **D** is an array of linear operators [see IV, §19.1.2] or zeros, thus:

$$\mathbf{D} = \begin{bmatrix} \dfrac{\partial}{\partial x} & 0 & 0 & \dfrac{\partial}{\partial y} & 0 & \dfrac{\partial}{\partial z} \\[2mm] 0 & \dfrac{\partial}{\partial y} & 0 & \dfrac{\partial}{\partial x} & \dfrac{\partial}{\partial z} & 0 \\[2mm] 0 & 0 & \dfrac{\partial}{\partial z} & 0 & \dfrac{\partial}{\partial y} & \dfrac{\partial}{\partial z} \end{bmatrix}. \quad (10.3)$$

The equilibrium equations (10.1) are true at any interior point; on the surface the special version needed can be written as

$$[\mathbf{D}n]\boldsymbol{\sigma} = \boldsymbol{\phi}, \tag{10.4}$$

where the surface force components per unit area are assembled as the vector

$$\boldsymbol{\phi}^T = [\phi_x\, \phi_y\, \phi_z]$$

and n is the local coordinate normal to the surface. The expression $\mathbf{D}n$ is taken to mean that *all* the operators in (10.3) act upon the variable n. The resulting quantities $\partial n/\partial x, \partial n/\partial y, \partial n/\partial z$ are simply the *direction cosines* [see V, §13.3.7] of the surface normal, and they arise because it is necessary to resolve the various stress components in equilibrium with the three components of $\boldsymbol{\phi}$. If strains were not small then none of the fundamental equations (10.1), (10.2) and (10.4) would be linear.

In many problems—particularly those concerned with structures which are complex and demand subtle idealizations (like shells)—it is often easier to derive equilibrium or compatibility equations *indirectly* using *virtual work* or *minimum energy* arguments (Davies, 1982). For example, one may readily discard small energy terms since energy is a scalar, but errors in the vector equations (10.1), (10.2) and (10.4) are more difficult to discuss. Also, when using *approximate methods of analysis* it is found that virtual work or extremum energy methods are a powerful tool, as we shall discuss later.

The *principle of virtual work* simply equates the product of force times displacement for internal forces to that for external forces. Thus

$$\int_V \boldsymbol{\sigma}^T \boldsymbol{\varepsilon}\, dV = \int_V \boldsymbol{\omega}^T \mathbf{u}\, dV + \int_{S_\phi} \boldsymbol{\phi}^T \mathbf{u}\, ds, \tag{10.5}$$

where V denotes the entire volume and S_ϕ that part of the surface where the applied pressures $\boldsymbol{\phi}$ are given [see IV, §§6.3 and 6.4]. (The other part of the surface S_u may have the displacements constrained and the consequent supporting reactions would not then be known pressures.) Equation (10.5) is used in two distinct forms: first, the *principle of virtual displacements* (PVD) and strains (denoted as $\bar{\mathbf{u}}$ and $\bar{\boldsymbol{\varepsilon}}$) and second, the *principle of virtual forces* (PVF) ($\bar{\boldsymbol{\omega}}$, $\bar{\boldsymbol{\phi}}$, $\bar{\boldsymbol{\sigma}}$). Before amplifying the reason for selecting these curious hypothetical virtual forces, stresses, strains and displacements, we mention that the equivalent extremum form of (10.5) provides no more information but does require a prior definition of certain energies, namely,

(i) The (usual) external potential energy:

$$U_e = -\int_V \boldsymbol{\omega}^T \mathbf{u}\, dV - \int_{S_\phi} \boldsymbol{\phi}^T \mathbf{u}\, ds.$$

(ii) The internal potential energy:

$$U_i = \int_V \int \boldsymbol{\sigma}^T \, d\boldsymbol{\varepsilon} \, dV.$$

The internal (strain) energy involves an inner integral from an initial strain state to a final one, and can only be evaluated if the stress–strain law is defined; however, this can be circumvented by writing the *principle of minimum potential energy* as

$$\delta(U_i + U_e) = 0, \tag{10.6}$$

which becomes

$$\int_V \boldsymbol{\sigma}^T \, d\boldsymbol{\varepsilon} \, dV = \int_V \boldsymbol{\omega}^T \, d\mathbf{u} \, dV + \int_{S_\phi} \boldsymbol{\phi}^T \, d\mathbf{u} \, ds,$$

which is identical to the PVD version of (10.5) when we put $d\boldsymbol{\varepsilon} = \bar{\boldsymbol{\varepsilon}}$ and $d\mathbf{u} = \bar{\mathbf{u}}$. Similar *complementary energies* have to be defined to give an extremum version of the PVF. In applying extremum principles like (10.6) we have to use *variational methods* [see IV, Chapter 12] in seeking stress, strain and displacement fields which minimize a functional like $U_e + U_i$. Such arguments can be avoided by treating (10.5) as a virtual work theorem and seeing what information is delivered when the theorem is applied to a particular structure. This is readily demonstrated by writing down a particular form of the *Gauss–Ostrogradsky theorem* [see IV, Theorem 6.4.2], namely,

$$\int_V (\boldsymbol{\sigma}^T \mathbf{D}^T \mathbf{u} + \mathbf{u}^T \mathbf{D} \boldsymbol{\sigma}) \, dV = \int_S \mathbf{u}^T [\mathbf{D}n] \boldsymbol{\sigma} \, ds,$$

which is true for *any* continuously differentiable functions $\boldsymbol{\sigma}$ and $\mathbf{u}$ [see IV, §5.3] and any matrix of linear operators—not just $\mathbf{D}$ in fact. $S = S_\phi + S_u$ is the entire surface. On substituting this identity into (10.5) we obtain, first, as the PVD,

$$\int_V [\boldsymbol{\sigma}^T(\bar{\boldsymbol{\varepsilon}} - \mathbf{D}^T \bar{\mathbf{u}}) - \bar{\mathbf{u}}^T(\mathbf{D}\boldsymbol{\sigma} + \boldsymbol{\omega})] \, dV$$

$$+ \int_{S_\phi} \bar{\mathbf{u}}^T \{[\mathbf{D}n]\boldsymbol{\sigma} - \boldsymbol{\phi}\} \, ds + \int_{S_u} \bar{\mathbf{u}}^T [\mathbf{D}n]\boldsymbol{\sigma} \, ds = 0. \tag{10.7}$$

We can at last take advantage of the hypothetical nature of the virtual displacements and conclude that for (10.7) to be zero everywhere, the coefficients of $\bar{\mathbf{u}}^T$ must be zero everywhere, except on S_u where $\bar{\mathbf{u}} = 0$, *provided we eliminate* $\bar{\boldsymbol{\varepsilon}} - \mathbf{D}^T \bar{\mathbf{u}} = 0$. Thus the PVD will deliver the equations of equilibrium $\mathbf{D}\boldsymbol{\sigma} + \boldsymbol{\omega} = 0$ inside V and $[\mathbf{D}n]\boldsymbol{\sigma} = \boldsymbol{\phi}$ on the surface S_ϕ, but we must insist that the virtual displacements satisfy the compatibility equations. Since we therefore have to find $\bar{\boldsymbol{\varepsilon}} = \mathbf{D}^T \bar{\mathbf{u}}$ for the structure, it is common practice to put the real strains $\boldsymbol{\varepsilon} = \mathbf{D}^T \mathbf{u}$ also, and thence on using the stress–strain law the PVD is now a *pure displacement method* never utilizing stresses or equilibrium equations directly.

Similarly the PVF version of (10.5) will deliver

$$\int_V [\bar{\sigma}^T(\epsilon - \mathbf{D}^T\mathbf{u}) - \mathbf{u}^T(\mathbf{D}\bar{\sigma} + \bar{\omega})]\, dV$$
$$+ \int_{S_\phi} \mathbf{u}^T\{[\mathbf{D}n]\bar{\sigma} - \bar{\phi}\}\, ds + \int_{S_u} \mathbf{u}^T[\mathbf{D}n]\bar{\sigma}\, ds = 0. \qquad (10.8)$$

This must be zero for any values of $\bar{\sigma}$. Thus, provided we satisfy equilibrium equations (and therefore eliminate the second and third terms above), we obtain indirectly the equations of compatibility $\epsilon = \mathbf{D}^T\mathbf{u}$ in V and $\mathbf{u} = \mathbf{0}$ on S_u. It is common to obtain a *pure force solution* without displacements by selecting $\bar{\phi} = \mathbf{0}$ and $\bar{\omega} = \mathbf{0}$ in (10.5), i.e. $\mathbf{D}\bar{\sigma} = \mathbf{0}$ in V and $[\mathbf{D}n]\bar{\sigma} = \mathbf{0}$ on S_ϕ. The stress system $\bar{\sigma}$ is *self-equilibrating* and is simply a *complementary function* [see IV, §7.3.2] solution of (10.1).

We have now summarized the four analytical requirements of structural mechanics: idealization of a stress or displacement field, the stress–strain law and the equations of equilibrium and compatibility, the last two being derivable indirectly by energy/work arguments if necessary. The main difficulty in solving linear small-displacement structural problems lies not in the relatively simple governing equations but in the geometrical shape of typical structures. The boundaries and interiors of many structures are very complicated to describe and consequently approximate numerical solutions to the equations are sought, rather than closed analytical forms. For this reason we have divided structural mechanics into two sections—*analytical* and *numerical*—both with their rather different approaches. Nevertheless, both approaches start with the object of satisfying the governing equations (10.1) and (10.2), although finite element numerical methods start with the equivalent (10.5) (and sometimes equations 10.7 and 10.8).

10.2 Analytical linear problems with no gross deformations

Analytical solutions for *three-dimensional* structures are not possible except for one or two special cases where symmetry reduces the number of independent variables (Timoshenko and Goodier, 1951). We therefore restrict our discussion to two-dimensional cases where no variation in loading or geometry occurs in one direction (plane strain) or, more usually, where one structural dimension is so small compared to others that the stress or displacement field may be assumed through that small thickness; e.g. beams, tubes, thin plates and shells. All such linear two-dimensional problems turn out to be *boundary-value problems for elliptic differential equations* [see IV, §8.5] or simplified forms of such. However, the various types of structure lead to special forms of these equations and it is possible to take advantage of these special forms such as the *biharmonic equation* [see III, §9.5], *Laplace's equation* [see IV, §8.2], etc.

Beams

Starting with the first—and easiest—a *beam* is characterized by the two following assumptions:

1. It is slender (in the z direction) with a small taper, and the deformations in the xy plane of its cross-section can be ignored.
2. Cross-sections normal to the z axis remain plane after bending. This assumption also relies on the beam being so slender that bending deformations dominate over shear deformations.

For illustration, in the simplest case of a beam section symmetrical about the y axis and loaded by a force $p(z)$ per unit length in this direction, it will deflect $v(z)$ in the same direction. For small rotations $(dv/dz \ll 1)$, assumption 2 enables us to write (see Figure 10.1) the axial displacement as

$$w(y, z) = -yv'(z),$$

where $(' = d/dz)$ [see IV, §3.1.1]. Substituting $v(z)$, $w = -yv'(z)$ and $u = 0$ into (10.2) we obtain only $\varepsilon_{zz} = -yv''$, and so, assuming Hooke's law $\sigma_{zz} = E\varepsilon_{zz}$, the PVD version of (10.5) becomes

$$\int_0^L \int \int Ey^2 v'' \bar{v}'' \, dx \, dy \, dz = \int_0^L p(z)\bar{v} \, dz, \tag{10.9}$$

whence on *integrating by parts* [see IV, §4.3] (the one-dimensional form of

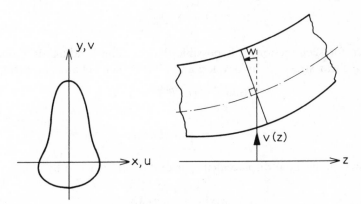

Figure 10.1: Simple beam deformation.

Gauss' theorem [IV, Theorem 6.4.2])

$$\int_0^L [(EIv'')'' - p]\bar{v}\, dz + [EIv''\bar{v}']_0^L - [(EIv'')'\bar{v}]_0^L = 0$$

$$\text{for arbitrary } \bar{v}(z), \qquad (10.10)$$

where

$$I(z) = \int\int y^2\, dx\, dy$$

and the x axis passes through the centroid in order that

$$\int\int \sigma\, dx\, dy = 0.$$

Equation (10.10) is the equivalent of (10.7). The equation of equilibrium we extract as

$$(EI(z)v'')'' = p(z), \qquad (10.11)$$

which is a *linear fourth-order differential equation* [see IV, §7.3] with four required boundary conditions which may be kinematic (say v and v' zero if the end is clamped) or, from (10.9), $EIv'' = 0$ and $(EIv'')' = 0$ if the end is free to rotate ($\bar{v}' \neq 0$) or free to displace ($\bar{v} \neq 0$).

The *twisting of slender beams* is not so simply described—see *two-dimensional torsion and flexure*—although an equation for the twist $\theta(z)$ analogous to (10.11) can be derived:

$$(GJ(z)\theta')' = q(z), \qquad (10.12)$$

where G is the shear modulus, $q(z)$ is the applied torque per unit length and J is a section constant (equal to the polar second moment of area for circular sections only). For *thin-walled sections* whose thickness $t(s)$ varies around the section and is small compared to the entire periphery S, J is approximately

$$\tfrac{1}{3}\int_0^S t^3\, ds;$$

however, for such sections it is possible that (torsion-bending' or 'differential-bending' shear stresses may develop and (10.12) then acquires an extra term:

$$(GJ\theta')' - (E\Gamma\theta'')'' = q(z). \qquad (10.13)$$

Here

$$\Gamma = \int_0^S w^2 t\, ds,$$

where $w(s)$ is the axial displacement per unit θ' and is given by

$$w(s) = -\int p\, ds,$$

$p(s)$ being the offset of a tangent to the section wall from the centre of twist, the limits in the last integral being chosen to make

$$\int_0^s w \, ds = 0.$$

Equation (10.13) may be deduced in exactly the same way as (10.11) (Davies, 1982).

If the beam is subjected to a compressive axial force P then it may buckle. We may predict the *onset of buckling* by adding to the PVD the work done by P during lateral deflections $v(z)$. We noted that the governing equations of compatibility are only linear if insignificant changes in geometry take place, and clearly this is not true for buckling from (say) an initially straight configuration. However, we may sidestep this difficulty by applying the PVD to a *small* virtual displacement $\delta v(z)$ from the deflected shape $v(z)$. (This is the first time we have had to make our virtual displacements small compared to the real ones!) It may be shown that, if pure bending displacements $v(z)$ occur, the beam shortens by an amount

$$\Delta = \int_0^L \tfrac{1}{2}(v')^2 \, dz.$$

(Notice that Δ is a non-linear function of v.) The virtual work done by P is equal to

$$P \, \delta\Delta = P \int_0^L v' \, \delta v' \, dz.$$

Adding this term to the right-hand side of (10.9) we obtain in the same way as (10.10)

$$\int_0^L [(EIv'')'' + Pv'' - p] \, \delta v \, dz + [EIv'' \, \delta v']_0^L - [(EIv'')' + Pv']_0^L \, \delta v = 0. \tag{10.14}$$

Notice that both the equation of equilibrium and one boundary condition (end free to displace) are modified by the axial load P. Suppose we solve the equation of equilibrium in the simplest case of a uniform section:

$$EIv'''' + Pv'' = p(z). \tag{10.15}$$

The solution can be written as

$$v(z) = C_1 + C_2 z + C_3 \sin \alpha z + C_4 \cos \alpha z + f(z),$$

where $\alpha^2 = P/EI$ and $f(z)$ is the *particular integral* [see IV, §7.3.2]. When this solution is inserted into the four boundary conditions, we obtain four equations for the constants $\mathbf{C} = [C_1 \ C_2 \ C_3 \ C_4]^T$ in the form

$$\mathbf{AC} = \mathbf{F}. \tag{10.16}$$

A contains α and F contains the particular integral which of course is a function of the lateral loading $p(z)$. Inverting (10.16),

$$C = \frac{A_{adj}}{|A|} F \tag{10.17}$$

[see I, §6.12.1]. Normally the solution for C (and therefore $v(z)$) is finite and a linear function of p (contained in F) but a non-linear function of P (contained in A). Moreover, if $|A| \to 0$ we see that the deflections become infinitely large—i.e. the beam buckles. The *critical value* of α is therefore given by the smallest positive root of $|A| = 0$. (Exactly the same type of equation as (10.16) arises if the beam has initial deformations rather than a lateral load $p(z)$.) If the beam is not laterally loaded and is perfectly straight, then (10.16) becomes $AC = 0$, from which we conclude that $C = 0$, i.e. no deflections *unless* $|A| = 0$. Consequently, we deduce initial buckling loads for perfect beams also by setting $|A| = 0$, but the solution for C now is indeterminate; in other words, initial buckling is a state of *neutral equilibrium*.

Similar transcendental equations occur if we seek the natural frequencies of freely vibrating beams: we simply replace $p(z)$ by $-m(\partial^2 v/dt^2)$ (the reversed inertia 'force') and assuming $v(z, t) = v(z) e^{i\Omega t}$ we then obtain

$$EIv'''' - m\Omega^2 v = 0, \tag{10.18}$$

which has a mixed trigonometric–hyperbolic solution [see IV, §§2.12 and 2.13].

Many buildings, of course, are complex three-dimensional frameworks consisting of a myriad collection of beams, both vertical and horizontal, connected together at the frameworks joints. Nevertheless, the PVD still holds provided we remember that the integral is over the entire volume, i.e. we sum equation (10.9) over all beams. It is usual to solve such structures by setting $p(z) = 0$ (the framework is loaded only at its joints) or perhaps $p(z) = $ constant. In both cases $v(z)$ for a single beam can be solved as a simple cubic or quartic [see III, Chapter 5] in terms of its end rotations and deflections, whence we can sum all the boundary terms in (10.10) as coefficients of the discrete values of $\bar{v}$ and $\bar{v}'$ of all joints. The resulting equations for the joint displacements and rotations for the complete framework are best assembled using the technique described in 'the finite element' method of approximate solutions (Section 10.6 and Chapter 15), although here, of course, each beam 'element' is solved exactly rather than approximately.

Plates

The *bending* of a thin flat plate in the xy plane whose surfaces are $z = \pm t/2$ is assumed to be a simple two-dimensional extension of beam theory. There is one restriction to this simple model, namely that the deflections $w(x, y)$ are not in excess of t, otherwise further membrane strains are induced—a phenomenon not present in simple beams. If we assume again that $u = -y(\partial w/\partial x)$ and $v = -y(\partial w/\partial y)$ and apply the PVD, we obtain a fourth-order differential equation (Timoshenko, 1940) for w, which if the plate is isotropic becomes the well-known *biharmonic equation* [see III, §9.5]

$$D\nabla^4 w(x, y) = p(x, y), \qquad (10.19)$$

where $p(x, y)$ is now the pressure, the flexural rigidity $D = Et^3/12(1 - v^2)$, $\nabla^4 = \nabla^2\nabla^2$ and ∇^2 is the *Laplace or harmonic operator* $\partial^2/\partial x^2 + \partial^2/\partial y^2$ [see also IV, §8.2]. If in-plane edge stresses σ_{xx}, σ_{yy} and σ_{xy} are applied which might cause buckling, terms similar to the beam's Pv'' have to be added (Mansfield, 1964) to (10.19), namely

$$t\sigma_{xx}\frac{\partial^2 w}{\partial x^2} + t\sigma_{yy}\frac{\partial^2 w}{\partial y^2} + 2t\sigma_{xy}\frac{\partial^2 w}{\partial x\,\partial y}.$$

It so happens that the *stretching* of plates is also governed by the same equation and we should therefore consider both together. The stresses in the plate σ_{xx}, σ_{yy} and σ_{xy} are assumed constant through the thickness (plane stress), and the components σ_{zz}, σ_{zy} and σ_{zx} are ignored, although strictly they are only zero on the stress-free surfaces $z = \pm t/2$. We could convert the strains in the compatibility equations (10.2) to stresses, using the linear stress–strain–temperature law $\varepsilon_{xx} = (1/E)(\sigma_{xx} - v\sigma_{yy}) + \alpha T$, etc., and on substituting into the equilibrium equations (10.1), obtain second-order differential equations in the displacements u and v (Love, 1944). However, it is more convenient to introduce a *stress function* $F(x, y)$ defined so as automatically to satisfy (10.1). Thus

$$\sigma_{xy} = \frac{-\partial^2 F}{\partial x\,\partial y}; \qquad \sigma_{xx} = \frac{\partial^2 F}{\partial y^2} + V; \qquad \sigma_{yy} = \frac{\partial^2 F}{\partial x^2} + V, \qquad (10.20)$$

where V is another body force potential defined by $\omega_x = -\partial V/\partial x$, $\omega_y = -\partial V/\partial y$. Eliminating u and v from (10.2) and inserting the stress–strain law for an isotropic material, the compatibility equation is expressed in terms of stress alone, which when using (10.20) becomes

$$\nabla^4 F = -\nabla^2[E\alpha T + (1 - v)V]. \qquad (10.21)$$

The known right-hand sides of (10.19) and (10.21) present no problem when finding *particular integrals*; we therefore concentrate on finding *complementary functions* of the biharmonic which then enable us to satisfy boundary conditions. Particular integrals and complementary functions of partial differential equations are analogous to the ordinary case [see IV, §7.3.2]. In the case of plate bending, a common ruse is to *separate variables* [see IV, §8.2] in a *partial Fourier Series* fashion like

$$w = \sum_n f_n(y) \sin\left(\frac{n\pi x}{a}\right) \qquad \text{for a rectangular plate of width } a,$$

or

$$w = \sum_n f_n(r) \sin\left(\frac{n\pi\theta}{a}\right) \qquad \text{for a sector bounded by edges } \theta = 0, a$$

[see IV, §20.5]. This technique works when the assumed function is constant on the boundary; in the above example the assumed sine terms satisfy the simply supported edge conditions. *A double Fourier series* [see IV, §20.7]

$$w = \sum_n \sum_m f_{nm} \sin\left(\frac{n\pi x}{a}\right) \sin\left(\frac{m\pi y}{b}\right)$$

is possible for a rectangular simply supported plate of sides a and b, and if the right-hand side of the equation is similarly expanded as a Fourier series and coefficients of like terms on both sides are equated, we obtain the values of f_{nm}. Solutions for regions bounded by strips, rectangles, sectors and circles are well tabulated (Timoshenko, 1940).

Separation of variables is not as successful for the plane stress problem, mostly because the boundary conditions are not simply on $F(x, y)$ or one derivative; and of course if the boundary is not straight or circular then $F(x, y)$ and $w(x, y)$ become almost intractable. One direct and elegant method is to transform to the *complex variable* $z = x + iy = re^{i\theta}$ [see I, §2.7] and $\bar{z} = x - iy = re^{-i\theta}$. The interested reader should consult the standard texts (Milne-Thomson, 1960; Muskhelishvili, 1953) but briefly the method rests on the fact that the harmonic operator ∇^2 becomes $4[\partial^2/(\partial z\, \partial\bar{z})]$, and so the complementary function of (10.21) can be immediately integrated as

$$2F(z, \bar{z}) = z\overline{\phi(z)} + \bar{z}\phi(z) + \int \psi(z)\, dz + \int \overline{\psi(z)}\, d\bar{z}. \tag{10.22}$$

(The curious '2' and use of integrals of $\psi(z)$ in (10.22) is merely to conform with the notation of the definitive text; see Muskhelishvili, 1953.) The need for solving a differential equation has therefore been replaced by a search for *analytic functions* [see IV, §9.1] which satisfy the boundary conditions. The stresses (10.20) depend on the derivatives of $\phi(z)$ and $\psi(z)$ which is why these functions must be analytic—i.e. possess unique derivatives $\partial\phi/\partial z$, $\partial\psi/\partial z$, irrespective of the direction of the vector dz. When finding the stresses from $\phi(z)$ and $\psi(z)$ it is much more convenient to form the following combinations, particularly so if we seek stress components in a normal (n) and tangential (s) direction at angles α to the y and x axes (see Figure 10.2):

$$\sigma_{nn} + \sigma_{ss} = \sigma_{xx} + \sigma_{yy} = 2[\phi'(z) + \overline{\phi'(z)}], \tag{10.23}$$

$$\sigma_{nn} - \sigma_{ss} + 2i\sigma_{sn} = (\sigma_{yy} - \sigma_{xx} + 2i\sigma_{xy})e^{2i\alpha} = 2[\bar{z}\phi''(z) + \psi'(z)]e^{2i\alpha}. \tag{10.24}$$

The displacement vector $u + iv$ is given by

$$2G(u + iv) = \kappa\phi(z) - z\overline{\phi'(z)} - \overline{\psi(z)}, \tag{10.25}$$

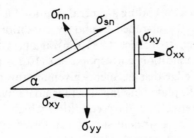

Figure 10.2: Stresses on inclined
surface.

where $\kappa = (3 - v)/(1 + v)$ for plane stress and $3 - 4v$ for plane strain. If the displacements are prescribed on a boundary, then (10.25) is the boundary condition, and if the edge stresses are given then we form (10.23) and (10.24), i.e. $2(\sigma_{nn} + i\sigma_{sn})$. However, a much more convenient stress boundary condition not involving the tiresome angle α is obtained if we integrate the known edge stresses to find the total *force* $X + iY$ acting on a boundary from some arbitrary starting point to the point z, viz.

$$\phi(z) + z\overline{\phi'(z)} + \overline{\psi(z)} = -i(X + iY). \tag{10.26}$$

Clearly (10.26) and (10.25) are almost identical.

For a singly connected circular region—i.e. a disc, or a circular hole in a very large plate—it is only necessary to assume a *Laurent series* [see IV, §9.7] for $\phi(z)$ and $\psi(z)$, and to then substitute into (10.26) (or 10.25), expand the right-hand side of this equation as a simple *complex Fourier series* [see IV, §20.5.6] and equate like terms in $z^{\pm n}$. This technique is little more than a standard Fourier analysis, but the method becomes more powerful for *non-circular* regions which are mappable onto the interior (or exterior) of a unit circle $|\zeta| = 1$ by a *conformal transformation* [see IV, §9.7.2] $z = \Omega(\zeta)$. The boundary condition for the potentials $\phi(z) = \phi[\Omega(\zeta)] = \phi_1(\zeta)$ and $\psi_1(\zeta)$ is identical in form to (10.26) or (10.25) except for the term

$$\overline{z\phi'(z)} = \frac{\Omega(\zeta)\overline{\phi_1'(\zeta)}}{\overline{\Omega'(\zeta)}}.$$

It can be shown that the quotient $\Omega(\zeta)/\overline{\Omega'(\zeta)}$ is always expandable as a convergent series in ζ; therefore equating coefficients of Laurent expansions for $\phi_1(\zeta)$ and $\psi_1(\zeta)$ still delivers $\phi_1(\zeta)$, but this time it is necessary to use the Cauchy integral formula [see IV, Theorem 9.6.1] to extract $\psi_1(\zeta)$ in closed form from the boundary condition. (Actually, because $\phi_1(\zeta)$, $\psi_1(\zeta)$ and $\Omega(\zeta)$ have poles both inside and outside the unit circle, it is necessary to use *generalized* Cauchy integrals; see Muskhelishvili, 1953.)

All the above remarks for finding a general solution for $F(z, \bar{z})$ are equally true of $w(z, \bar{z})$. In bending, the clamped boundary condition is $\partial w / \partial z = 0$ for any direction z, and is seen to reduce to (10.26) again. The edge force/moment prescribed boundary condition corresponds to (10.25) (Savin, 1961)—i.e. the roles of edge forces or edge displacements given are reversed in the plane stress problem compared with plate bending. There is no equivalent in plane stress to the simply supported edge boundary condition which is very lengthy (Mansfield, 1964, p. 52), but for a *straight* simply supported edge it simplifies to

$$\nabla^2 w = 0$$

or

$$\phi'(z) + \overline{\phi'(z)} = -\nabla^2 w_{\mathrm{p}},$$

where w_{p} is the particular integral.

Finally, it should be honestly admitted that the use of complex potentials simply converts a conventional boundary-value problem to another one of finding *conformal mapping functions* [see IV, §9.12]. One of the most useful is the *Schwarz–Christoffel* which maps n-sided polygons having exterior angles α_k onto circular boundaries by the following products:

For exteriors:
$$\frac{\partial z}{\partial \zeta} = \text{constant} \times \prod_{k=1}^{k=n} \left(1 - \frac{\sigma_k}{\zeta}\right)^{-\alpha_k/\pi}$$

and

For interiors:
$$\frac{\partial z}{\partial \zeta} = \text{constant} \times \prod_{k=1}^{k=n} \left(1 - \frac{\zeta}{\sigma_k}\right)^{\alpha_k/\pi}$$

where σ_k is a point on the unit circle $\zeta = \sigma = e^{i\theta}$. By expanding the terms in the above to any chosen extent we may produce a 'rounded polygon' approximating any shape we wish, and—equally important—ensure that the roots of $\partial z / \partial \zeta = 0$ lie just outside the boundary where they will do no harm—inside they render the mapping non-conformal.

Torsion and bending of solid sections

Earlier we outlined the simple bending and twisting of beams. This analysis is adequate provided that the beam or tube has a thin-walled section for which it is permissible to assume that the shear stress lies parallel to the median line everywhere—and not just on the surface where of course it must be (i.e. $\sigma_{nz} = 0$). Fortunately, most beams and tubes are indeed thin walled but some are not— e.g. aerofoil sections in turbine blades, helicopter blades, solid wings and controls in missiles, and camshafts of all shapes. For these sections, the shear stress distribution under pure torque has to be solved from the two-dimensional field equations.

It is possible to satisfy equilibrium from a stress function $F(x, y)$ again by this time putting

$$\sigma_{xz} = G\theta' \frac{\partial F}{\partial y}, \qquad \sigma_{yz} = G\theta' \frac{\partial F}{\partial x}$$

and on the boundary, of course,

$$\sigma_{nz} = G\theta' \frac{\partial F}{\partial s} = 0,$$

whence F is a constant on the boundary, which can be made zero for singly connected regions. On substituting into the compatibility equations we find $\nabla^2 F = -2$, and it is usual to remove the particular integral by putting

$$\psi(x, y) = F(x, y) + \tfrac{1}{2}(x^2 + y^2)$$

so that we now have

$$\nabla^2 \psi = 0; \qquad \psi = \tfrac{1}{2}z\bar{z} \qquad \text{on the boundary.} \qquad (10.27)$$

Equation (10.27) is the standard *Dirichlet problem* [see IV, §8.2] and can be integrated immediately to give $\psi = f(z) + \overline{f(z)}$ or, in the notation of the only text devoted entirely to this problem (Milne-Thomson, 1962),

$$\psi(z, \bar{z}) = -\frac{i}{2}\omega(z) + \frac{i}{2}\overline{\omega(z)},$$

whence

$$\sigma_{xz} - i\sigma_{yz} = G\theta'(\omega'(z) - i\bar{z}).$$

The boundary condition in (10.27) gives $\omega(z)$ immediately using *Cauchy integrals* [see IV, Theorem 9.6.1], and even if we have to map the section, using $z = \Omega(\zeta)$, onto the unit circle $\zeta = \sigma$, we find

$$\omega(z) = \omega[\Omega(\zeta)] = \frac{1}{2\pi} \oint \frac{\Omega(\sigma)\overline{\Omega(\sigma)}\, d\sigma}{\sigma - \zeta}. \qquad (10.28)$$

The torsional rigidity

$$J = -\frac{i}{4} \oint z\bar{z}^2\, dz - \tfrac{1}{2} \oint \omega(z)\, d(z\bar{z}) \qquad (10.29)$$

is best evaluated, knowing $\omega(z)$, using *Cauchy's residue theorem* [see IV, Theorem 9.9.1].

If the section is subjected to a constant shear force Q, giving direct stresses

$$\sigma_{zz} = \frac{Qzy}{I},$$

then these known stresses modify (through the equations of equilibrium) the previous stress field. Because this new term is known—like body forces—it simply produces a right-hand side to (10.28). It is possible to find the position through which Q must act so that the bending problem and torsion problem are completely decoupled (Milne-Thomson, 1962).

Finally, it should be mentioned that another method of solving two-dimensional elasticity problems, of some limited use in particular regions like strips, discs, half-planes, etc., is by the use of *integral transforms* [see IV, §13.1] such as *Fourier, Laplace, Hankel, Mellin, Stieltjes*, etc. (Flügge, 1962a).

Shells

It is logical to carry the methods used in solving one-dimensional beams and two-dimensional plates through to three-dimensional shells, but in so doing we have to make some fairly restrictive assumptions, or otherwise we are faced with the general three-dimensional equations (10.1) and (10.2) to which there is no general analytical solution. Fortunately, most shells are very thin, the value t/r (r being a typical radius of curvature) lying between 0.01 for robust pressure vessels and other industrial plant to 0.001 for aircraft, cooling towers and ships and even down to 0.0001 for some space vehicles. If we are able to assume $t/r \ll 1$, then the problem becomes two dimensional; we can describe the middle surface of a shell by two curvilinear coordinates and make the same simplifying assumption that we made in plates, namely that the displacement varies linearly through the thickness. If $t/r \ll 1$ then a linear variation in strain follows this kinematic assumption. If we further assume that displacements are as small as order 't' then the final equations will be linear. This is not unreasonable for shells which are, as we shall see, much more efficient structures than flat plates. We could now proceed to set up the governing differential equations for arbitrary shells, but these are so complex and have so few analytical solutions that the interested reader is best referred to the single text (Goldenweizer, 1961) on shells which is completely general. The complexities arise solely from the geometry of the surface and are considerably simplified if we are able to use, as surface coordinates, the orthogonal lines of *principal curvature*. The nature of these lines, together with the Gauss–Codazzi conditions, which are necessary prerequisites to a treatment of shells, follow from the *differential geometry of surfaces* [see V, Chapter 12]. If there is an analytical solution to a shell problem, then separation of variables in the two surface coordinates is invariably necessary, in which case the above restriction means that the shell boundary conditions have to be applied along a line of curvature; fortunately most shells are supported in this way. In fact, most shells are surfaces of revolution or a part of such a surface. There are several texts (Kraus, 1967; Novoshilov, 1959) which base their analysis on this assumption and the interested reader is referred to them; suffice it to say here that the expected six equations of equilibrium and six equations of compatibility are reduced to eleven if we ignore normal deformations through the thickness.

Instead of quoting the general equations we will first of all examine the simplest of all shells—the axisymmetrical cylinder loaded by an axisymmetrical pressure distribution $p(x)$—as there are several peculiar features of thin shells which are present even in this simplest case. The governing differential equation for the normal deflection $w(x)$ is very similar to the beam equation (10.15). Thus

$$D\frac{d^4w}{dx^4} - N\frac{d^2w}{dx^2} + \frac{Etw}{r^2} = p(x) - \frac{vN}{r},\tag{10.30}$$

where D is defined in equation (10.19). This shows that pressure is resisted by three mechanisms: first, by the flexural term $D(d^4w/dx^4)$ present in beams and plates; second, by $N(d^2w/dx^2)$ which is the effective normal component of axial tension (N) per unit length and which is similar to the destabilizing force Pv'' on struts (10.15); and third, by Etw/r^2 which looks like a spring stiffness proportional to w. This last term is a normal component of the circumferential *membrane* force σt equal to $(\sigma t)(1/r) = (Ew/r)(t/r)$, but this term arises from the undeformed geometry $1/r$ and not the induced curvature d^2w/dx^2. The term 'membrane' is coined to distinguish this stress, uniform through the thickness, from the linearly varying bending stresses which would be negligible in an infinitesimally thin membrane. In the general case, with *two* lines of curvature, we would have $N_1/r_1 + N_2/r_2$; this mechanism is very efficient and explains why shells are much stiffer than flat plates. A pure membrane shell with no bending stresses is a highly desirable objective, but not always attainable as we shall see. The last term on the right-hand side of (10.30) is a Poisson contraction due to the longitudinal stress N/t.

The second term in (10.30) is of particular interest when the longitudinal stress N/t is compressive, since it is the destabilizing term. For example, if we take a simply supported shell of length L and test (10.30) for neutral equilibrium (initial buckling) by assuming $w(x) = A\sin(m\pi x/L)$, we find that the smallest buckling stress is

$$\frac{N}{t} = \frac{E}{\sqrt{3(1-v^2)}}\frac{t}{r}$$

where $m = \sqrt{2}\beta/\pi$ and $\beta^4 = 3(1-v^2)/r^2t^2$. The half-wavelength of the buckles, $L/m = 0.17\sqrt{rt}$, is clearly very small compared to the radius and shows that boundary conditions are unimportant. This initial buckling stress is, however, totally unreliable, because the post-buckling behaviour—occurring when $w > t$ and hence is non-linear—involves non-axisymmetrical modes occurring at applied stresses less than one half of the above for practical shells with some *initial imperfections* (Thompson and Hunt, 1973). However, if the shell is in tension, the second term in (10.30) is now stabilizing and *reduces* bending stresses roughly by an amount $1/\sqrt{1 + \gamma}$, where γ is the ratio of the applied stress to the critical buckling stress given above. In all texts, and in what follows, this stabilizing term is ignored—hopefully optimistically—since its inclusion would in general involve the non-linear product of an unknown membrane stress with an unknown change in curvature. Equation (10.30) therefore may be written

$$\frac{d^4w}{dx^4} + 4\beta^4 w = \frac{p}{D} - \frac{vN}{Dr},\tag{10.31}$$

where $\beta^4 = \dfrac{3(1 - v^2)}{r^2 t^2}$ as before.   (10.32)

In view of the desirability of a pure membrane solution, it is of interest to note when this might happen, i.e. when we may put $D = 0$ in (10.30). First, if we discard high-order derivatives in a differential equation we must expect possible violation of boundary conditions, and if the simple membrane solution does violate boundary conditions there will be an error which can only be removed by the *complementary function* [see IV, §7.3.2] of (10.31) which is

$$w(x) = e^{-\beta x}(k_1 \sin \beta x + k_2 \cos \beta x) + e^{\beta x}(k_3 \sin \beta x + k_4 \cos \beta x).\tag{10.33}$$

If we use (10.33) to satisfy boundary conditions at $x = 0$ (say) then we must discard positive exponents since, at some typical distance L from the end, βL is of order $(L/r)\sqrt{r/t}$ which is $\gg 1$. The remaining solution therefore represents a bending *boundary layer* of thickness order $1/\beta$ or $\sqrt{rt}$ which is small compared with r. Away from this thin bending layer a membrane solution is possible provided the *particular integral* [see IV, §7.3.2] of (10.31) is given simply by

$$4\beta^4 w = \frac{p}{D} - \frac{vN}{Dr}.$$

This will be true provided the variation in $p(x)$ is not as strong as $e^{\beta x}$ which, in view of (10.32), in practice means *any* loading with the exception of concentrated forces.

This specific behaviour of thin cylinders enables us to judge better the solution of *any* thin axisymmetric shell loaded axisymmetrically, since we might expect a membrane solution to be adequate everywhere away from concentrated loads and discontinuities in geometry. We refer the interested reader to the books by Kraus (1967) or Flügge (1962b) for details, but briefly the equations are most concisely summarized in terms of the normal shear force Q per unit length and the rotational displacement χ of the meridional line. Denoting r_1 and r_2 as the meridional and circumferential principal radii of curvature [see V, §12.4.3] and putting $U = r_2 Q$, we can derive a pair of simultaneous differential equations of second order for the bending solution (assuming the particular integral has been taken care of by the membrane solution):

$$L(\chi) - \frac{v}{r_1}\chi - \frac{U}{D} = 0, \tag{10.34}$$

$$L(U) + \frac{v}{r_1}U + Et\chi = 0, \tag{10.35}$$

where

$$L(\ldots) = \frac{r_2}{r_1}\frac{(\ldots)''}{r_1} + \left[\frac{r_2}{r_1}\cot\phi + \left(\frac{r_2}{r_1}\right)'\right]\frac{(\ldots)'}{r_1} - \frac{r_1}{r_2}\cot^2\phi\frac{(\ldots)}{r_1} \tag{10.36}$$

and ϕ is the angle between a normal and the shell axis; primes denote differentiation with respect to ϕ. Either variable can be eliminated from (10.34) and (10.35). For example,

$$LL(U) + vL\left(\frac{U}{r_1}\right) - \frac{v}{r_1}L(U) - \frac{v^2}{r_1^2}U + \frac{Et}{D}U = 0. \tag{10.37}$$

This equation is intractable analytically unless the two middle terms cancel, which can happen if:

(i) $v = 0$. This is reasonable for reinforced concrete shells.
(ii) $r_1 = $ constant. This is true for cylinders, cones, spheres and toroids.

Equation (10.37) can then be factorized as

$$(L + i\mu^2)(L - i\mu^2)U = 0, \tag{10.38}$$

where $\mu^4 = \dfrac{Et}{D} - \dfrac{v}{r_1^2} \approx \dfrac{Et}{D}$ for $t/r \ll 1$.

Solutions of $(L \pm i\mu^2)U = 0$ are possible in series form. For cylinders we have already seen trigonometric and hyperbolic terms; for cones we have *Bessel's equation* and *Kelvin functions* [see IV, (10.4.5)]; for spheres it becomes *hypergeometric* [see IV, §7.6.1].

For axisymmetrical shells loaded in an arbitrary fashion we may take advantage of the necessary cyclic nature of the solution and expand both displacements and applied forces as unknown and known *Fourier series* [see IV, §20.5] in the polar coordinate θ. Upon equating coefficients of like harmonics we obtain a set of *ordinary* differential equations in the meridional variable ϕ. However, the algebraic complexity is fearsome and a much simpler approximation due to Geckeler (see Kraus, 1967) is possible for very thin shells in which (10.37) is replaced by a condensed version, which can then be extended if required as a Fourier series to cope with non-axisymmetrical loads. We take advantage of the expected boundary layer in which all bending takes place and assume that all derivatives with respect to ϕ are an order of magnitude $\sqrt{r/t}$

greater than the function differentiated. Equation (10.37) then simply becomes

$$Q'''' + 4\beta^4 Q = 0, \tag{10.39}$$

where now
$$\beta^4 = 3(1 - v^2)\left(\frac{r_1}{r_2}\right)^2\left(\frac{r_1}{t}\right)^2, \tag{10.40}$$

and, because $\beta \gg 1$, solutions of (10.39) behaving like $e^{-\beta\phi}$ decay so rapidly that variations in $r_1(\phi)$ and $r_2(\phi)$ may be much simplified or even ignored altogether in the boundary layer. This latter ruse is equivalent to approximating the geometry of such a layer as a a short conical frustum. The Geckeler approximation is not valid near $\phi = 0$ since the cot ϕ terms in L in (10.36) prevent us from ignoring all but the highest derivative. Fortunately a special 'shallow shell theory' is available for this region, the consequence of which is a pair of equations like the separate flat plate equations of bending and stretching (10.19 and 10.21) but with limited coupling due to the curvature $1/r$. Thus

$$D\nabla^4 w + \frac{1}{r}\nabla^2 F = p - D\alpha(1 + v)\nabla^2 T \tag{10.41}$$

and

$$\nabla^4 F - \frac{Et}{r}\nabla^2 w = -\nabla^2[E\alpha T + (1 - v)V]. \tag{10.42}$$

The use of complex variables is not as productive as it was in (10.19) and (10.21); nor is it necessary since the shallow shell solution is mostly used to blend into the shell proper a short distance away from $\phi = 0$ around a simple circular contour. In the axisymmetric case the above equations are *homogeneous* and can be integrated [see IV, §8.2]; otherwise we rewrite them as follows. First, we nondimensionalize

$$\bar{F} = \frac{F}{a^2 N}$$

and

$$\bar{w} = \frac{wEt^2}{12(1 - v^2)a^2 N},$$

where N/t is a reference membrane stress and a is a typical shallow shell dimension $(a \ll r)$ [see I, §17.8]. We then form $\phi = \bar{w} - i\bar{F}$ and combine the equations thus to find the complementary function

$$\nabla^4\phi + i\beta^2\nabla^2\phi = 0; \qquad \beta^2 = \frac{a^2}{rt}[12(1 - v^2)]^{1/2},$$

whence $\phi = \phi_1 + \phi_2$ where $\nabla^2\phi_1 = 0$ and $\nabla^2\phi_2 + i\beta^2\phi_2 = 0$; ϕ_1 is the familiar harmonic solution and ϕ_2 is expressible in terms of *Hankel functions* [see IV, (10.4.12)].

The Geckeler approximation is a fairly crude reduction which nevertheless gives answers to within a few percent even for t/r as large as 0.1. If the reader desires a closer approximation, the method of *asymptotic integration* [see IV, §7.6.4] may be used. This technique also exploits the large parameter μ^2 in (10.38). As a first step the operator $L(\ldots)$ in (10.36) is converted to *normal form* by a suitable change of variable

$$Y = U \left[\frac{r_2 \sin \phi}{r_1 t} \right]^{1/2}$$

so that (10.38) becomes of the form

$$Y'' + [f(\phi) + i\beta^2 g(\phi)]Y = 0$$

(cf. equation 10.39). This equation may then be solved exactly as $\beta \to \infty$ by an asymptotic solution which may still be adequate even for moderate β, depending on the nature of $f(\phi)$ and $g(\phi)$ for the particular variations $\Gamma_1(\phi)$ and $r_2(\phi)$.

Finally, we return to the *membrane solution* of shells for two reasons. First, as we have seen, it is convenient because it is simply a particular integral of the full equations in equilibrium with the applied loads, provided the loading is reasonably 'smooth'. Second—a much better reason—a pure membrane is an efficiently stressed structure. If we examine the equations of equilibrium with bending moments and shears removed, three unknown membrane stress components are left in the three equations. Therefore the solution is statically determinate and many solved examples for various shapes are available in texts (Flügge, 1962b, Gould, 1977; Kraus, 1967). We content ourselves here with a look at the form of the solution; thus if we choose to eliminate two stress components the final equation in (say) the meridional stress component N/t

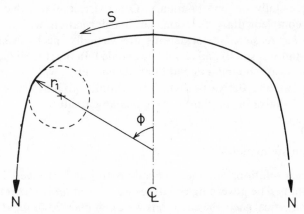

Figure 10.3: Membrane stresses in axisymmetrical shell.

looks like

$$\frac{\partial^2 N}{\partial s^2} + \frac{1}{r_1 r_2 \sin^2 \phi} \frac{\partial^2 N}{\partial \theta^2} + \text{terms in first and zeroth derivatives.}$$

The change of variable from ϕ to s ($ds = r_1 \, d\phi$) is necessary because ϕ may increase or decrease as we move a distance s along a meridian, and this clouds the behaviour of the above equation. We see that the sign of the second term depends on the *gaussian curvature* [see V, §12.4.4] $1/r_1 r_2$, and this dictates whether the equation is *elliptical* (a sphere, $r_1 = r_2$), *parabolic* (a cone, $r_1 = \infty$) or *hyperbolic* (a saddle-type surface). The behaviour of *elliptic* and *hyperbolic* equations is quite different [see IV, §8.5] and the engineer should be aware of these differences. In elliptic equations any discontinuity in N on a boundary will diffuse into the interior in a characteristic *diffusion length*—this is particularly important if a shell rests on discrete supports—whereas in a shell with negative gaussian curvature such a boundary force may propagate into the shell along its *characteristics* [see III, §9.1] and appear at another boundary where no equilibrating reaction is available. It is therefore much more difficult to design the edges of such shells so that the membrane solution satisfies the equilibrium boundary conditions and also the kinematic conditions, as we shall see. Without going into details, once we have solved the membrane stresses it is possible to insert them in the compatibility equations as strains and then integrate to find displacements. In fact, this procedure can be made algebraically identical to the solution of the stresses from equilibrium equations (Novoshilov, 1959), and the previous remarks apply concerning the elliptic or hyperbolic nature of the equations. The displacements will contain a particular integral from the known stresses and a complementary function consequent upon integrating. To preserve a membrane solution it is necessary to hold the boundaries in the correct fashion. In the case of hyperbolic shells—with two edges—this can only be done successfully on one boundary. Otherwise it is possible to have an unbounded complementary function, i.e. a deformation without membrane stress. Such *inextensional* deformations are virtually mechanisms (since the bending resistance is small) and should be avoided. In practice, if the membrane solution violates any boundary conditions, then bending will take place and the higher order 'bending' derivatives which then come into play are always elliptic in nature irrespective of the sign of the gaussian curvature.

10.3 Approximate methods

All analytical solutions obtained in Sections 10.1 and 10.2 were for relatively simple structures. The governing equations were linear and yet, even in the case of the axisymmetrical shell, the equations were complicated to handle for purely geometrical reasons. For arbitrary shells or three-dimensional problems the geometry usually forces us to seek an *approximate solution* in which some assumption is made for the displacement or stress fields.

A common ruse, much used before the advent of accessible computers, is the Krylov and Kantorovitch (1964) technique in which we *assume* a likely variation of the displacement field (or stress field) in one or two independent variables—satisfying prescribed boundary conditions of course—and then substitute into the PVD (or PVF) version of (10.5) and obtain *ordinary* differential equations in the remaining independent variable. There are several ways of doing this: e.g. in two-dimensional plate-bending (or stretching) problems we might use the PVD and assume a displacement variation (or use the PVF and a stress function variation) of the form

$$w(x, y) = \phi_1(x)f_1(y) + \phi_2(x)f_2(y) + \cdots,$$

where $f_n(y)$ are assumed. We obtain, by putting the coefficients of the virtual $\bar{\phi}_n(x)$ to zero, a set of equations which can be integrated with respect to y, leaving ordinary differential equations and boundary conditions in $\phi_n(x)$. If the answers are questionable then the solved $\phi_n(x)$ can be used as a basis for then solving new $f_n(y)$, and so on. This procedure can be carried a stage further and functions of *all* independent space variables can be assumed, in which case only the *magnitudes* of the variations remain as unknowns and the problem of solving a differential equation has been completely bypassed. This technique is known as the *Rayleigh–Ritz method* [see III, §9.8] when using form (10.5) or the *Galerkin method* when using (10.7) or (10.8). Provided the *necessary* boundary conditions (kinematic for equation 10.7 and equilibrium for equation 10.8) are satisfied, both the Rayleigh–Ritz and Galerkin methods yield the same equations. The assumed variations are taken as some convenient *complete series* such as a polynomial (Higgins, 1977) or Fourier series [see IV, §20.6.4]. For instance, if the plate-bending problem was a simply supported rectangular plate of sides a and b it could be solved using the PVD by assuming

$$w(x, y) = \sum_n \sum_m A_{mn} \sin\left(\frac{n\pi x}{a}\right) \sin\left(\frac{m\pi y}{b}\right).$$

This satisfies the necessary kinematic boundary conditions and—as a bonus—the edge moments are zero also. Using the PVD (10.5) and making the plate-bending assumptions of Section 10.2 ($u = -y(\partial w/\partial x)$; $v = -y(\partial w/\partial y)$), we find after integrating through $-t/2 < y < t/2$ that

$$\int_A \left[\left(\frac{\partial^2 w}{\partial x^2} + v\frac{\partial^2 w}{\partial y^2}\right)\frac{\partial^2 \bar{w}}{\partial x^2} + \left(\frac{\partial^2 w}{\partial y^2} + v\frac{\partial^2 w}{\partial x^2}\right)\frac{\partial^2 \bar{w}}{\partial y^2} + 2(1-v)\frac{\partial^2 w}{\partial x \partial y}\frac{\partial^2 \bar{w}}{\partial x \partial y} \right.$$
$$\left. -\frac{p(x, y)}{D}\bar{w} \right] dA = 0.$$

On now substituting for w (x, y) and setting to zero the coefficients of each $\bar{A}_{mn}$ term, we obtain a set of simultaneous equations in the unknown A_{mn}.

The Rayleigh–Ritz procedure therefore converts a *continuous* field problem to a search for *discrete* unknowns; however the use of assumed 'shape functions' spanning the entire structure is too limiting for real structural shapes, and further types of discretization are necessary in practice. The overwhelmingly popular technique in structural mechanics is the *finite element method* of Chapter 15, which is simply a piecewise Rayleigh–Ritz using localized shape functions, but before expanding upon this we mention two other forms of discretization.

The *finite difference method* [see III, §9.1] replaces a continuous function by its discrete (unknown) values at selected mesh or nodal points. The next step is usually to approximate the various derivatives in the governing differential equation and boundary conditions throughout the region at all nodes. With linear equations of compatibility or equilibrium, a set of linear simultaneous equations emerges. Alternatively, the same approximations can be inserted into the PVD or PVF, with the computational advantage that a *positive-definite matrix* [see I, §6.7(v)] of the coefficients of the discrete unknowns can be guaranteed and 'natural' boundary conditions satisfied implicitly. The difference between the finite element and finite difference techniques is not as great as their historical development would imply, but it is probably true that higher order approximations and very complex boundaries are better solved using finite elements.

The *boundary integral equation method* has been applied successfully in both fluid (see Section 2.4) and structural mechanics, amongst other field problems. It relies upon obtaining, from the differential equation, a singular solution corresponding to some sort of 'source' on the boundary. The boundary condition is then integrated piecewise using segments containing source magnitudes, or some equivalent measure, as the discrete unknowns. The problem is therefore only discretized on the surface boundary and errors are confined to a surface layer since the method is applied to elliptic boundary-value problems only. Although the number of unknowns is much smaller than in the finite element or difference methods (particularly in three dimensions) the matrix to be inverted is fully populated. The method has found limited favour in structural mechanics. However, it has advantages for regions which are semi-infinite and is usefully employed, for example, in determining local stress fields around surface cracks and evaluating stress intensity factors.

The finite element method

The finite element method has long passed into common usage in all branches of applied structural mechanics and there exists a staggering amount of capital investment in the many commercial finite element packages ranging from the small to the very large, from the specialized to the general purpose, such as NASTRAN, ASKA, etc. A brief account follows in order to highlight the various mathematical requirements, but a more general treatment is given in Chapter 15.

The method can be visualized as two distinct operations: the methodical evaluation of a complete volume and surface integral (10.5) over the structure, by summing all the integrals over small elements, together with an approximation of the field variables over these small 'finite' elements. Historically the first operation was originally applied to *frameworks* whose component parts are exact finite elements and need no approximation, and it was only later that the approximate finite element idealization was introduced to enable arbitrary structural configurations to be analysed. An element is assumed to be small enough so that the variation of a field over it can be decently represented by (usually) a polynomial. Thus there are three possible finite element formulations.

(i) *Displacement models.* Assume displacement fields. Satisfy compatibility exactly ($\varepsilon = \mathbf{D}^T \mathbf{u}$) and use the PVD to satisfy equilibrium approximately.

(ii) *Equilibrium models.* Assume stress fields which satisfy equilibrium exactly ($\mathbf{D}\sigma + \omega = 0$) and use the PVF to satisfy compatibility approximately.

(iii) *Mixed models.* Assume *both* stress and displacement fields and use a mixed formulation of both PVF and PVD to satisfy equilibrium and compatibility in an approximate (weighted residual) fashion.

By far the most popular is the displacement method, which we will consider alone. Option (ii) involves *integrating* a field equation ($\varepsilon = \mathbf{D}^T \mathbf{u}$ is much easier) whilst option (iii) avoids this but eventually leads to sets of equations which are not positive definite.

The gth element of volume V_g and surface S_g has its displacement field approximated by a series of *interpolation polynomials* $\mathbf{w}(x, y, z)$ of magnitude ρ_g. Thus

$$\mathbf{u} = \mathbf{w}\rho_g \tag{10.43}$$

where the ρ_g are usually the displacements or equivalent measures at selected nodal points in the gth element. For example, ρ_g may be displacements at the nodes of a rectilinear mesh in the xyz planes and ω the product of the well-known *Lagrange polynomials* [see III, §2.3.4] in x, y and z. Alternatively, there are computational advantages in avoiding nodes *inside* an element and just using nodes on s_g alone; in that case to allow the displacement field to embrace high-order polynomials we may select both displacements and various derivatives in ρ_g and construct $\mathbf{w}$ from hermitian polynomials [see III, §6.5.3]. It is not appropriate to dwell here on the choice of $\mathbf{w}$ save to say that $\mathbf{w}\rho_g$ should contain a linear field at least to ensure convergence as $V_g \rightarrow 0$ (see Chapter 15). It should also preferably be *isotropic* to avoid directional accuracy, i.e. the field should contain exactly the same polynomial powers after a rotational transformation of the coordinate system.

Substituting $\mathbf{u} = \mathbf{w}\boldsymbol{\rho}_g$ into (10.2) we obtain

$$\boldsymbol{\varepsilon} = \mathbf{D}^T \mathbf{w}\boldsymbol{\rho}_g$$

or

$$\boldsymbol{\varepsilon} = \boldsymbol{\alpha}\boldsymbol{\rho}_g$$

where $\boldsymbol{\alpha} = \mathbf{D}^T \mathbf{w}$. Now write the linear stress–strain law as

$$\boldsymbol{\sigma} = \boldsymbol{\kappa}\boldsymbol{\varepsilon}$$

where $\boldsymbol{\kappa}$ is a symmetrical material stiffness matrix. For example, in the case of an *isotropic* material, this stiffness matrix is given in terms of the familiar constants E and v:

$$\boldsymbol{\kappa} = \frac{E}{(1-2v)(1+v)} \begin{bmatrix} 1-v & v & v & 0 & 0 & 0 \\ & 1-v & v & 0 & 0 & 0 \\ & & 1-v & 0 & 0 & 0 \\ & \text{Symmetric} & & \frac{1}{2}-v & 0 & 0 \\ & & & & \frac{1}{2}-v & 0 \\ & & & & & \frac{1}{2}-v \end{bmatrix}$$

The PVD version of (10.5) then becomes, summing over all n elements,

$$\sum_{g=1}^{n} (\boldsymbol{\rho}_g^T \mathbf{k}_g \bar{\boldsymbol{\rho}}_g - \mathbf{P}_g^T \bar{\boldsymbol{\rho}}_g) = 0, \tag{10.44}$$

where

$$\mathbf{k}_g = \int_{V_g} \boldsymbol{\alpha}^T \boldsymbol{\kappa} \boldsymbol{\alpha} \, dV$$

and

$$\mathbf{P}_g^T = \int_{V_g} \boldsymbol{\omega}^T \mathbf{w} \, dV + \int_{S_g} \boldsymbol{\phi}^T \mathbf{w} \, ds$$

taking the element g to have a surface, part of which (S_g) has on it the surface forces $\boldsymbol{\phi}$ prescribed. Equation (10.44) can now be solved by connecting elements together simply by writing a *global* list of displacements $\mathbf{r}$ and selecting for one element $\boldsymbol{\rho}_g = \mathbf{a}_g \mathbf{r}$, where $\mathbf{a}_g$ is a *Boolean* matrix [see I, chapter 16] of ones and zeros which extracts the values of $\boldsymbol{\rho}_g$ from the complete list. Thus

$$\left(\mathbf{r}^T \sum_g \mathbf{a}_g^T \mathbf{k}_g \mathbf{a}_g - \sum \mathbf{P}_g^T \mathbf{a}_g \right) \bar{\mathbf{r}} = \mathbf{0}.$$

Since all $\bar{\mathbf{r}}$ are arbitrary, all coefficients of all components of $\bar{\mathbf{r}}$ must vanish separately, whence

$$\mathbf{K}\mathbf{r} = \mathbf{R}, \tag{10.45}$$

where $\mathbf{K} = \sum_g \mathbf{a}_g^T \mathbf{k}_g \mathbf{a}_g$

and   (10.46)

$$\mathbf{R} = \sum_g \mathbf{a}_g^T \mathbf{P}_g.$$

The set of simultaneous equations (10.45) has to be solved for the global list of displacements, $\mathbf{r}$, after which the local element information can be extracted like

$$\boldsymbol{\rho}_g = \mathbf{a}_g \mathbf{r} \qquad \text{and} \qquad \boldsymbol{\sigma} = \kappa \alpha \boldsymbol{\rho}_g$$

The *global stiffness matrix* $\mathbf{K}$ in (10.46) is simpler than it appears. It may readily be demonstrated that if $\mathbf{a}_g$ is a Boolean matrix which selects two local displacements, say ρ_i and ρ_j, from the list $\mathbf{r}$ as r_I and r_J, then the *congruent transformation* [see I, (9.1.6)] $\mathbf{a}_g^T \mathbf{k}_g \mathbf{a}_g$ will simply place the element stiffness k_{ij} in the global stiffness at K_{IJ}. Equation (10.46) is, of course, a summation so k_{ij} from several elements may contribute to K_{IJ}.

The *physical* explanation of our approximation becomes clear when we imagine the previous steps applied to the alternative PVD equivalent equation (10.7). Thus the term

$$\int_{V_g} (\mathbf{D}\boldsymbol{\sigma} + \boldsymbol{\omega}) \bar{\mathbf{u}} \, dV$$

shows that we will satisfy equilibrium, $\mathbf{D}\boldsymbol{\sigma} + \boldsymbol{\omega} = \mathbf{0}$, only *in the mean* through each V_g, the residual error becoming hopefully small as we increase the number of degrees of freedom in the weight $\bar{\mathbf{u}} = \mathbf{w}\boldsymbol{\rho}_g$.

Now equation (10.7) was derived from the PVD using the Gauss theorem, which is valid only when the integrand possesses continuous first derivatives. Suppose then that displacements are assumed which have continuous forms (strains) inside the elements but which may have discontinuous gradients across the interfaces. Then in applying Gauss' theorem separately *to each element*, the summation (10.44) contains different contributions from adjacent elements (labelled '1' and '2' in Figure 10.4) which are grouped thus:

$$\int_{S_g} [(\mathbf{D}n)\boldsymbol{\sigma}_1 - (\mathbf{D}n)\boldsymbol{\sigma}_2] \mathbf{w}\bar{\boldsymbol{\rho}}_g \, ds = 0.   \tag{10.47}$$

The differing signs in the integrand in (10.47) arise since n is in opposite senses for elements 1 and 2 at their common interface. The equation assumes that $\mathbf{u} = \mathbf{w}\boldsymbol{\rho}_g$ is continuous across S_g, in which case we see that it is enforcing *in the mean* continuity of tractions $(\mathbf{D}n)\boldsymbol{\sigma}$. Thus displacement fields should be continuous but stress fields need not be. The connectivity equation $\boldsymbol{\rho}_g = \mathbf{a}_g \mathbf{r}$ ensures continuity at those nodal points on interfaces, whilst the interpolation function $\mathbf{w}$ should then ensure continuity between interface nodes (although it must be admitted that very successful 'non-conforming' elements do exist which violate this requirement; see Zienkiewicz, 1977).

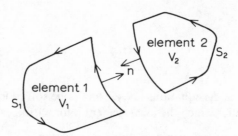

Figure 10.4: Contiguous finite elements.

The global stiffness matrix **K** in (10.45) and (10.46) is clearly symmetrical; it is also *positive definite* [see I, §6.7(v)], as may also be seen if we note that the stored strain energy is $\frac{1}{2}\mathbf{R}^T\mathbf{r} = \frac{1}{2}\mathbf{r}^T\mathbf{K}\mathbf{r}$ and must be positive for any **r**. The stiffness **K** is also likely to have its non-zero elements clustered along the leading diagonal if we order the node numbering suitably. Consider the physical nature of element K_{ij}. If we impose a unit displacement $r_j = 1$ and *all others zero*, then K_{ij} is the *i*th row of the necessary forces **R** required to maintain this displacement pattern. But such forces are certain to be grouped locally around node *j*, and so if the node-numbering scheme is suitable, the necessary forces **R** will be grouped about R_j—i.e. **K** will have a finite diagonal bandwidth. These two characteristics make the solution of (10.45) fairly straightforward—even for large **K**—by the *Cholesky decomposition method* [see III, §4.2], for example, in which decomposition can be performed entirely within the bandwidth and it is not necessary to hold all of **K** simultaneously in the main store. Inversion of **K** is rarely attempted since the decomposition technique is perfectly equipped to cope even with several loading cases **R**. Also, $\mathbf{K}^{-1}$ will be fully populated. *Substructuring* [see III, §4.3] techniques can also be used for large **K** and the Irons (1970) *frontal solution* is popular; it is an extension of the substructuring idea applied at the element level with the frontal nodes arranged so that the 'substructures' are series-linked. *Ill conditioning* [see III, §3.5] is not usually a problem, even for large **K**, although it can occur artificially by bad node numbering. Some packages evaluate *conditioning numbers* [see III, 4.7.34] such as the ratio of the largest to smallest *eigenvalues* [see III, 4.7.37] of **K** but this is expensive. *Iterative methods* [see III, §3.4] are an alternative to elimination; indeed, precomputer solutions were found by *relaxation,* and *dynamic relaxation,* which is numerically equivalent to an improved *Jacobi iterative method* [see III, §3.4.4], is sometimes used (Otter, 1966; Young, 1974).

The method used for the integral evaluation of the element stiffness matrix $\mathbf{k}_g$ in (10.44) depends on the form of **w**. Lagrange polynomials are easily generated but inconvenient to integrate in closed form and so *Gaussian quadrature* [see III, §7.3] is mostly used. This is particularly true of the popular *isoparametric* elements in which both displacements *and* geometry are written as $\mathbf{u} = \mathbf{w}\boldsymbol{\rho}$ and

$x = wx_n$, where $x^T = [x\ y\ z]$ and x_n are chosen generalized coordinates at, say, the nodes of a simple rectilinear mesh $(\zeta_1\ \zeta_2\ \zeta_3)$ on which the element is mapped. It is necessary to express the operators in $\varepsilon = D^T u$ in terms of $\partial/\partial\zeta$ using J, the *Jacobian* [see IV, Example 5.12.1], which will also occur in k_g when integrating with respect to $d\zeta_1\ d\zeta_2\ d\zeta_3$ (see Zienkiewicz, 1977, for a more detailed explanation).

Non-linear problems

It is fortunate that most structural behaviour is linear, or nearly so, but sometimes non-linearities in either the stress–strain law or the equilibrium and compatibility equations cannot be ignored. The solution of both types of non-linear problem is rarely possible analytically except for the simplest possible configurations, and so the freedom that the finite element (or difference) method allows is doubly welcome. Before describing the common approach we should say that one form of inelastic behaviour—*linear viscoelasticity*—can be integrated, and is a reasonable approximation to the time-dependent behaviour (creep) of hot metals and some polymers. The analysis of viscoelastic problems is described in detail in the next chapter.

The treatment of non-linearities, whether due to gross deformations or to plasticity, is the same. We consider small increments in the loads dR producing small displacements dr, and establish an instantaneous or tangential stiffness K_T relating the two:

$$dR = K_T\, dr.$$

K_T is a function of past history in stresses, displacements, or both, as we briefly demonstrate, first considering the effect of gross displacements but still small strains. We recall that $\varepsilon = D^T u = D^T w\rho = \alpha\rho$ for an element, but now admit that the changes in element geometry are so significant that α is no longer constant. We therefore apply the PVD using *small* virtual displacements $d\bar{\rho}$ and correspondingly small changes in element loads dP, geometry $d\alpha$ and stress $d\sigma$. Equation (10.44) for a single element becomes

$$d\bar{\rho}^T(P + dP) = d\bar{\rho}^T \int_{V_g} (\alpha + d\alpha)^T(\sigma + d\sigma)\, dV$$

whence, substracting (10.44) and ignoring the second-order product,

$$dP = \int_{Y_g} (\alpha^T\, d\sigma + d\alpha^T\sigma)\, dV. \tag{10.48}$$

The first term in the integrand is the usual elastic stiffness due to $d\sigma = \kappa_E\, d\varepsilon$ (the suffix E is for elastic), producing

$$\int_V \alpha^T K_E \alpha\, dV\, d\rho = k_E\, d\rho.$$

The second term is a stiffness due to existing stresses $\boldsymbol{\sigma}$ being realigned in a changing element field ($d\boldsymbol{\alpha}$). We can write

$$\int_V d\boldsymbol{\alpha}^T \boldsymbol{\sigma} \, dV = \int_V \sum_{i=1}^{6} \sigma_i \, d\alpha_i^T \, dV = \sum_i \int_V \sigma_i \frac{\partial \alpha_i^T}{\partial \boldsymbol{\rho}} \, dV \, d\boldsymbol{\rho} = \mathbf{k}_G \, d\boldsymbol{\rho},$$

where $\mathbf{k}_G$ is known as the *element geometric stiffness*. Then

$$d\mathbf{P} = (\mathbf{k}_E + \mathbf{k}_G) \, d\boldsymbol{\rho} = \mathbf{k}_T \, d\boldsymbol{\rho}. \tag{10.49}$$

Notice that $\mathbf{k}_G$ is a function of geometry *change* $d\boldsymbol{\alpha}$, as is $\mathbf{k}_E$ a function of $\boldsymbol{\alpha}$. At this stage we note an immediate use for $\mathbf{k}_G$ not involving cumulative history. 'Initial' buckling concerns stresses $\boldsymbol{\sigma}$ in the undeformed state due to $\mathbf{R}$. Let $\lambda \mathbf{R}$ therefore cause $\lambda \boldsymbol{\sigma}$. In this case, from (10.48) summed over the entire structure,

$$d\mathbf{R} = (\mathbf{K}_E + \lambda \mathbf{K}_G) \, d\mathbf{r},$$

where

$$\mathbf{K}_G = \sum \mathbf{a}_g^T \mathbf{k}_G \mathbf{a}_g.$$

Initial buckling is a state of *neutral equilibrium* in which case $d\mathbf{r}$ exists for $d\mathbf{R} = 0$ $= (\mathbf{K}_E + \lambda \mathbf{K}_G) \, d\mathbf{r}$. Hence $|\mathbf{K}_E + \lambda \mathbf{K}_G| = 0$, which is an *eigenvalue problem* [see III, Chapter 4], the lowest value of λ giving the initial buckling stress. Similar eigenvalue problems occur in any structure in which a destabilizing force exists whose magnitude is proportional to deformations, such as *shaft whirling* (the eigenvalue is rotational speed) or *aeroelasticity* (the eigenvalue is divergence speed).

Second, we consider *plasticity* in the shape of the *Prandtl–Reuss incremental approach* applied to an element. The incremental strain $d\boldsymbol{\varepsilon}$ consists of two parts $d\boldsymbol{\varepsilon}_E$, the elastic recoverable, and $d\boldsymbol{\varepsilon}_p$, the permanent strains which give rise to no volume change and are proportional to stress deviators $\mathbf{s} = \boldsymbol{\sigma} - \sigma_m \mathbf{e}$, where σ_m is the mean hydrostatic stress equal to $\frac{1}{3}(\sigma_{xx} + \sigma_{yy} + \sigma_{zz})$ and $\mathbf{e}^T = [1, 1, 1, 0, 0, 0]$. Then

$$d\boldsymbol{\varepsilon}_p = \frac{3}{2E_p} \frac{d\bar{\sigma}}{\bar{\sigma}} \mathbf{s} \qquad \text{when plastic,}$$

or

$$d\boldsymbol{\varepsilon}_P = \mathbf{0} \qquad \text{when elastic.}$$

Here $\bar{\sigma}$ is the 'equivalent' stress given by

$$\bar{\sigma}^2 = \tfrac{3}{2} \mathbf{s}^T \mathbf{s}$$

and E_p given by

$$\frac{1}{E_p} = \frac{1}{E_T} - \frac{1}{E}.$$

Substituting into $d\boldsymbol{\varepsilon} = d\boldsymbol{\varepsilon}_E + d\boldsymbol{\varepsilon}_p$ and inverting we find now that

$$d\boldsymbol{\sigma} = (\boldsymbol{\kappa}_E - \boldsymbol{\kappa}_p) \, d\boldsymbol{\varepsilon}; \qquad \boldsymbol{\kappa}_p = \frac{9G^2 \mathbf{s}\mathbf{s}^T}{(3G + E_p)} \bar{\sigma}^2.$$

We take $\kappa_P = 0$ unless yielding takes place, i.e. $\bar{\sigma} = Y$ for $d\bar{\sigma} > 0$. If $\bar{\sigma} < Y$ or $d\bar{\sigma} < 0$ (unloading) the behaviour is elastic. Summarizing,

$$k_T = k_E - k_p,$$

where
$$k_p = \int_{V_g} \alpha^T \kappa_p \alpha \, dV.$$

In general therefore, (10.49) becomes

$$d\mathbf{P} = (k_E + k_G - k_p) \, d\rho = k_T \, d\rho$$

and

$$K_T = \sum a_g^T k_T a_g.$$

How now do we construct the $\mathbf{R} \sim \mathbf{r}$ relationship?

There are various numerical and iterative procedures for solving such non-linear problems as $\mathbf{K}_s(\mathbf{r})\mathbf{r} = \mathbf{R}$, where $\mathbf{K}_s(\mathbf{r})$ is called the *secant* stiffness matrix, such as the *Newton–Raphson* method and its *modification* [see III, §5.4]. However, the terms in $\mathbf{K}_s(\mathbf{r})$ are usually not known unless the rotations are small and strains can be related to the original configuration in terms of total displacements (a Lagrange formulation). If deformations are gross, then we must resort to the incremental approach and tangent stiffness relationship $\mathbf{K}_T(\mathbf{r}) \, d\mathbf{r} = d\mathbf{R}$. One simple solution technique is to select convenient incremental values of load $d\mathbf{R}$ and at each step evaluate

$$d\mathbf{r} = [\mathbf{K}_T(\mathbf{r})]^{-1} \, d\mathbf{R}$$

to prepare the next step based on $\mathbf{K}_T(\mathbf{r} + d\mathbf{r})$. This method is simple, without iteration, but tends to drift above the correct solution for softening structures and below for hardening; the step size should therefore be small. However, larger load increments can be applied if iteration is used within each step to reduce the error, using one of the *predictor–corrector* techniques [see III, §8.3.1]. A fairly rapidly converging method (Argyris and Chan, 1972) is to evaluate $\mathbf{K}_T(\mathbf{r} + d\mathbf{r})$ and then return to the beginning of the step and find a new

$$d\mathbf{r}^* = [\mathbf{K}_T^*(\mathbf{r})]^{-1} \, d\mathbf{R},$$

where $\quad \mathbf{K}_T^* = \frac{1}{2}[\mathbf{K}_T(\mathbf{r}) + \mathbf{K}_T(\mathbf{r} + d\mathbf{r})]$.

This is repeated until successive $d\mathbf{r}^*$ differ by an acceptably small amount—usually three or four iterations suffice. The new position of $(\mathbf{R}, \mathbf{r})$ will be exact for a quadratic relationship and so fairly large step sizes are permissible for most relationships. All methods fail when $\mathbf{K}_T$ is singular [see I, Definition 6.4.2] and if this is due to plastic collapse or buckling, then the desired critical value of $\mathbf{R}$ has been found. However, there may be stable post-buckling or snap-through problems to which we do need the answer after $\mathbf{K}_T$ ceases to be singular. Such problems are best solved dynamically; indeed, the real problem may be influenced by dynamic behaviour. Even if the problem is static and monotonic, the dynamic algorithm can be as efficient as static incremental loading since inversion of $\mathbf{K}_T$ can be avoided, as we shall see in the next section.

10.4 Dynamics

One of the most powerful applications of discretized models, such as the finite element one, is to dynamic problems where the techniques applicable to one degree of freedom, r, can often be carried over to many degrees, $\mathbf{r}$. We shall only briefly describe the techniques here; the interested reader should consult one of the standard texts such as Hurty and Rubenstein (1964) or Clough and Penzien (1975).

Consider our previous finite element models modified to include dynamic terms; this can be done in a quasi-static fashion simply by incorporating inertia terms $-\rho\ddot{\mathbf{u}}$ as body forces (ρ = density) in the PVD (10.5) whence, instead of obtaining simply (10.45), we find

$$\mathbf{M}\ddot{\mathbf{r}} + \mathbf{K}\mathbf{r} = \mathbf{R}, \tag{10.50}$$

where the *mass matrix*

$$\mathbf{M} = \sum_g \mathbf{a}_g^T \mathbf{m}_g \mathbf{a}_g \tag{10.51}$$

and the gth element mass matrix

$$\mathbf{m}_g = \int_{V_g} \mathbf{w}^T \rho \mathbf{w} \, dV. \tag{10.52}$$

A viscous damping term can be incorporated in exactly the same way, starting from a linear viscoelastic material stiffness matrix; however, in practice the addition of viscous damping is rarely based on such fundamental knowledge and for most structures it is common to extend (10.50) as

$$\mathbf{M}\ddot{\mathbf{r}} + \mathbf{C}\dot{\mathbf{r}} + \mathbf{K}\mathbf{r} = \mathbf{R}(t), \tag{10.53}$$

where the damping matrix $\mathbf{C}$ is inferred from experience. It is often distributed as a purely diagonal array [see I, (6.7.3)] the elements of which are guessed or found experimentally and are usually some small fraction of critical damping. Another common assumption is that of *proportional*, or Rayleigh, damping

$$\mathbf{C} = \beta\mathbf{M} + \gamma\mathbf{K}.$$

This is a reasonable approximation to many structures and has the major analytical advantage that the *eigenvectors* [see I, §7.1] of the undamped system are unchanged by such damping.

Dynamic problems can be classified as:

(i) Steady-state vibration in which periodic—not necessarily simple harmonic—excitation is applied over a time long enough for transients to decay.

(ii) Excitation from rest in which transients are important.

Considering the latter problem, there are two basic approaches: *direct integration* of the equations of motion in all the discrete variables $\mathbf{r}$ or *modal synthesis* in a much reduced number of generalized coordinates $\boldsymbol{\phi}$. Modal synthesis is probably the most widely used method, so to this end let us first look at the undamped *free vibration* analysis of (10.50) and assume that $\mathbf{r}(t) = \mathbf{r}_0\, e^{i\omega t}$; thus

$$[-\mathbf{M}\omega^2 + \mathbf{K}]\mathbf{r}_0 = \mathbf{0}. \tag{10.54}$$

This is a standard *eigenvalue problem* [see III, Chapter 4] with the *eigenvalues* $\lambda_i = \omega_1^2$ being the natural frequencies and the *eigenvectors* the corresponding mode shapes. In many cases this information suffices if all we wish to know are the exciting resonant frequencies which we must avoid, and perhaps where nodes and antinodes occur. If, however, we wish to know the response to *any* excitation, we first write the solutions of the above equation as

$$-\mathbf{M}\mathbf{q}\lambda + \mathbf{K}\mathbf{q} = \mathbf{0},$$

where $\lambda = \lfloor \lambda_1\, \lambda_2 \cdots \lambda_i \cdots \lambda_n \rfloor$ is a diagonal matrix of the eigenvalues and $\mathbf{q} = [\mathbf{q}_1\, \mathbf{q}_2 \cdots \mathbf{q}_i \cdots \mathbf{q}_n]$ is a matrix of the eigenvectors. Suppose we order the eigenvalues from the lowest λ_1 (fundamental) to a 'sufficiently high' value λ_m since for large $\mathbf{K}$ the *very high* natural frequencies are of no interest. We now expand the displacements in terms of the restricted set $\mathbf{q} = [\mathbf{q}_1 \cdots \mathbf{q}_m]$; thus

$$\mathbf{r} = \mathbf{q}\boldsymbol{\phi} \tag{10.55}$$

and on substituting into (10.50) and premultiplying by $\mathbf{q}^T$,

$$\mathbf{q}^T\mathbf{M}\mathbf{q}\ddot{\boldsymbol{\phi}} + \mathbf{q}^T\mathbf{K}\mathbf{q}\boldsymbol{\phi} = \mathbf{q}^T\mathbf{R}(t).$$

But the *eigenvectors are orthogonal* [see III, §4.6] with respect to both $\mathbf{M}$ and $\mathbf{K}$; consequently $\mathbf{q}^T\mathbf{M}\mathbf{q} = \mathbf{m}$ and $\mathbf{q}^T\mathbf{K}\mathbf{q} = \mathbf{k}$ are diagonal matrices and we have *decoupled* the equations of motion into a set of equations in single degrees of freedom. If there are *rigid body modes* ($\lambda = 0$), we partition the eigenvectors accordingly, $\mathbf{q} = [\mathbf{q}_0 \vdots \mathbf{q}_d]$ where $\mathbf{q}_d$ are the deforming modes. Similarly, $\mathbf{m} = \lceil \mathbf{m}_0\, \mathbf{m}_d \rfloor$ and $\mathbf{k} = \lceil \mathbf{0}\; \mathbf{k}_d \rfloor$. The rigid body movement is the solution of

$$\mathbf{m}_0\ddot{\boldsymbol{\phi}}_0 = \mathbf{q}_0^T\mathbf{R}(t),$$

and if $\boldsymbol{\phi}_0 = \dot{\boldsymbol{\phi}}_0 = 0$ at $t = 0$, then

$$\boldsymbol{\phi}_0 = \int_0^t \int_0^t \mathbf{m}_0^{-1}\mathbf{q}_0^T\mathbf{R}(\tau)\, d\tau\, d\tau. \tag{10.56}$$

The deforming modes are solutions of

$$\ddot{\boldsymbol{\phi}}_d + \lambda \boldsymbol{\phi}_d = \mathbf{m}_d^{-1} \mathbf{q}_d^T \mathbf{R}(t)$$

and if, say, $\boldsymbol{\phi}_d = \mathbf{0}$ at $\mathbf{t} = 0$, the solution can be written in terms of the *convolution* or *Duhamel integral*

$$\boldsymbol{\phi}_d = (\lambda^{1/2} \mathbf{m}_d)^{-1} \int_0^t \sin\left[\omega(t - \tau)\right] \mathbf{q}_d^T \mathbf{R}(t)\, d\tau, \qquad (10.57)$$

where $\sin \omega t = \lceil \sin \omega_1 t \quad \sin \omega_2 t \cdots \sin \omega_m t \rfloor$.

Thus,

$$\mathbf{r} = \mathbf{q}_0 \boldsymbol{\phi}_0 + \mathbf{q}_d \boldsymbol{\phi}_d \qquad (10.58)$$

The inclusion of proportional damping is straightforward since the damped eigenvectors are the same as the undamped. In general, though, (10.53) can be solved by rewriting as

$$\begin{bmatrix} \mathbf{O} & \mathbf{M} \\ \mathbf{M} & \mathbf{C} \end{bmatrix} \begin{bmatrix} \ddot{\mathbf{r}} \\ \dot{\mathbf{r}} \end{bmatrix} + \begin{bmatrix} -\mathbf{M} & \mathbf{O} \\ \mathbf{O} & \mathbf{K} \end{bmatrix} \begin{bmatrix} \dot{\mathbf{r}} \\ \mathbf{r} \end{bmatrix} = \begin{bmatrix} \mathbf{O} \\ \mathbf{R}(t) \end{bmatrix}$$

and proceeding as before for the new vector $\{\dot{\mathbf{r}}\ \mathbf{r}\}$, although now the matrices are no longer *symmetric and positive definite* [see I, §6.7(v)] and therefore the *eigenvalues and vectors can be complex* [see I, Theorem 7.8.1].

In the case of periodic excitation we may simply use a *Fourier transform* [see IV, §13.2] over the period T_0 of the forcing functions $\mathbf{R}(t)$ to decompose them into frequency components $2\pi n / T_0$ and solve the equations at each frequency. This approach can be extended to forcing which is not repetitive by extending T_0 to infinity, putting n/T_0 to $d\omega$ and replacing the summation by an integral so that the representation of $\mathbf{R}(t)$ becomes a *Fourier integral* [cf. IV, §20.5.6]. The response is thus analysed initially in the *frequency domain* and

$$\mathbf{R}(t) = \frac{1}{2\pi} \int_{\omega = -\infty}^{\infty} \mathbf{f}(\omega)\, e^{i\omega t}\, d\omega,$$

where $\mathbf{f}(\omega) = \int_{t = -\infty}^{\infty} \mathbf{R}(t)\, e^{-i\omega t}\, dt.$

In practice $\mathbf{R}(t)$ is evaluated numerically over a range considered adequate, and both integrals are evaluated as discrete summations. By far the most efficient algorithm for doing this is the *fast Fourier transform* (Brigham, 1974; Chari and Silvester, 1980).

In large systems it may be necessary to reduce the number of unknowns, not only by choosing a restricted set of eigenvectors, but *before* modal synthesis since eigenvalue analysis can be expensive. This is done by selecting 'master' degrees of freedom $\mathbf{r}_m$ in $\mathbf{r} = \{\mathbf{r}_m \vdots \mathbf{r}_s\}$ and condensing out the 'slave' variables $\mathbf{r}_s$.

This is often done quite crudely by using 'lumped' masses only at r_m and assuming no significant inertia or damping forces at r_s, so *static condensation* is justifiable; thus

$$\begin{bmatrix} \mathbf{K}_{mm} & \mathbf{K}_{ms} \\ \mathbf{K}_{sm} & \mathbf{K}_{ss} \end{bmatrix} \begin{bmatrix} \mathbf{r}_m \\ \mathbf{r}_s \end{bmatrix} = \begin{bmatrix} \mathbf{R}_m \\ \mathbf{O} \end{bmatrix}$$

and

$$\mathbf{K}_m \mathbf{r}_m = \mathbf{R}_m,$$

where $\mathbf{K}_m = \mathbf{K}_{mm} - \mathbf{K}_{ms} \mathbf{K}_{ss}^{-1} \mathbf{K}_{sm}.$

Even without 'lumped' mass idealization, static condensation is tolerable for lower resonant frequencies whose mode shapes are governed largely by stiffness, but a more systematic way of using condensation without lumping is the Guyan (1965) reduction. *Dynamic condensation*—including $\mathbf{M\ddot{r}}$—would be more accurate but this would presuppose knowledge of the frequency response and is really feasible only when simple harmonic excitation is forced.

Turning now to methods of evaluating eigenvalues and eigenvectors, it is clear that we may have to deal with large matrices, unless drastic condensation is employed. However, we should be able to take advantage of the expected small diagonal bandwidth of $\mathbf{K}$ and $\mathbf{M}$ [see III, §4.9]. Furthermore, the complete set of eigenvalues is unlikely to be required since the high frequencies arise solely because of the large number of discrete variables, $\mathbf{r}$, necessary to find element stresses accurately, and they play no part in the dynamic response. A family of suitable matrix *iteration* schemes exists [see III, §4.8] of which the Stodola is the simplest. A fundamental mode is assumed and by successive improvement the mode and *lowest* eigenvalue can be found by finding the highest eigenvalue $1/\lambda$ of $\mathbf{K}^{-1}\mathbf{M}$. Knowing this, the next eigenvalue and mode can be found. Unfortunately $\mathbf{K}^{-1}\mathbf{M}$ is expensive to find and store, and for fully populated matrices *transformation methods* [see III, §4.11] are preferred such as **LR**, **QR**, with Householder's reduction (see also Wilkinson, 1965). Also purifying solved vectors from the trial solution is somewhat expensive and so *subspace* or *simultaneous vector iteration* (Parlett, 1980) which involves all the chosen modes is common in many dynamic finite element packages. The user decides on how many modes are considered adequate and then chooses a larger number (such as twice) for iteration to hasten convergence (the lowest frequency modes usually converge faster).

A rather specialized technique based on the *Sturm sequence* property [see I, §14.8] of the principal minors of $|\mathbf{A} - \lambda\mathbf{I}| = 0$ is useful in obtaining eigenvalues by *bisection* [see III, §4.9], particularly for stiffness matrices of 'exact' beam–column elements where the elements of $\mathbf{K}$ are transcendental functions of λ (Williams and Wittrick, 1970). The *Sturm sequence* method can be used to find eigenvalues and then inverse iteration to find the vector [see III, §4.8].

Finally, we return to direct step-by-step integration methods applied to

(10.50) or (10.53). This technique is necessary if the problem is non-linear—such as dynamic plastic collapse, the dynamics of very flexible structures, earthquake foundation problems, etc. (It also has adherents even in linear static cases since it explicitly avoids inversion of the stiffness matrix. As mentioned before, this approach is known as *dynamic relaxation*.) For a fuller discussion of the merits of various idealizations and algorithms for step-by-step integration, the reader is best referred to the standard texts (Mitchell and Griffiths, 1980; Smith, 1978), but briefly, the equations in time are discretized over small time intervals Δt and some approximation is made in which the values of $\mathbf{R}$ and $\mathbf{r}$ (and $\mathbf{K}$, $\mathbf{M}$, $\mathbf{C}$ in non-linear problems) are deduced at some time $t + \Delta t$ in terms of those at time t. The use of 'marching' finite *difference* schemes—as opposed to finite *element* techniques—is preferred because there are no tortuous geometric problems in the $\mathbf{r} - t$ planes.

Of all *explicit* (Clough and Penzien, 1975) schemes, in which $\ddot{\mathbf{r}}$ is solved from (10.50) using $\mathbf{r}$ and $\dot{\mathbf{r}}$ from the previous step, the central difference operator of order two has the maximum stability limit, low storage requirements and is accurate provided Δt is small compared with the smallest eigenperiod $T_{\min}$ (which may of course be minute for large order $\mathbf{K}$). As $\Delta t \approx 0.1\, T_{\min}$ is advisable, larger values would lead to inaccuracies or unstable error growth. A more stable algorithm is based on *linear acceleration* using current and future values, and, with iteration within the time step, values of $\Delta t \approx 0.2\, T_{\min}$ are possible for accurate and stable results. For (10.53) the inclusion of damping improves convergence. Higher order schemes, such as a *cubic* variation in $\mathbf{M}\ddot{\mathbf{r}}$ (λ), can be used with larger time steps, less drift and more accurate response to impact loading (Argyris and Chan, 1972). In all these schemes $\mathbf{K}$ is not inverted, only $\mathbf{M}$ which is easier; hence the attraction for non-linear problems. It is rare for deformations to be so large that $\mathbf{M}$ has to be updated.

There are whole families of *unconditionally stable algorithms* which allow much larger time steps. They all include a form of numerical damping which irons out the embarrassing high frequency modes but which unfortunately modifies the amplitude eventually and elongates the period of the low frequencies. The *Houbolt* algorithm (Houbolt, 1950) is the most stable with highest damping, but it involves computationally unattractive backward differences. The *Newmark β method* (Newmark, 1959) uses forward differences and is very popular for both non-linear and linear problems, and can be used with Δt up to one-hundredth of the expected response period. The *Wilson θ method* (Wilson, 1968) has similar damping to Newmark (with $\theta = 1.4$), but is difficult to use with iteration in non-linear problems.

A completely different way of avoiding unstable growth is simply to integrate using a reduced set of the eigenvectors as variables instead of the complete displacement vector $\mathbf{r}(t)$. The smallest natural period is that of the largest retained eigenvalue and even for an ambitiously large set this is probably an order of magnitude increase upon the smallest eigenperiod of $\mathbf{M}^{-1}\mathbf{K}$.

10.5 Optimum design

Finally, mention should be made of the impact of *mathematical programming* methods on structural analysis [see I, Chapters 11 and 12, IV, Chapters 15 and 16], where we take the name to mean techniques of selecting (many) sets of variables to minimize (or maximize) a merit function like weight (or the load-carrying capacity) subject to various inequality constraints upon these variables. For example, *linear programming* is a well-posed problem with a satisfactory solution based upon the *simplex algorithm* [see I, §11.4] and is in common use, for example, for resource-allocation problems in operational research. We may simplify the problem of optimum structural design, with some justification, to that of selecting member sizes in a structural configuration that has already been chosen. However, even this problem is non-linear, except for very special cases such as frameworks with idealized plastic collapse as the failure criteria. The use of *non-linear programming* [see IV, Chapter 15] methods has met with only limited success, however, partially because many design constraints are not easy to quantify and partially because such programming methods as do exist cannot guarantee to find a *global* optimum, unless a good initial guess can be made. For some structural models *geometric programming* [see IV, §15.6.6] is possible, and this refines the quest for a global optimum. The development of finite element software has provided a great incentive for optimum-design packages which iterate upon the answer delivered by a first approximation to **K**. The general topic of optimum system design is treated in Chapter 13.

G.A.O.D.

References

Argyris, J. H., and Chan, A. S. L. (1972). Applications of finite elements in space and time, *Ing. Archiv.*, **41**.

Brigham, E. O. (1974). *The Fast Fourier Transform*, Prentice-Hall.

Chari, M. V. K., and Silvester, P. P. (eds.) (1980). *Finite Elements in Electrical and Magnetic Field Problems*, pp. 145–160 and references therein, Wiley.

Clough, R. W., and Penzien, J. (1975). *Dynamics of Structures*, McGraw-Hill.

Davies, G. A. O. (1982). *Virtual Work in Structural Analysis*, Wiley.

Flügge, W. (1962a). *Handbook of Engineering Mechanics*, McGraw-Hill.

Flügge, W. (1962b). *Stresses in Shells*, Springer-Verlag.

Goldenweizer, A. L. (1961). *Theory of Elastic Thin Shells*, Pergamon.

Gould, P. L. (1977). *Static Analysis of Shells*, Lexington Books.

Guyan, R. J. (1965). Reduction of stiffness and mass matrices, *A.I.A.A. Jnl.* **3**.

Higgins, J. R. (1977). *Completeness and Basic Properties of Sets of Special Functions*, Cambridge University Press.

Houbolt, J. C. (1950). A recurrence matrix solution for the dynamic response of elastic aircraft, *J. Aero. Sci.*, **17**, 540–550.

Hurty, W. C., and Rubenstein, M. F. (1964). *Dynamics of Structures*, Prentice-Hall.

Irons, B. M. (1970). A frontal solution program for finite element analysis, *Int. J. Numerical Meth. in Eng.*, **2**.

Knott, J. (1973). *Fundamentals of Fracture Mechanics*, Butterworths.

Kraus, H. (1967). *Thin Elastic Shells*, Wiley.

Krylov, V. I., and Kantorovitch, L. (1964). *Approximate Methods of Higher Analysis*, Interscience, New York.

Love, A. E. H. (1944). *The Mathematical Theory of Elasticity*, Dover, New York.

Mansfield, E. H. (1964). *The Bending and Stretching of Plates*, Pergamon.

Milne-Thomson, L. M. (1960). *Plane Elastic Systems*, Springer-Verlag.

Milne-Thomson, L. M. (1962). *Antiplane Elastic Systems*, Springer-Verlag.

Mitchell, A. R., and Griffiths, D. F. (1980). *The Finite Difference Method in Partial Differential Equations*, Wiley.

Muskhelishvili, N. I. (1953). *Some Basic Problems in the Mathematical Theory of Elasticity*, P. Noordhoff.

Newmark, N. M. (1959). A method of computation for structural dynamics, *J. Eng. Mech. Div. ASCE*, 67–94.

Novoshilov, V. V. (1959). *Thin Shell Theory*, P. Noordhoff.

Otter, J. R. M. (1966). Dynamic relaxation, *Proc. Inst. Civ. Eng.*, **35**.

Parlett, B. N. (1980). *The Symmetric Eigenvalue Problem*, Prentice-Hall.

Savin, G. N. (1961). *Stress Concentrations around Holes*, Pergamon.

Smith, G. D. (1978). *Numerical Solution of Partial Differential Equations: Finite Difference Methods*, 2nd edn, Clarendon Press.

Sneddon, I. N., and Lowengrub, M. (1969). *Crack Problems in the Classical Theory of Elasticity*, Series in App. Maths, Wiley.

Thompson, J. M. T., and Hunt, G. W. (1973). *A General Theory of Elastic Stability*, Wiley.

Timoshenko, S. (1940). *Theory of Plates and Shells*, McGraw-Hill.

Timoshenko, S., and Goodier, J. N. (1951). *Theory of Elasticity*, McGraw-Hill.

Wilkinson, J. H. (1965). *The Algebraic Eigenvalue Problem*, Clarendon Press, Oxford.

Williams, F. W., and Wittrick, W. H. (1970). An automatic computational procedure for calculating natural frequencies of skeletal structures, *Int. J. Mech. Sci.*, **12**.

Wilson, E. L. (1968). *A Computer Program for the Dynamic Stress Analysis of Underground Structures*, SESM Report No. 68-1, Division of Structural Engineering and Structural Mechanics, University of California, Berkeley.

Young, D. (1974). Iterative methods for solving partial differential equations of elliptic type, *Trans. Am. Math. Soc.*, **76**.

Zienkiewicz, O. C. (1977). *The Finite Element Method*, McGraw-Hill.

11

Polymer Engineering

11.1 Introduction

The great upsurge in the use of polymers for engineering applications over recent decades has been accompanied by an expansion of the theory necessary to characterize their behaviour. This expansion has taken place not only in the fields of polymer chemistry and physics, but also in applied mechanics, in which the mechanical behaviour of polymers is studied; it is this last discipline which is discussed in this chapter. The motivation for seeking analyses which are specifically relevant to polymers is that their behaviour differs, sometimes significantly, from that of the more traditional materials when subjected to similar loading conditions. The objective, then, is to examine the differences between the behaviour of polymers and those materials for which the classical theories are appropriate. It will soon become apparent that a major activity is to determine constitutive relations which both give realistic representations of behaviour and also are suitable for incorporation into a general analysis.

The most significant way in which solid polymers differ from metals at normal working temperatures is that polymers exhibit *time-dependent behaviour* and, as such, are said to be *viscoelastic*. The incorporation of time is a radical departure from conventional elasticity theory and so the section on viscoelasticity represents the bulk of this chapter. It is worth pointing out that the development of the theory puts a great reliance on *integral transform techniques* [see IV, Chapter 13] and some familiarity with these methods, particularly *Laplace transforms* [see IV, §13.4], though not essential, is advisable. Whilst the treatment is directed to solids, some of the analysis is also applicable to fluids. For more information on elastic liquids, see Lodge (1964).

The second section gives a very brief account of finite elasticity, indicating that the approximations made in infinitesimal theory and the assumptions that a structure does not significantly change shape may be unreasonable. This is because polymers have a low modulus and also because large deformations may take place in the solid forming of components. A constitutive stress–strain relationship suitable for such applications is given and there is little mathematical involvement.

The final section considers some aspects of the analysis which is relevant to a study of the operations involved in producing polymer components. The subject of polymer processing is altogether too vast to allow much more than an extremely brief and selective presentation, and the books by Fenner (1979), Pearson (1966) or Tadmor and Gogos (1979) should be consulted by interested readers. Nevertheless, it is intended that by considering the important process of single screw extrusion, appropriate mathematical methods are indicated. Because of the complexity of the analysis, these techniques are, necessarily, numerical in nature. Once again, an important feature is the adoption of a suitable constitutive equation since polymers in molten form do not usually conform to classical Newtonian representations.

11.2 Time-dependent behaviour

The nature of viscoelasticity

The theory of elasticity is the cornerstone of the engineering treatment of solid mechanics and deals with materials which are considered to be perfectly elastic. These materials deform instantaneously in response to a given applied load and do not deform further unless the load is altered. Additionally, they have the capacity to store mechanical energy without loss or dissipation, so that all the energy stored is recoverable. Generally, Hooke's law is held to apply so that stress is proportional to strain and the behaviour is linear.

In the study of fluid mechanics, the classical theory deals with the behaviour of perfectly viscous fluids, for which the stress depends on the rate of strain but not the strain itself, and the fluid has a capacity for dissipating mechanical energy but not for storing it. For a Newtonian viscous fluid, the stress is proportional to the strain rate and so the response to an applied stress is a steady flow. (Dynamic effects can be ignored in this context.)

Both of these treatments, therefore, are based on idealizations, not only because the assumed relationships between stress and strain or strain rate may not always be realistic, but also because the assumptions regarding the rôle of time categorize the material as either a solid without any liquid characteristics whatsoever, or vice versa. In a vast number of practical engineering applications, this approach is perfectly sensible, but in dealing with polymers and other materials under certain conditions behaviour which has elements associated with both solids and liquids acting simultaneously is encountered. These materials have the capacity both to store and to dissipate mechanical energy, and the response to a constant applied stress is an instantaneous deformation followed by a flow process. Such materials are said to be *viscoelastic* and if the strain is proportional to the stress, then the material is *linearly viscoelastic*. The strain response of a viscoelastic material to a constant applied stress is illustrated in Figure 11.1, which shows that the strain increases with time, but at a continuously decreasing rate. This behaviour is known as *creep*.

If after a period of time τ, say, the applied stress is instantaneously and completely removed, then there will be an instantaneous but not total drop in strain followed by a period of strain decay, as illustrated in Figure 11.2. This is *creep recovery* and shows that a viscoelastic material can experience strain in the absence of stress. In contrast, when a viscoelastic material is subjected to an instantaneous constant applied strain, there is an instantaneous stress response followed by stress decay, as illustrated in Figure 11.3. This is known as *relaxation*.

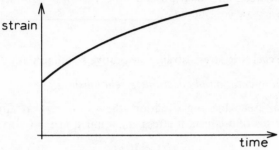

Figure 11.1: Schematic creep curve of a viscoelastic material.

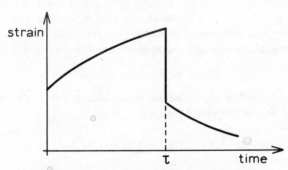

Figure 11.2: Schematic creep recovery curve.

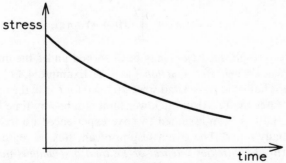

Figure 11.3: Schematic relaxation curve for viscoelastic material.

The phenomena of creep and relaxation are basic characteristics of viscoelastic materials and can be explained in terms of the molecular structure. However, in an engineering continuum analysis, microscopic considerations are ignored and a macroscopic approach, whereby the bulk material can be characterized by measurable parameters, is adopted. The successful description and prediction of viscoelastic behaviour then relies on the development of sound constitutive relationships between stress, strain and time via these parameters, and these relationships are now considered.

11.3 Linear viscoelastic stress–strain constitutive relationships

The integral representation of constitutive relationships

In a uniaxial creep loading situation, the strain, $\varepsilon(t)$, at any time can be expressed in terms of a constant stress, σ_0, applied at $t = 0$, by

$$\varepsilon(t) = J(t)\sigma_0, \tag{11.1}$$

where $J(t)$ is the *creep compliance function* of the material and is experimentally determined. If the applied stress remains unchanged, then this expression will predict the strain response for all times, but if at some time τ the stress changes by $\Delta\sigma$, then by the *Boltzmann superposition principle* the subsequent strain response is computed as though the applied stresses were acting separately and

$$\varepsilon(t) = J(t)\sigma_0 + J(t - \tau)\,\Delta\sigma. \tag{11.2}$$

For a series of loading steps, each of $\Delta\sigma_i$ applied at $t = \tau_i$, the strain response is obtained by *summation* [see IV, §1.7], so that

$$\varepsilon(t) = J(t)\sigma_0 + \sum_{i}^{n} J(t - \tau_i)\,\Delta\sigma_i \tag{11.3}$$

and for a *continuous* stress variation, the total strain can be expressed by the *integral* representation [see IV, §14.8]:

$$\varepsilon(t) = J(t)\sigma_0 + \int_{0}^{t} J(t - \tau)\,d\sigma(\tau). \tag{11.4}$$

The pure creep response, $J(t)\sigma_0$, may be absorbed under the integral sign by use of the *Heaviside unit step function* [see IV, Example 4.8.1] (as may any subsequent step change in stress) and since $J(t) = 0$ for $t < 0$, the upper limit can be any time greater than t. Also, the lower limit can be any time less than zero since the material has been assumed to have experienced no load up till zero time or is 'initially quiet'. The strain response can thus be represented by the *convolution* form of a *Stieltjes integral or Faltung or Duhamel integral*, and as such has the properties of *commutivity, associativity* and *distributivity* [see I, §2.3] as established by Gurtin and Sternberg (1962), whose work should be consulted for a rigorous mathematical approach to the subject. If the stress history is *differentiable* [see IV, §5.3], then equation (11.4) can be written as

$$\varepsilon(t) = J(t)\sigma_0 + \int_0^t J(t - \tau) \frac{\partial \sigma(\tau)}{\partial \tau} \, d\tau \tag{11.5}$$

and so, if the applied stress is known as a function of time, the complete strain response can be determined. Other useful expressions can be obtained by a *change of variables* and *integration by parts* [see IV, §4.3], and in particular

$$\varepsilon(t) = J(0)\sigma(t) + \int_0^t \sigma(t - \tau) \frac{\partial J(\tau)}{\partial \tau} \, d\tau. \tag{11.6}$$

Therefore, if the strain response is known, then the stress input can, in principle, be determined by recognizing this expression as a *Volterra integral equation of the second kind* [see III, §10.3]. Alternatively, it may be that as a result of relaxation tests the stress is known in terms of an applied strain, ε_0, by a relationship $\sigma(t) = E(t)\varepsilon_0$, where $E(t)$ is the *relaxation modulus*. Then the procedure outlined above can be reproduced to obtain a stress response to a strain history in the form:

$$\sigma(t) = \int_0^t E(t - \tau) \frac{\partial \varepsilon(\tau)}{\partial \tau} \, d\tau. \tag{11.7}$$

$E(t)$ is therefore a function corresponding to the elasticity modulus of Chapter 10 and, indeed, if $E(t)$ is taken to be a constant with a step load of strain, Hooke's law is retrieved. Similarly, the constitutive equation for a Newtonian fluid is obtained by setting $E(t) = \eta \, \delta(t)$, where η is the fluid viscosity and $\delta(t)$ is the *Dirac delta function* [see IV, §8.2].

It follows that if both stress and strain are obtainable either from equations involving the creep compliance or the relaxation modulus, then there must be a relationship between the two material functions. This relationship is most conveniently deduced by applying *Laplace transforms* [see IV, §13.4] to the stress–strain relationships. Denoting $\bar{f}(p)$ as the Laplace transform of $f(t)$, then by the *convolution property or Borel's theorem* [see IV, §13.4.4]

$$\bar{f}(p)\bar{g}(p) = \int_0^\infty e^{-pt} \, dt \int_0^t f(t - \tau)g(\tau) \, d\tau = \int_0^\infty e^{-pt} \, dt \int_0^t f(\tau)g(t - \tau) \, d\tau. \tag{11.8}$$

By use of the *Laplace transform of derivatives* [see IV, §13.4.3], then (11.7) becomes

$$\bar{\sigma}(p) = \bar{E}(p)p\bar{\varepsilon}(p). \tag{11.9}$$

Therefore, $\bar{\varepsilon}(p)$ can be found by rearranging and $\varepsilon(t)$ can be found by *inversion* [see IV, §13.4.2]. Thus there is a convenient technique for solving the integral equation (11.7) if $\sigma(t)$ is known. A similar expression is obtained from the stress–strain relationship in terms of creep compliance:

$$\bar{\varepsilon}(p) = \bar{J}(p)p\bar{\sigma}(p) \tag{11.10}$$

and so $p\bar{E}(p) = (1/p)\bar{J}(p)$ and the p multiplied Laplace transforms, or *Carson transforms*, of the relaxation modulus and creep compliance show the same reciprocal interdependence as the elasticity modulus and compliance. This relationship can be inverted to give:

$$\int_0^t E(t - \tau)J(\tau)\, d\tau = \int_0^t E(\tau)J(t - \tau)\, d\tau = t. \tag{11.11}$$

If a fading memory hypothesis is assumed which states that the more recent the input, the stronger the response, then it can also be assumed that the creep compliance is a *monotonically increasing function* [see IV, §2.7] of time and the relaxation modulus is a monotonically decreasing one; thus it can be argued (e.g. Pipkin, 1972) that $J(t)E(t) \leqslant 1$. Further, a consequence of the *initial- and final-value theorems* of the Laplace transforms (Spiegel, 1965) is that the product of these functions is always equal to unity in the limiting cases of zero and infinite time. In fact, for all practical cases, the reciprocal relationship between creep compliance and relaxation modulus is approximately observed throughout the time range.

The integral representation described can therefore be used to determine both stress and strain responses for any uniaxial loading history, given that the creep compliance or relaxation modulus are known. Other representations of the constitutive relationships can be used, however, but before proceeding with a consideration of these, it is worth pointing out that while many texts dealing with viscoelasticity commence with Boltzmann's superposition principle others, which include the texts by Christensen (1971) and Lockett (1972) and are excellent books on linear and non-linear viscoelasticity respectively, do not. Instead, the starting point is to assume that, at some spacial point, there is a law which *transforms* an admissible strain history to an admissible stress history. This can be expressed mathematically by

$$\sigma(t) = \underset{\tau=-\infty}{\overset{\tau=t}{\psi}} \, [\varepsilon(\tau)],$$

where ψ is an experimentally determined *linear continuous functional* [see IV, §19.1.2]. Restriction to hereditary materials and use of the *Riesz representation theorem* (Riesz and Nagy, 1955) [see IV, Theorem 19.2.2] allows the functional to be expressed as a Stieltjes integral and the remaining analysis follows. Gurtin and Sternberg (1962) also adopt this approach.

Differential operator form of constitutive relationship

The integral form of stress–strain relationship can be converted to a form involving *differential operators* [see IV, §8.1] by, once again, the use of the Laplace transformation as follows. Let the p-multiplied transform of the creep compliance be written as the ratio of two *polynomials* [see I, §14.1.1] so that

$$\frac{1}{p\bar{E}(p)} = p\bar{J}(p) = \frac{\bar{P}(p)}{\bar{Q}(p)} \tag{11.12}$$

where $\qquad \bar{P}(p) = \sum_{i=0}^{N} a_i p^i \qquad$ and $\qquad \bar{Q}(p) = \sum_{i=0}^{N} b_i p^i.$

Then $\qquad\qquad\qquad \bar{P}(p)\bar{\sigma}(p) = \bar{Q}(p)\bar{\varepsilon}(p). \tag{11.13}$

By the Laplace transform of derivatives, this can be inverted to give a *linear ordinary differential equation*:

$$P(D)\sigma(t) = Q(D)\varepsilon(t), \tag{11.14}$$

where $D \ (= d/dt)$ is the D *operator* [see IV, §7.3], provided the following requirement on initial conditions is met:

$$\sum_{r=i}^{N} a_i \sigma^{(r-i)}(0) = \sum_{r=i}^{N} b_i \varepsilon^{(r-i)}(0); \qquad \text{for } r = 1, 2, \ldots, N, \tag{11.15}$$

where $\sigma^{(r-i)}(0)$ means that $(r-i)$ order derivative of $\sigma(t)$ evaluated at $t = 0$. This is not a severe restriction as in most cases the material will be initially quiet and all derivatives of stress and strain will be zero at zero time. The differential form then relates current values of stress and strain and their derivatives. The earliest formulation of differential constitutive relationships stemmed from the use of physical models using spring and dashpot elements. Quite simply, the idea behind this approach is that if elastic behaviour is represented by a spring so that $\sigma = Ee$ and viscous behaviour is represented by a dashpot with $\sigma = \mu(d\varepsilon/dt)$, then viscoelastic behaviour can be represented by some physical combination of the two. The two most simple combinations are the Maxwell model which has the elements in series and the Voigt or Kelvin model which has them in parallel (Figure 11.4).

The constitutive relationship for a Maxwell model is easily obtained by analysis of Figure 11.4 and is

$$\dot{\varepsilon} = \frac{\dot{\sigma}}{E} + \frac{\sigma}{\eta}, \tag{11.16}$$

where $\dot{\sigma} = d\sigma/dt$ and $\dot{\varepsilon} = d\varepsilon/dt$, which is a form of the general relationship with $a_0 = 1/\eta$, $a_1 = 1/E$, $b_0 = 0$, $b_1 = 1$, and all other coefficients zero. Considering either creep or relaxation inputs, then the responses are easily obtained by solving a *first-order differential equation* [see IV, §7.2]. In fact, the Maxwell

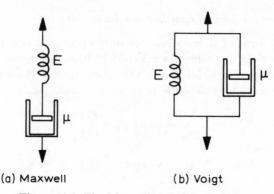

(a) Maxwell   (b) Voigt

Figure 11.4: The Maxwell and Voigt models.

model gives a poor description of creep behaviour (which has a linear dependence of strain on time) but reasonable relaxation characteristics, given by

$$\sigma(t) = E\varepsilon_0 \, e^{-Et/\eta} = E\varepsilon_0 \, e^{-t/\tau_R}, \qquad (11.17)$$

where τ_R is the relaxation time and is the time taken for the model to relax completely if it maintained its initial relaxation rate. The relaxation modulus for a Maxwell material is therefore Ee^{-t/τ_R} and this can be directly obtained by substitution of the appropriate coefficients in equation (11.12) and inverting. The use of *Heaviside's expansion formula* (Davis, 1978) is helpful in performing this operation. The constitutive equation for a Voigt model is

$$\eta\dot{\varepsilon} + E\varepsilon = \sigma, \qquad (11.18)$$

so that $a_0 = 1$, $b_0 = E$, $b_1 = \eta$, and all other coefficients of the differential operator polynomials are zero. This model shows no time-dependent relaxation due to the dominance of the spring, but creep is given by

$$\varepsilon = \frac{\sigma_0(1 - e^{-t/\tau_c})}{E} \qquad (11.19)$$

where τ_c is η/E as before but is now the retardation time. The use of Maxwell and Voigt models is convenient for considering relaxation and retardation spectra. The compliance and relaxation functions obtained from the analysis of these models have only a single retardation or relaxation time; since real materials do not exhibit behaviour which can be characterized by single retardation or relaxation times a more realistic representation is to consider a chain of individual models, each having different ratios of modulus to viscosity. In particular, a chain of N Maxwell models connected in parallel leads to

$$\sigma(t) = \varepsilon_0 \sum_i^N E_i \, e^{-t/\tau_R}, \qquad (11.20)$$

thus exhibiting a discrete relaxation spectrum, whilst for an infinite series of models

$$\sigma(t) = \varepsilon_0 \int_0^\infty E(\tau_R)\, e^{-t/\tau_R}\, d\tau_R = \varepsilon_0 E(t) \tag{11.21}$$

and $E(\tau_R)$ is the relaxation spectrum. Retardation spectra are obtained by considering a chain of Voigt models in series. Ferry (1970) gives the experimentally determined relaxation and retardation spectra of both cross-linked and uncrosslinked polymers but indicates that it is more convenient to use a logarithmic time scale and introduce a new function as the continuous spectrum.

Since the Maxwell and Voigt models show deficiencies in creep and relaxation respectively, other combinations of elements can be selected to give more realistic behaviour. The most common of these is the standard linear solid which has a spring in series with a Voigt component and Burgers model which is a series combination of a Maxwell and Voigt model. The characteristic parameters of these and other models are given in several texts (see Bland, 1960, and Flugge, 1975) and a most convenient method for solving the governing differential equations is, once again, by application of Laplace transforms, rearrangement using *partial fractions* [see I, §14.10] and inverting. Constant applied stress or strain rate can be expressed using the *Heaviside* and *Dirac delta functions* [see IV, Example 4.8.1 and §8.2] which are easily handled within the transform technique. All such physical models lead to relationships in which the time dependences are expressed in terms of exponential decay and all can be described by the generalized differential relationship. Indeed, since the models have no greater significance than the equations which describe them, some workers find their use objectionable and unnecessary, although Bland (1960) developed his theme on viscoelasticity in terms of spring and dashpot combinations. Physical models can help in interpretation in a qualitative manner, however, provided they are simple enough to visualize and the elementary models are widely referred to and used, even in the most sophisticated analyses.

Complex modulus formulation of constitutive equations

A practical situation in which the response of a viscoelastic material can show marked differences to that of an elastic counterpart is steady-state oscillating loading. If a viscoelastic component is subjected to a strain history which is a *harmonic function* [see IV, Definition 8.2.1] of time so that $\varepsilon(t) = \varepsilon_0\, e^{-i\omega t}$ (where $i = \sqrt{-1}$, according to *complex number* notation) [see I, (2.7.6)] then the steady-state stress response, after transient effects have died down, must be of similar form and

$$\sigma(t) = E^*(i\omega)\varepsilon_0\, e^{i\omega t}, \tag{11.22}$$

where $E^*(i\omega)$ is the *complex modulus* which can be written in terms of its *real and imaginary parts* [see I, §2.7.1] as

$$E^*(i\omega) = E_1(\omega) + iE_2(\omega), \tag{11.23}$$

where $E_1(\omega)$ is the in-phase or storage modulus and is associated with stored energy and $E_2(\omega)$ is the out-of-phase or loss modulus and is associated with dissipated energy. An alternative expression for the complex modulus is

$$E^*(i\omega) = E_1(\omega)[1 + i \tan \delta(\omega)], \tag{11.24}$$

where $\tan \delta(\omega) = E_2(\omega)/E_1(\omega)$ and is known as the loss tangent. The angle $\delta(\omega)$ is the loss angle and gives the extent to which the response is out of phase with the input. The stress response due to the harmonic strain input can be determined by the usual use of the convolution integral, but in order to avoid transient terms in the analysis and obtain true steady-state response, it is best to consider that the input has been applied for a long time and let the upper limit approach infinity. It is also helpful to express the relaxation modulus $E(t)$ as the sum of a constant term E_∞ and a time-dependent term $E'(t)$ which tends to zero as t tends to infinity, so that

$$E(t) = E_\infty + E'(t). \tag{11.25}$$

Procedures of this type are necessary to avoid *divergence* and ensure that the *limit of the integral* exists [see IV, §4.6]. The stress response can now be confidently determined and equated to the harmonic response of equation (11.22) to give the two relationships:

$$E_1(\omega) = E_\infty + \omega \int_0^\infty [E(t) - E_\infty] \sin \omega t \, dt, \tag{11.26a}$$

$$E_2(\omega) = \omega \int_0^\infty [E(t) - E_\infty] \cos \omega t \, dt, \tag{11.26b}$$

and from (11.26a), $E_\infty = E_1(0)$. These expressions are *Fourier sine and cosine transforms* [see IV, §13.3], which may be inverted to give expressions for the relaxation modulus in terms of the loss and storage moduli. Gross (1953) gives the relationship between loss and storage moduli which must exist since there are two expressions relating these moduli to the relaxation modulus.

The energy dissipated per unit volume under harmonic loading can be calculated by integrating the stress with respect to strain. The result is that the rates of dissipated energy to the maximum (elastic) energy which can be stored is equal to $2\pi \sin \delta$ and is thus only dependent on loss angle. This is the damping energy.

For general loading, the convolution integrals can be subjected to *Fourier transformations* [see IV, §13.2] to give a form of stress–strain constitutive relationship comparable to the Laplace transformed expressions. Denoting the

Fourier transform of $f(t)$ as $\bar{f}(\omega)$, such that

$$\bar{f}(\omega) = \int_{-\infty}^{\infty} f(t) e^{-i\omega t} \, dt. \tag{11.27}$$

(Note that the transform and its inverse are often normalized with a constant $\sqrt{2\pi}$.)

Then using the *convolution property of Fourier transforms* [see IV, §13.2.3], the integral relationships become

$$\bar{\sigma}(\omega) = i\omega\bar{E}(\omega)\bar{\varepsilon}(\omega) = E^*(i\omega)\bar{\varepsilon}(\omega), \tag{11.28a}$$

$$\bar{\varepsilon}(\omega) = i\omega\bar{J}(\omega)\bar{\sigma}(\omega) = J^*(i\omega)\bar{\sigma}(\omega), \tag{11.28b}$$

where $J^*(i\omega)$ is the complex compliance, and the following relationships between moduli are established:

$$E^*(i\omega) = \frac{1}{J^*(i\omega)}, \tag{11.29a}$$

$$1 + \omega^2\bar{J}(\omega)\bar{E}(\omega) = 0. \tag{11.29b}$$

Thus, a compact relationship between stress and strain can be expressed in terms of Fourier transforms and complex moduli in a similar way to relationships involving Laplace transforms, and since the formal substitution of $i\omega = p$ in the Fourier transform gives the *two-sided Laplace transform* [see IV, §13.4.10], such similarities are not too surprising. However, since many functions do not have a Fourier transform because of divergence of the integral at its infinite limits and more functions can be handled analytically and conveniently using the Laplace transformation, the latter is generally much easier to use.

Fourier transformations can also be performed on the general differential operator form of representation (equation 11.13) to give

$$P(i\omega)\bar{\sigma}(i\omega) = Q(i\omega)\bar{\varepsilon}(i\omega) \tag{11.30}$$

and

$$E^*(i\omega) = \frac{Q(i\omega)}{P(i\omega)} = \frac{1}{J^*(i\omega)}. \tag{11.31}$$

Relationships between experimentally determined material functions

The foregoing relationships show that, provided one material function is known, the others can be derived from that function. However, the use of these techniques is based on the assumption that the known material function is in a suitable analytical form. Since the material function is deduced from experiments, this is rarely possible without approximation. As will be seen later, there are techniques of approximation which give forms suitable for analysis as well as methods for handling the inversion of transforms, which is usually the most difficult operation. Leaving aside such techniques for the time being, as they have wider applicability, then material functions can be calculated from an experimentally determined function by *numerical methods of integration* [see III, Chapter 7] and the *solution of integral equations* [see III, Chapter 10]. For example, Hopkins and Hamming (1965) give an efficient procedure for obtaining creep compliance from relaxation modulus, or vice versa, by a step forward integration using the *trapezoidal rule* [see III, §7.1.3] and the *mean value theorem* [see IV, §4.5], which is as accurate as the obtainable data. Other interrelations can be calculated by methods which involve *expansion into eigenfunctions*, using *Fourier series* [see IV, §7.8] and *iterative techniques* [see III, §10.5] and are discussed by Ferry (1970), who also gives further references. On the other hand, particularly if a material is to be subjected to a single mode of loading, it may be preferable to conduct experiments to determine each specific, relevant function. There are arrangements designed to measure creep compliance and relaxation modulus as well as dynamic experiments to determine complex moduli from forced or free vibration tests or from stress waves in bars (e.g. Kolsky, 1967). These latter may involve the use of *Fourier analysis* [see IV, Chapter 20] to determine the required functions.

One form of approximate material representation which is easy to determine and lends itself to analytical treatment is the power law approximation. If the strain response from a creep test is plotted against time on a logarithmic basis (i.e. logarithmic strain versus logarithmic time), then in many cases a linear dependence is revealed for much of the range. This suggests that the creep compliance can be expressed in the form $J(t) = J_0 t^n$, where n is the slope of the line and is usually less than 0.1. This representation tends to break down as t becomes very small, but it is suitable for many practical viscoelastic situations. The Laplace transform of power law compliance is given by

$$J_0 \int_0^\infty t^n e^{-pt} \, dt = \frac{J_0 n!}{p^{n+1}} = \frac{J_0 \Gamma(n+1)}{p^{n+1}}, \tag{11.32}$$

where $\Gamma(n)$ is the *gamma or factorial function* [see IV, §10.2.1]. Use of the relationship between transformed moduli (equation 11.10) and the *reflection formula* [see IV, (10.2.12)] gives

$$J_0 E_0 = \frac{\sin(n\pi)}{n\pi}; \qquad E(t) = E_0 t^{-n}. \tag{11.33}$$

The product of creep compliance and relaxation modulus is very nearly unity for small n and Pipkin (1972) discusses relationships between functions for power law materials and gives approximate forms for complex compliance and loss angle.

The power law representation can also be usefully applied to the relaxation spectra and this leads to a modified form of relaxation modulus which still has power law dependence but does not have any difficulties as t becomes very small. Unfortunately, such a representation leads to complicated transforms involving incomplete *beta functions* [see IV, §10.2.1] (Williams, 1964).

Extension to three dimensions

In describing three-dimensional viscoelasticity, *cartesian tensor (suffix) notation* [see V, §7.1] is usually used. Tensor notation is compact and transposition to *matrix notation* [see I, §6.2] is straightforward. Stresses and strains are usually split into their dilatational and deviatoric components so that Hooke's law of elasticity for isotropic materials (see also Section 10.3) can be rewritten as

$$\sigma_{ij} = \delta_{ij}\lambda\varepsilon_{kk} + 2\mu\varepsilon_{ij} \tag{11.34}$$

and the deviatoric components are

$$s_{ij} = 2\mu e_{ij}. \tag{11.35a}$$

The dilational stress is

$$\sigma_{kk} = 3K\varepsilon_{kk}. \tag{11.35b}$$

Therefore the deviatoric stress and strains are

$$s_{ij} = \sigma_{ij} - \tfrac{1}{3}\delta_{ij}\sigma_{kk}; \qquad s_{ii} = 0, \tag{11.36a}$$

$$e_{ij} = \varepsilon_{ij} - \tfrac{1}{3}\delta_{ij}\varepsilon_{kk}; \qquad e_{ii} = 0. \tag{11.36b}$$

Notice that both summations s_{ii} and e_{ii} are zero. Also, λ and μ are the Lamé constants (μ is the shear modulus), δ_{ij} is the *Kronecker delta* [see I, §6.2.8] and the *summation convention* [see V, §7.1] is understood to apply; K is the bulk modulus and $\varepsilon_{ij} = \tfrac{1}{2}(u_{i,j} + u_{j,i})$, where ',$j$' denotes $\partial/\partial x_j$ [see IV, Definition 5.3.3]. Equation (11.34) is a special case of the anisotropic relationship

$$\sigma_{ij} = C_{ijkl}\varepsilon_{kl}, \tag{11.37}$$

where C_{ijkl} is a fourth-order tensor describing the stress and strain independent elastic moduli which, in general, has eighty-one elements but because of symmetry of the stress and strain tensor has thirty-six at most.

For linear viscoelasticity, by proceeding in an analogous way to the formulation of the one-dimensional theory, the general stress is given by

$$\sigma_{ij}(t) = \int_{-\infty}^{t} C_{ijkl}(t - \tau)\dot{\varepsilon}_{kl}(\tau)\,d\tau, \tag{11.38}$$

where C_{ijkl} are now stress relaxation functions which, for anisotropy, have a maximum of thirty-six independent functions. Anisotropy is not discussed further here, and Shu and Onat (1967) and Gurtin and Herrera (1965) give greater elaboration. For isotropic materials, equation (11.38) reduces to

$$\sigma_{ij}(t) = \delta_{ij} \int_{-\infty}^{t} \lambda(t - \tau)\dot{\varepsilon}_{kk}(\tau)\, d\tau + 2 \int_{-\infty}^{t} \mu(t - \tau)\dot{\varepsilon}_{ij}(\tau)\, d\tau \qquad (11.39)$$

and

$$s_{ij}(t) = 2 \int_{-\infty}^{t} \mu(t - \tau)\dot{e}_{ij}\, d\tau, \qquad\qquad\qquad (11.40a)$$

$$\sigma_{kk}(t) = 3 \int_{-\infty}^{t} K(t - \tau)\dot{e}_{kk}\, d\tau, \qquad\qquad\qquad (11.40b)$$

where $\lambda(t)$, $\mu(t)$ and $K(t)$ are now time-dependent functions. The similarity both with the uniaxial approach and the corresponding elastic relationships is apparent. Applying Laplace and Fourier transformations [see IV, (13.1.3) and (13.1.5)] to equation (11.39) underlines the similarity and yields respectively:

$$\bar{\sigma}_{ij}(p) = \delta_{ij} p\bar{\lambda}(p)\bar{\varepsilon}_{kk}(p) + 2p\bar{\mu}(p)\bar{\varepsilon}_{ij}(p), \qquad\qquad (11.41)$$

$$\bar{\sigma}_{ij}(\omega) = \delta_{ij} i\omega\bar{\lambda}(i\omega)\bar{\varepsilon}_{kk}(i\omega) + 2i\omega\bar{\mu}(i\omega)\bar{\varepsilon}_{ij}(i\omega)$$

$$= \delta_{ij}\lambda^*(i\omega)\bar{\varepsilon}_{kk}(i\omega) + 2\mu^*(i\omega)\bar{\varepsilon}_{ij}(i\omega), \qquad (11.42)$$

where the asterisk means that $\lambda^*(i\omega) = i\omega\bar{\lambda}(i\omega)$ is now a complex function. The various relationships between viscoelastic functions are obtained, as before, from the corresponding elastic relations and replacing the moduli or Poisson's ratio by the p-multiplied equivalent functions or the complex functions, so that, for example:

$$E(p) = \frac{9\mu(p)K(p)}{\mu(p) + 3K(p)}; \qquad E^*(i\omega) = \frac{9\mu^*(i\omega)K^*(i\omega)}{\mu^*(i\omega) + 3K^*(i\omega)}. \qquad (11.43)$$

The differential form can be similarly extended to three dimensions by considering the deviatoric and dilational parts separately to give

$$P_1 s_{ij}(t) = Q_1 e_{ij}(t), \qquad\qquad\qquad (11.44a)$$

$$P_2 \sigma_{kk}(t) = Q_2 \varepsilon_{kk}(t), \qquad\qquad\qquad (11.44b)$$

where P and Q are polynomials of the differential operator as before. Time-dependent functions are obtained from the elastic relationships by the substitutions $2\mu = Q_1/P_1$ and $3K = Q_2/P_2$.

11.4 Stress analysis

The solution of any stress analysis problem requires the satisfaction of equilibrium, compatibility and constitutive relationships, together with the boundary conditions which specify the particular problem to be considered. A viscoelastic stress analysis, therefore, differs only from any other stress analysis by the constitutive relations which have been discussed previously, and so a particular problem can, in principle, be solved simply by their introduction into the problem specifications. Many problems can be readily tackled in this way by incorporating the integral or differential forms of stress–strain laws and solving directly. The similarity with elasticity is evident in such analyses and the replacement of elastic expressions with the relevant convolution integral or differential operator polynomials converts an elastic situation to a viscoelastic one. Solution may then involve standard *analytical* [see IV, Chapters 7 and 8] *or numerical* [see III, Chapters 8, 9 and 10] *techniques for the solution of differential, integral or integrodifferential equations.* In many cases, the use of transform methods, particularly Laplace transforms, gives the most convenient approach. There is a wide range of problems for which the transform technique offers an even greater facility for solving viscoelastic problems when the corresponding elastic solution is known. This involves the *elastic–viscoelastic correspondence principle* and since it takes slightly different forms for different situations it is considered in various categories.

Quasi-static, isothermal viscoelastic problems

The restrictions to this class of problem are that dynamic and temperature effects are negligible but may include situations where loads may move provided that inertia terms can be ignored in the force–equilibrium equations, viz.

$$\sigma_{ij,j} + \rho F_i = 0, \tag{11.45}$$

where ρ is density and F_i are body forces. If this relationship as well as the expressions for compatibility and boundary conditions are subjected to either Laplace [see IV, (13.3)] or Fourier [see IV, (13.1.5)] transforms, then the transformed relationships correspond exactly with the original expressions with time-dependent stress, strain, surface tractions, displacement, etc., replaced by their transforms. As shown previously, the transformed constitutive stress–strain law corresponds to the elastic expression with transformed stress and strain replacing stress and strain and the elastic properties replaced by p-multiplied transforms or complex moduli, depending on whether a Laplace or Fourier transform is applied. Therefore, transformed relationships are directly equivalent to their elastic counterparts and so the solutions must be directly equivalent also. From this follows the correspondence principle for quasi-static viscoelastic problems which is that the viscoelastic solution is obtained from the corresponding elastic solution by substitution of the loading or displacements by their transforms and the elastic constants by their p-multiplied Laplace transforms or complex functions and by inverting the expression so obtained. This is subject to the limitation on boundary conditions that, if over some part of the boundary tractions are prescribed and over the remainder of the boundary displacements are specified, then the interface of these regions must not move with time. This means that at any point on the boundary the conditions must always be of the same *type*. Failing this restriction, transform techniques can still be used but *Weiner–Hopf methods* (Davis, 1978; Hochstadt, 1973) will be involved (Pipkin, 1972) and the solution becomes more complicated. Examples which do not conform to the requirements for the correspondence principle to apply include indentation by a rigid curved plunger, which gives a stress-free surface up to the time that the indenter contacts the surface and a displacement condition subsequently, and moving cracks. Graham (1968) has extended the correspondence principle subject to other limitations and Schapery (1975) has applied this to fracture problems, pointing out that some relaxation of these limitations is allowable.

Subject to the foregoing restriction on boundary conditions, therefore, if the elastic solution is known, the corresponding viscoelastic solution can be obtained. The solution is even more straightforward if a structure subject to proportional loading is considered. In this case, the stress or strain throughout the structure can be expressed as a function of position multiplied by a function of time throughout the event period. This is the *separation of variables* condition [see IV, §8.2] and the solutions consist of a spatial term which is the same as for the elastic solution multiplied by a temporal term.

It is useful to illustrate the application of the correspondence principle by the simple example of a cantilever subjected to a moving load. In this case, separation of variables is not allowed because the loading is not proportional due to the fact that the load moves. Nevertheless, the analysis is straightforward. The problem considered is that of a cantilever of length L and moment of inertia I which is initially quiet. At time $t = 0$, a load of magnitude W moves from the built-in end at a constant speed V and the deflection $\delta_0(t)$ of the end of the cantilever is required both for the time when the load is in contact with the beam and when it has dropped off the end, which is for $t > t_0$ where $t_0 = L/V$. For this example, a power law material, for which the relaxation modulus $E(t) = E_0 t^{-n}$, is considered.

First, the elastic solution is recalled, bearing in mind that the solution must be given for all times greater than zero. This is:

$$\delta_0(t) = \frac{WL^3}{EI} \left\{ \left[\frac{1}{2} \left(\frac{Vt}{L} \right)^2 - \frac{1}{6} \left(\frac{Vt}{L} \right)^3 \right] [1 - H(t - t_0)] \right\}$$

$$= \frac{PL^3}{EI} Y(t), \text{ say,} \tag{11.46}$$

where $H(t)$ is the *Heaviside step function* [see IV, Example 4.8.1]. Then by the correspondence principle, the transform of the viscoelastic solution is given by

$$\bar{\delta}(p) = \frac{WL^3}{p\bar{E}(p)} \bar{Y}(p). \tag{11.47}$$

In order to evaluate $\bar{Y}(p)$, it is necessary to deal with the Heaviside function using the *shift theorem of Laplace transforms* [see IV, (13.14.15)]; to do so, time terms, which are functions of t, have to be expressed as functions of $(t - t_0)$. This is best done by expansion of t^2 and t^3 in equation (11.46) by a *Taylor series* about t_0 [see IV, §3.6]. Laplace transformation can now be easily applied to give

$$\bar{\delta}_0(p) = \frac{\Gamma(1 + n)WL^3}{Ip^{1+n}E(p)} \left[\left(\frac{V}{pL}\right)^2 - \left(\frac{V}{pL}\right)^3 + \left(\frac{V}{pL}\right)^3 \left(1 - \frac{t_0^2 p^2}{2} - \frac{t_0^3 p^3}{2}\right) e^{-t_0 p} \right].$$

(11.48)

Using the properties of the gamma function that $\Gamma(1 + n) = n\Gamma(n)$ [see IV, (10.2.2)], this expression can then be rearranged so that terms are in standard transforms of the type $\Gamma(a + n)/p^{a+n}$ (a is an integer) which can be inverted to give

$$\delta_0(t) = \frac{WL^3}{E_0 I} \left\{ \left(\frac{Vt}{L}\right)^2 \frac{t^n}{(1 + n)(2 + n)} \left[1 - \left(\frac{Vt}{L}\right) \frac{1}{3 + n}\right] \right.$$
$$+ \left(\frac{Vt_0}{L}\right)^3 \left[\frac{(t/t_0 - 1)^{3+n}}{(1 + n)(2 + n)(3 + n)} - \frac{(t/t_0 - 1)^{1+n}}{2(1 + n)}\right.$$
$$\left. \left. - \frac{(t/t_0 - 1)^{3+n}}{3}\right] t^n H(t - t_0) \right\}.$$

(11.49)

This is the same expression given by Williams (1980) obtained by direct integration of the convolution integral.

Dynamic, isothermal viscoelasticity

When inertia terms are significant, they must be included in the force equilibrium equations which then become

$$\sigma_{ij,j} + \rho F_i = \rho \frac{\partial^2 u_i}{\partial t^2},$$

(11.50)

where u_i are displacements. Since time-dependent derivatives are introduced, the correspondence principle which was applicable for the quasi-static case can no longer be used. There is a correspondence principle, but for dynamic problems the substitution for the elastic constants must be made in the *transform of the elastic solution* and then the inversion is made to yield the viscoelastic solution. This has the effect of making dynamic problems less tractable than the quasi-static case. Inversion involves *branch points* [see IV, Definition 9.17.1] in the *contour integration process of inversion* and is much more difficult. Recourse to the elastic solution may not be particularly helpful and a direct approach may be preferable. Fourier transforms are often used in dealing with dynamic problems as the most appropriate material data are in complex modulus form and although analytical inversion can be impossible, inversion by *numerical integration* [see III, Chapter 7] is usually practicable. Christensen (1971) considers dynamic response problems and wave propagation and gives further relevant references on this topic.

Thermal effects

Thermal problems can be treated in the same way as isothermal problems, provided that the material properties depend only on some reference ruling temperature and not on any variations from this temperature. In this case, the correspondence principle is applied as before to the thermoelastic solutions to obtain thermoviscoelastic solutions using the appropriate form, depending on whether the problem is quasi-static or dynamic, and so there is no inherent difficulty in dealing with these situations.

A somewhat different thermal problem is the effect of temperature on the constitutive equations. Since viscoelastic materials are dissipative to some degree, then there are likely to be temperature rises and these can have a quite dramatic effect on viscoelastic properties. Experiment has shown that the viscoelastic behaviour of many materials at high temperatures and high strain rates is similar to that at low temperatures and low strain rates. Such materials are called 'thermorheologically simple' and for these materials the effect of time and temperature can be combined into a single parameter by the 'time–temperature superposition principle'. This effectively means that the effect of changing the temperature is the same as changing the real time for events to occur. Considering a relaxation modulus, a relationship exists of the form:

$$E(T, t) = E(T_0, t_R), \tag{11.51a}$$

$$t_R = \frac{t}{a_T(T)}, \tag{11.51b}$$

where T is the temperature, T_0 is a reference temperature and t_R is the 'reduced time', related to real time t by the temperature shift factor $a_T(T)$. The most common expression used to relate the shift factor and temperature is the WLF equation, after Williams, Landel and Ferry (1955). This is

$$\log_{10} a_T(t) \equiv \log\left(\frac{t}{t_R}\right) = \frac{-k_1(T - T_0)}{k_2 + (T - T_0)}, \tag{11.52}$$

where k_1 and k_2 are material constants. The principle is more clearly illustrated by visualizing a series of plots of logarithmic modulus against logarithmic time, each for a different temperature. Then if each plot is shifted parallel to the time axes a certain distance, dependent on the temperature, the plots will coincide. Therefore, the temperature effect on properties can be encompassed using a change in time scale.

General principles

The theory of linear viscoelasticity is well developed and techniques and theorems applicable to classical elasticity have been extended to linear viscoelasticity. The classical *variational theorems* [see IV, §12.3] (also see Chapter 10) of elasticity have several different generalizations to viscoelasticity and the use of these is a much more rigorous procedure than the use of integral transform techniques with the correspondence principle. Once again, Christensen (1971) can be consulted for more detailed consideration and compendium of variational principles, together with *minimum theorems* [see IV, §12.4], *uniqueness theorems* and a reciprocal theorem in which transform methods are avoided because of the associated limitations of their use in establishing general theorems. Additional references can also be found therein.

11.5 The solution of practical problems—approximate and numerical methods

The body of linear viscoelastic theory which has been briefly summarized affords a framework for the prediction of time-dependent responses of polymeric materials. If the creep compliance or relaxation modulus is in a sufficiently simple and suitable form, then stress analysis is for most cases not too much more difficult than an elastic analysis. The use of transforms [see IV, Chapter 13], whether in conjunction with a correspondence principle or as a tool for solving differential or integral equations [see III, §7.7.14], is a powerful technique, but other methods for the solution of these equations can be just as appropriate. Nevertheless, unless the material behaviour can be characterized by equations which describe the simpler models, such as the Voigt and Maxwell equations, the performance of the operations required to obtain analytical solutions can be extremely difficult—if not impossible. The restriction to simple material functions is often helpful to obtain some indication of the likely behaviour of a time-dependent polymer, however, and an assessment of viscoelastic response in complex problems is often made with their use. Moving on to the treatment of real materials, it has been seen that a power law material can give good representation of behaviour and is suitable for analysis. Another approximation is to treat the relaxation modulus as a series of exponential decays, so that

$$E(t) = E_0 + \sum E_k e^{-t/t_K}. \tag{11.53}$$

This is a *Dirichlet or Prony series* and the Laplace transform of such a series is straightforward. The constants in the series (E_k) can be determined by a *collocation method* proposed by Schapery (1962) by splitting the range of experimental data into N decades of time such that a value of t_K with an associated $E(t_K)$ lies in each decade. Adding to these points the modulus for the instantaneous response, $E(o)$, a set of *linear algebraic equations* is obtained [see

I, §5.7] from which the E_k can be deduced [see III, Chapter 3]. Schapery points out that the matrix representing these equations is essentially triangular and therefore the computation can be performed quickly and efficiently by use of a calculator, even for data which are given over a very wide time range.

The power of the correspondence principle lies in the fact that if the elastic solution is known, whether it is in suitable analytical or purely numerical form, then the transformed viscoelastic solution is also known. The stumbling block is the difficulty of inverting these solutions. If the transformed equation can be arranged into suitable standard form, then inversion can take place by reference to any set of tables of Laplace transforms. Alternatively, the *complex inversion formula* can be used [see IV, §13.4.2] which usually means that *contour integration* involving *branch points* and *residue theory* will be required [see IV, §§9.3 and 9.9 and Definition 9.17.1]. Approximate inversions can be achieved by *saddle-point integration* (Pipkin, 1972), or *asymptotic expansion* [see IV, §2.15] can be used but this may be laborious. Fortunately, there are approximate inversion techniques which, provided care is taken in using them for appropriate circumstances, can be successfully employed in inverting transformed equations, even if data are available in numerical form only. Two of the most useful are due to Schapery (1962). The first is a direct approximation which is valid when the time-dependent functions are nearly linear in log t and is

$$f(t) = [p\bar{f}(p)]_{p=t/2} \tag{11.54}$$

Schapery's second method, which is more generally applicable, is based on the *principle of least squares* [see III, §6.1] and expresses the solution in the form of a Dirichlet series as follows. If a function $F(t)$ having a Laplace transform $F(p)$ is approximated by

$$F_A(t) = \sum_{i=1}^{N} A_i e^{-t/t_i}, \tag{11.55}$$

then the coefficients A_i can be determined by minimizing the total square error between $F(t)$ and $F_A(t)$ to give

$$\int_0^\infty [F(t) - F_A(t)] e^{-t/t_i} = 0, \tag{11.56}$$

which is the Laplace transform of $[F(t) - F_A(t)]$ with $1/t_i$ replaced by p. Using the series approximation (11.55) gives

$$\left[\sum_{j=1}^{N} \frac{A_j}{p + (1/t_j)} \right]_{p \to 1/t_i}, \tag{11.57}$$

which is a series of algebraic equations from which the coefficients of the series can be determined in a similar way to the collocation technique mentioned previously. Cost (1964) gives both of these methods and others which have been suggested for approximate transform inversion. Approaches such as these greatly extend the range of practical problems which can be tackled by the correspondence principle and transform techniques.

Solution by use of transforms can be avoided, however, particularly when the correspondence principle is not applicable. The integral equation approach often leads to a relatively simple numerical scheme since the convolution integral can be solved by a *finite difference method of integration* [see III, §7.7.1] and yields a triangular matrix from which an approximate solution is readily obtained (Lee and Rogers, 1961).

In a general approach to a particular analysis, it is advisable to break the problem down as far as possible into a series of subproblems and finally use linear superposition. Proportional loading and step function inputs are easy to handle and if the situation can be treated at least in part or approximately in these terms, so much the better. In some circumstances, a *quasi-elastic approach* can be adopted which essentially reduces the problem to one of elasticity and dramatically eases the usual difficulties. In these cases, the moduli in the elastic solution are simply replaced by the time-dependent moduli. This elementary method should not be underestimated for assessing the effects of time dependence in many circumstances. For example, Williams and Marshall (1975) have used quasi-elastic solutions which give a good description of the experimental data in studies on the fracture of polymers. Naturally, there will be many classes of problem for which this approach will lead to large inaccuracies and solutions so obtained should be subjected to scrutiny and tested as to their meaning.

There will, of course, be problems for which the correspondence principle does not apply or an elastic solution does not exist and analytical methods cannot cope with the complexity, so that computer methods are needed. As with structural mechanics in Chapter 10, the most popular is the *finite element method* (see Chapter 15), although some work has been done using *boundary integral equations* (e.g. Prideleanu and Screpel, 1980). The problem for real materials is the vast amount of storage required to evaluate the convolution integrals. This can be reduced by taking advantage of the fading memory behaviour and provided that early events do not include relatively large deformations, then integrals may be truncated. Alternatively, the relaxation functions can be expressed as a Dirichlet series so that summation is replaced by a *recurrence relation* [see I, §14.12], as shown by Zak (1967). Zienkiewicz (1971) uses the differential form so that relationships are in terms of current stress and strain and their time derivatives. The success of these approaches depends on how few terms can be used in approximating the material functions.

11.6 Non-linear viscoelasticity

Many polymers, even when subjected to small deformations, show deviations from linearity; i.e. a proportional change in strain causes a different proportional change in stress. Such materials are substantially more difficult to describe mathematically than their linear counterparts. A great deal of theory has been developed on the subject of non-linear viscoelasticity but, at this stage, it is usually complex and difficult to apply to practical situations. Constitutive equations can be represented, as in the linear case, in integral, differential or complex form, although the latter has had limited application because of the difficulty of obtaining non-linear dynamic data. The integral formulation for *uniaxial* non-linear viscoelasticity now becomes, via *functional analysis* [see IV, Chapter 19], a *multiple integral* representation [see IV, Chapter 6] of the form:

$$\varepsilon(t) = \int_0^t J_1(t - \tau_1)\dot{\sigma}(\tau_1) \, d\tau_1 + \int_0^t \int_0^t J_2(t - \tau_1, t - \tau_2)\dot{\sigma}(\tau_1)\dot{\sigma}(\tau_2) \, d\tau_1 \, d\tau_2$$

$$+ \int_0^t \int_0^t \int_0^t J_3(t - \tau_1, t - \tau_2, t - \tau_3)\dot{\sigma}(\tau_1)\dot{\sigma}(\tau_2)\dot{\sigma}(\tau_3) \, d\tau_1 \, d\tau_2 \, d\tau_3 + \cdots.$$

$$(11.58)$$

In order to use such equations, the kernel functions $J_1, J_2, J_3, \ldots$ must be obtained by experiment and their measurement is necessarily elaborate and involved. In the differential form, time derivatives should be frame indifferent and partial derivatives do not conform to this requirement. The formulation is therefore substantially more complex than the linear case and differential forms are usually used for fluids. Because of the complexity of these rigorous approaches, semi-empirical relationships have been proposed as a pragmatic means to characterize behaviour. Of these, Leadermann's (1943) relationship has received much attention and takes the form:

$$\varepsilon(t) = \int_{-\infty}^t J(t - \tau) \frac{\partial}{\partial \tau} [f(\sigma(\tau))] \, d\tau \qquad (11.59)$$

This reduces to the convolution integral form previously considered for linear viscoelasticity and has the advantage of being a single integral relationship. In dealing with stress analysis, the situation becomes even more complex and few problems can be fully solved. There is no correspondence principle to ease the burden and it would seem that computer approaches are required. For a more complete treatment, the texts by Lockett (1972) and Findley, Lai and Onaran (1976) should be consulted.

11.7 Finite elastic deformation

In the classical theory of elasticity, strains are considered to be infinitesimal and geometrical changes of structure under load are not taken into account. Such assumptions are often unrealistic for polymer engineering as deformations can be large and strains may be finite. In a pioneering work on finite elasticity, Murnaghan considered the classical theory to be a first approximation and indicated that more accurate predictions are obtained by retaining the *second-order terms* neglected in infinitesimal analysis (see Murnaghan, 1951). Strains may be expressed in terms of coordinate axes associated with either the undeformed or deformed body, giving Lagrangian or Eulerian viewpoints, and, for example, the Eulerian finite strain tensor is

$$\varepsilon_{ij} = \frac{1}{2}\left\{\frac{\partial u_i}{\partial x_j} + \frac{\partial u_j}{\partial x_i} - \frac{\partial u_k}{\partial x_i}\frac{\partial u_k}{\partial x_j}\right\} \tag{11.60}$$

The quadratic terms which introduce non-linearity are omitted in classical theory and the Lagrange and Euler representations become the same. Thus, for finite strain, geometrical non-linearity is observed even for elastic materials. By including second-order terms in the analysis, Murnaghan showed that an isotropic, compressible cylinder subjected to pure torsional loading would elongate by an amount proportional to the square of both the radius and the angle of twist, an experimentally observed fact. The angle of twist is related to the applied torsional couple, as in classical theory, but if this theory is applied to the deformation problem, the predicted elongation is zero. Since the relationship predicted by finite theory is second order, then, by definition, infinitesimal theory cannot even approximate such behaviour.

A further example of the effect of small-scale approximations on solutions is the important engineering problem of the bending of beams. If the thickness of a beam is small compared to its local radius of curvature R, then the strain distribution is linear, and if there is no elongation of the beam, then classical beam theory can be used to give the well-known expression $MR = EI$, where M is the local bending moment, E the elasticity modulus, I the moment of inertia of the beam and R is given by

$$R = \frac{[1 + (dv/dz)^2]^{3/2}}{d^2v/dz^2}, \tag{11.61}$$

where v is the deflection and z is the axial coordinate. In the usual theory, the second-order term is dropped to give $R = 1/(d^2v/dz^2)$ and a simple analysis ensues. However, if this approximation is not made, solutions are rather more difficult to obtain and are expressed in the form of *elliptic integrals of the first and second kind* [see IV, §§10.14 and 10.15], demonstrating non-linearity and, sometimes, instability. Rogers and Lee (1962), in considering the finite deformation of a viscoelastic cantilever, observed that elastic solutions in the form of elliptic integrals do not help in deducing viscoelastic solutions and so obtained the elastic solution numerically by *Picard's method of iteration* of integral equations (Kanwal, 1971). It can be seen that even with this simple example, the inclusion of second-order terms to obtain finite deformations greatly increases the complexity of the solution.

In a general treatment of large elastic deformations, conventional strain definitions are inconvenient and others must be sought. Additionally, Hooke's law may not be applicable and more general constitutive relationships are required. Since polymers demonstrate behaviour which is rubber-like in situations where they are likely to undergo large deformations, the approaches developed for rubber elasticity are usually adopted. Strain is expressed as an extension ratio, λ, or Green's strain, ε', which, in terms of the stretched and unstretched lengths of an element, l and l_0, are given by

$$\lambda = (1 + 2\varepsilon')^{1/2} = \frac{l}{l_0}. \tag{11.62}$$

Since for an elastic material all the work done in isothermal deformation is stored in the form of strain energy, the constitutive equations can be expressed in terms of the strain energy density, W. For an initially isotropic material, the strain energy density is a function of the *invariants* of the extension ratios I_1, I_2 and I_3, since it must not depend on the reference axes used to describe displacements. Restricting attention to the pure homogeneous deformation of a unit cube subjected to principal stresses σ_1, σ_2 and σ_3, acting in directions coincident with the axes of the cube, which support the extension ratios λ_1, λ_2 and λ_3, then

$$W = W(I_1, I_2, I_3) \tag{11.63}$$

where $I_1 = \lambda_1^2 + \lambda_2^2 + \lambda_3^2,$

$$I_2 = \lambda_1^2\lambda_2^2 + \lambda_2^2\lambda_3^2 + \lambda_3^2\lambda_1^2, \tag{11.64}$$

$$I_3 = \lambda_1^2\lambda_2^2\lambda_3^2.$$

The stresses can be derived by the principle of virtual displacement (see Chapter 10) which gives, using the *chain rule for partial differentiation* [see IV, §5.4],

$$\sigma_1 = \frac{1}{\lambda_2\lambda_3}\frac{\partial W}{\partial\lambda_1} = \frac{2}{I_3^{1/2}}\left(\lambda_1^2\frac{\partial W}{\partial I_1} - \frac{I_3}{\lambda_1^2}\frac{\partial W}{\partial I_2} + I_2\frac{\partial W}{\partial I_2} + I_3\frac{\partial W}{\partial I_3}\right) \tag{11.65}$$

with similar expressions for the other stresses.

A useful simplification, which is appropriate for most polymers in conditions when large deformations are experienced, is of incompressibility. In this case, $I_3 = 1$ and the stress is now given by

$$\sigma_1 = 2\lambda_1^2\frac{\partial W}{\partial I_1} - \frac{2}{\lambda_1^2}\frac{\partial W}{\partial I_2} + p. \tag{11.66}$$

The arbitrary hydrostatic pressure, p, is introduced because hydrostatic pressures cannot be defined by energy considerations of incompressible materials. This is the price paid for the simplification. It now remains to introduce a relationship between the strain energy density and the extension ratio invariants. It has been found that the simply expressed Mooney–Rivlin function gives good representation:

$$W = \frac{\phi}{2}(I_1 - 3) + \frac{\psi}{2}(I_2 - 3), \qquad (11.67)$$

where ϕ and ψ are experimentally determined constants. In practice, ψ is rarely greater than 0.2ϕ for polymers and setting ψ to zero gives the so-called 'neo-hookean' relationship. The stress–strain relationship for a Mooney–Rivlin material is therefore

$$\sigma_i = \phi\lambda_i^2 - \frac{\psi}{\lambda_i^2} + p. \qquad (11.68)$$

These relationships can now be incorporated with the equilibrium equations and boundary conditions to solve stress analysis problems. Further simplification can be obtained in considering two-dimensional problems either of plane strain or plane stress. In the former, there is a reciprocal relationship between the varying extension ratios and solutions may be obtained without detailed knowledge of the elastic constants; in the latter, p is obtained directly by setting the third stress equal to zero. Plane stress membrane problems have a particular application in the thermoformation of polymers, where the final shape is known, or can be assumed, to ease the solution. Williams (1980) considers some of these as well as large-strain thick-walled cylinder problems which have relevance in hose design, etc. Analytical solutions are usually too cumbersome or involved to be of much use and numerical solutions are advisable.

The description of the finite strain approach given has been, necessarily, restrictive and selective to indicate the salient features. More general treatments can be found in Green and Adkins (1970) and Atkin and Fox (1980).

11.8 Polymer processing—aspects of extrusion

Most polymer processing operations involve the conversion of feedstock in the form of powder or granules to a final product form via the operations of compaction; melting under the influence of pressure, temperature and mechanical work; melt pumping and shaping and, finally, cooling. The heart of the majority of processing activities is the screw extruder; therefore, the single-screw extrusion process is considered in order to indicate the techniques of analysis which may be implemented to solve polymer processing problems.

The single-screw extruder essentially consists of a screw rotating inside a heated barrel. Loose particulate feedstock enters through a hopper and is conveyed along the barrel. During this process, pressure is generated due to the frictional forces between barrel and feedstock and compaction takes place. Subsequently, melting is initiated and the screw section associated with further melting is tapered to give greater compression. Completely molten polymer is pumped and mixed in the final section of the screw and is eventually forced through the die to give the required shape. Therefore, there are three distinct operations: solids conveying, melting, and melt conveying and shaping. By far the greatest amount of research and analysis has been performed on the final section and much is appropriate to polymer melt flow in other processes; therefore this section is considered first.

Melt flow in extruders

The procedure for analysing polymer melt flow involves, first of all, the setting up of a mathematical model to describe the flow. This includes the selection of a suitable coordinate system, the derivation of the theoretical equations for the flow by application of the principles of continuum mechanics and the specification of the relevant boundary conditions. In principle, these *partial differential equations* can be solved [see III, Chapter 9] if a constitutive equation for the melt is known but, in practice, many simplifying assumptions are made in order to obtain solutions with a reasonable amount of effort. The nature of the assumptions will affect the form of the equations which finally have to be solved and the method of obtaining these solutions is based on this form.

The continuum mechanics equations, namely, the conservation of mass, momentum and energy, governing melt flow are expressed using *tensor notation* again [see V, Chapter 7]. The viscous stress and rate of deformation tensors are denoted by τ_{ij} and e_{ij}, where

$$e_{ij} = \frac{1}{2}\left(\frac{\partial V_i}{\partial x_j} + \frac{\partial V_j}{\partial x_i}\right) \tag{11.69}$$

and V_i is the local velocity component in the x_i direction. The polymer melt can be considered to be locally incompressible in extruders (for injection moulding, the pressure dependence of density may need to be included) and, therefore, conservation of mass gives

$$e_{ii} = 0. \tag{11.70}$$

The Reynolds number associated with melt flow is extremely small and such flows are laminar; inertia effects and body forces can be neglected in comparison with viscous and pressure forces, and the equation for the conservation of momentum becomes, therefore,

$$\frac{\partial p}{\partial x_i} = \frac{\partial \tau_{ij}}{\partial x_j}, \tag{11.71}$$

where p is the local hydrostatic pressure.

Steady flow is assumed and the thermal conductivity, k, and specific heat at constant pressure, C_p, are locally constant so that the conservation of energy equation becomes

$$\rho C_p V_i \frac{\partial T}{\partial x_i} = k \frac{\partial^2 T}{\partial x_i^2} + \tau_{ij} e_{ij} \tag{11.72}$$

where ρ is density and T is temperature.

In classical fluid mechanics, the fluid is assumed to be Newtonian. This assumption is not valid for polymer melts, although some Newtonian analyses are applied in the literature. The consequence of deviating from the Newtonian idealization, that viscosity is constant with respect to shear rate, is that relationships become non-linear and the superposition of flows is not possible. A good deal of work has been conducted on the study of the flow on non-Newtonian fluids and more information can be found in texts on the subject of 'rheology'; e.g. those of Astarita and Maracucci (1974) and Harris (1977). It is necessary, therefore, to adopt a constitutive equation which gives a reasonable characterization of behaviour, has readily measurable coefficients and which can be incorporated into the analysis with minimum difficulty. Fortunately, viscoelastic effects can be ignored, since in the extruder channels melts are subjected to relatively large deformations for relatively long times and so can be treated as inelastic viscous fluids. Although a number of constitutive relationships have been suggested (see, for example, McKelvey, 1962), it is usual to neglect cross-viscosity and adopt an empirical power law (Ostwald de Waele) model fluid for which the constitutive equation is

$$\tau_{ij} = 2\mu e_{ij}, \tag{11.73}$$

where the shear viscosity, μ, is given by

$$\mu = \mu_0 \left[\frac{(4I_2)^{1/2}}{\gamma_0} \right]^{n-1} \exp\left[-b(T - T_0) \right], \tag{11.74}$$

where n is the power law index which indicates the degree of non-newtonian behaviour, μ_0 is the effective viscosity at the reference shear rate γ_0 and temperature T_0, b is a temperature coefficient and I_2 is the second invariant of the rate of deformation tensor:

$$I_2 = \tfrac{1}{2} e_{ij} e_{ij}. \tag{11.75}$$

In extrusion, pressure variations are not particularly high and so the pressure dependence of viscosity can be ignored.

In order to obtain as much generality as possible from melt flow solutions and also to identify significant parameters characterizing flow behaviour, *dimensional analysis* is employed [see I, §17.8]. Flow rate and pressure drop are usually expressed as the dimensionless terms π_Q and π_p respectively. This assists in the comparison of similar processing machines of different size.

A further important simplifying assumption is the lubrication approximation, which is that local flow in a narrow gap can be replaced by uniform fully developed flow between plane parallel surfaces. This implies that changes in channel depth are gradual. Additionally, the restriction to narrow channel flow, i.e. flow in a channel whose depth H is substantially smaller than the other dimensions, is a further useful simplification which is appropriate in very many circumstances. Flow types which fall into this category are considered first.

Isothermal one-dimensional steady flow

This is the simplest type of flow. If the local velocity of the melt is w in the direction z, then the governing equations reduce to

$$\frac{dp}{dz} = \frac{d\tau_{zy}}{dy}; \qquad I_2 = \frac{1}{4}\left(\frac{dw}{dy}\right)^2 \qquad (11.76)$$

and the flow rate per unit width of channel, Q_z, is given by

$$Q_z = \int_0^H w \, dy. \qquad (11.77)$$

The flow rate for a given pressure gradient, or vice versa, is determined by the *integration* of the *ordinary first-order differential equations* [see IV, §7.2] and incorporating the boundary conditions which leads to a pair of *simultaneous non-linear algebraic equations* [see III, §5.6]. Analytical solutions are not possible, for general values of the power law index, if the coordinate position, y_0, of the surface at which the shear stress is zero (the 'stress-neutral surface') lies within the flow at an unknown point. An example for which an analytical solution may be obtained is for symmetrical flow, such as the flow due to pressure only (i.e. there is no drag flow) between parallel plates. The stress-free surface is then at the mid-point and the analysis can be completed. In the absence of this knowledge, a simple strategy is to calculate the pressure gradient and flow rate, separately, for selected values of y_0, thus generating a graphical plot giving the pressure gradient–flow rate relationship. Alternatively, a numerical solution can be obtained for y_0, using the known pressure gradient or flow rate, and this value is then used to determine the unknown quantity. An iterative approach using either *Newton's method* [see III, §§5.4.1 and 5.4.2] or an *interpolation technique* [see III, §2.3] gives rapid convergence to the required root of y_0.

Two-dimensional flow

The introduction of transverse (x direction) flow into a problem creates difficulties which are much greater than the one-dimensional case. The pressure gradient–shear stress relationships are

$$\frac{\partial p}{\partial x} = \frac{\partial \tau_{xy}}{\partial y}; \qquad \frac{\partial p}{\partial z} = \frac{\partial \tau_{yz}}{\partial y} \tag{11.78}$$

and the second invariant of the rate of strain tensor is

$$I_2 = \frac{1}{4}\left[\left(\frac{\partial w}{\partial y}\right)^2 + \left(\frac{\partial u}{\partial y}\right)^2\right]. \tag{11.79}$$

Integration of (11.78) gives relationships for the shear stresses in terms of two stress neutral surfaces, one for each flow direction. Combination with (11.79) and the constitutive equation gives expressions for the velocity gradients. These have been solved by successive substitution using *Runge–Kutta integration* [see III, §8.2.2] until the desired accuracy is obtained. Alternatively, and more efficiently, the boundary conditions can be incorporated to lead to four non-linear algebraic equations which can be solved by the *Newton–Raphson technique* [see III, §5.4.1], so the method is analogous to that for one-dimensional flow.

For flow in dies, solutions have been obtained by noting that the mean shear rate over some depth H is Q/H^2, where Q is the local flow rate and Q^2 is the sum of the squares of the flow rates in the x and y directions—Q_x^2 and Q_y^2. By introducing a stream function ψ such that $Q_x = \partial\psi/\partial y$ and $Q_y = -\partial\psi/\partial x$, continuity is automatically satisfied and the incorporation of the stream function into the analysis leads to

$$\frac{\partial}{\partial x}\left(\frac{\mu}{H^3}\frac{\partial\psi}{\partial x}\right) + \frac{\partial}{\partial z}\left(\frac{\mu}{H^3}\frac{\partial\psi}{\partial y}\right), \tag{11.80}$$

where for a power law material

$$\mu = \mu_0\left(\frac{Q}{\gamma_0 H^2}\right)^{n-1} \tag{11.81}$$

The mathematical behaviour of equation (11.80) is similar to that of a harmonic partial differential equation and it may be solved by *finite differences* [see III, Chapter 9] or, preferably, because of the complex geometries which may be encountered, by a *finite element technique* (see Chapter 15) with triangular elements.

For non-isothermal flow, the viscosity is dependent on the temperature and the energy equation must be used. For fully developed flow, in which there is no downstream variation of velocity or temperature, the energy equation has been solved simultaneously with the momentum equations using a *fourth-order predictor–corrector method* [see III, §8.3.1] started off by a Runge–Kutta technique. Fenner (1979) gives a strategy for solving developing flow problems, where the temperature varies in the downstream direction, in which the velocity profiles are determined for a known initial temperature profile by the Newton–Raphson method previously mentioned. The temperature analysis is performed by setting up a finite difference grid. It has been found that, in order to compute the next downstream temperature, *explicit methods* should be avoided as they may be unstable (Mitchell and Griffiths, 1980; Smith, 1978) [see §10.4] and a *Crank–Nicholson* method is recommended [see III, §9.4]. This approach is also suitable for fully developed flow, for which the computation is easier.

Flow in deep channels

When the assumption that the channel is narrow is no longer realistic, the approach must be modified and the momentum equation for isothermal downstream flow alone is

$$\frac{\partial p}{\partial z} = \frac{\partial \tau_{zx}}{\partial x} + \frac{\partial \tau_{yz}}{\partial y}. \tag{11.82}$$

The final partial differential equation can be solved using a finite difference technique and convergence is usually obtained by iteration using *successive over-relaxation*. Integration of the velocities to obtain the flow rates can be accomplished using *Simpson's rule* [see III, §7.1.5].

Finite element techniques may also be used, especially when the geometry is complicated. In these cases, a triangular element is used as it is much easier to fit the complex boundary shapes. The set of simultaneous linear algebraic equations obtained at the nodes can be solved using either *Gaussian elimination* [see III, §3.3.1] or *Gauss–Seidel methods* [see III, §3.4].

For non-isothermal, fully developed flow, finite difference techniques have been adopted to obtain numerical solutions.

Solids conveying

In the feed section, polymer in particulate form enters from the hopper and is compacted and transported along the extruder, and the frictional effect between polymer and barrel generates a pressure build-up. No true particulate analysis of this section has been attempted and the uncertainty of the relevant material data is such that a sophisticated analysis is unlikely to be justified. Therefore, great simplifying assumptions are made but analysis can be involved, nevertheless. Perhaps the simplest approach is to use the model of Darnell and Mol (1956). The particulate nature of the material is ignored and polymer is treated as a solid plug. A one-dimensional analysis is performed and it is assumed that transverse stresses are proportional to downstream stresses. This leads to the prediction of an exponential increase in pressure. Lovegrove and Williams (1973) have conducted a two-dimensional analysis in which pressure varies both in the downstream and transverse directions. By neglecting body forces and after lengthy analysis, a *hyperbolic partial differential equation* for pressure is obtained [see IV, §8.5], the solution for which is expressed in terms of exponential functions in both directions. These authors obtained numerical solutions including body forces using the *method of characteristics* [see III, §9.2] and showed that gravity and centrifugal forces can be significant and lead to pressure fluctuations which may affect the uniformity of the output.

Melting in single-screw extruders

After a number of turns, the compacted plug of polymers starts to melt at the barrel wall. This process continues under the influence of shear heat due to mechanical work and heat conducted from the barrel so that there is a solid plug of polymer surrounded by melt and, eventually, complete melting takes place. The modelling of this process is based on experimental observation and involves the consideration of polymer melt flows of differing types passing over faces of a rectangular solid plug. The heat transfer rate to the plug is equated to the melting rate via the latent heat of fusion and continuity of mass flow is observed. The equations of flow and heat transfer combine to give a complex description of the process and analytical methods are out of the question. Solutions which show good comparison with experiment have been obtained using finite differences and a Newton–Raphson technique. Far more information is presented by Fenner (1979) who gives a concise and well-organized account of the modelling and solution of melting problems as well as other polymer-processing phenomena.

D.P.I.

References

Astarita, G., and Maracucci, G. (1974). *Principles of Non-Newtonian Fluid Mechanics*, McGraw-Hill.

Atkin, R. J., and Fox, N. (1980). *An Introduction to the Theory of Elasticity*, Longman, London.

Bland, D. R. (1960). *The Theory of Linear Viscoelasticity*, Pergamon.

Christensen, R. M. (1971). *Theory of Viscoelasticity*, Academic Press, New York.

Cost, T. L. (1964). Approximate Laplace transform inversions in viscoelastic stress analysis, *AIAA J.*, **2**, 2157.

Darnell, W. H., and Mol, E. A. J. (1956). Solids conveying in extruders, *SPEJ.*, **12**, 20.

Davis, B. (1978). *Integral Transforms and their Applications*, Applied Mathematical Sciences, Vol. 25, Springer-Verlag.

Fenner, R. T. (1979). *Principles of Polymer Processing*, Macmillan, London.

Ferry, J. D. (1970). *Viscoelastic Properties of Polymers*, 2nd ed., Wiley.

Findley, W. N., Lai, J. S., and Onaran, K. (1976). *Creep and Relaxation of Non-Linear Viscoelastic Materials*, North-Holland, Amsterdam.

Flügge, W. (1975). *Viscoelasticity*, 2nd revised ed., Springer-Verlag, New York.

Graham, G. A. C. (1968). The correspondence principle of linear viscoelasticity theory for mixed boundary value problems involving time-dependent boundary regions, *Quart. Appl. Math.*, **26**, 167.

Green, A. E., and Adkins, J. E. (1970). *Large Elastic Deformations and Non-Linear Continuum Mechanics*, 2nd revised ed., Oxford University Press.

Gross, B. (1953). *Mathematical Structure of the Theories of Viscoelasticity*, Hermann, Paris.

Gurtin, M. E., and Herrera, I. (1965). On dissipation inequalities and linear visco-elasticity, *Quart. Appl. Math.*, **23**, 235.

Gurtin, M. E., and Sternberg, E. (1962). On the linear theory of viscoelasticity, *Arch. Ration. Mech. Anal.*, **11**, 291.

Harris, J. (1977). *Rheology and Non-Newtonian Flow*, Longman, London.

Hochstadt, H. (1973). *Integral Equations*, Wiley.

Hopkins, I. L., and Hamming, R. W. (1957). On creep and stress relaxation, *J. Appl. Phys.*, **28**, 906.

Kanwal, R. P. (1971). *Linear Integral Equations, Theory and Technique*, Academic Press.

Kolsky, H. (1967). Experimental studies of the mechanical behaviour of linear viscoelastic solids, in *Proc. Fourth Symp. Nav. Structural Mech.*, p. 357, Pergamon.

Leadermann, H. (1943). *Elastic and Creep Properties of Filamentous Materials*, Textile Foundation, Washington, D.C.

Lee, E. H., and Rogers, T. C. (1961). Solution of viscoelastic stress analysis problems using measured creep or relaxation functions, *Interim Tech. Rep. No. 1, No. 1892-E, Brown University*.

Lockett, F. J. (1972). *Non-Linear Viscoelastic Solids*, Academic Press.

Lodge, A. S. (1964). *Elastic Liquids*, Academic Press.

Lovegrove, J. G. A., and Williams, J. G. (1973). Solids conveying in a single-screw extruder, *J. Mech. Eng. Sci.*, **15**, 114.

McKelvey, J. M. (1962). *Polymer Processing*, Wiley.

Mitchell, A. R., and Griffiths, D. F. (1980). *The Finite Difference Method in Partial Differential Equations*, Wiley.

Murnaghan, F. D. (1951). *Finite Deformation of an Elastic Solid*, Wiley.

Pearson, J. R. A. (1966). *Mechanical Principles of Polymer Melt Processing*, Pergamon.

Pipkin, A. C. (1972). *Lectures on Viscoelasticity Theory*, Springer-Verlag, New York.

Prideleanu, M., and Screpel, J. (1980). Boundary integral equation method in thermo-viscoelasticity theory including crack problems, Addendum to *Proc. Second Int. Symp. on Innovative Num. Anal. in Appl. Eng. Sci.*, Montreal, **46**.

Riesz, F., and Nagy, B. (1955). *Functional Analysis*, Ungar, New York.

Rogers, T. C., and Lee, E. H. (1962). On the finite deflection of a viscoelastic cantilever, in *Proc. Fourth US Nat. Congr. Appl. Mech.*, **1962**, 977.

Schapery, R. A. (1962). Approximate methods of transform inversion for viscoelastic stress analysis, in *Proc. Fourth US Nat. Congr. Appl. Mech.*, **1962**, 1075.

Schapery, R. A. (1975). A theory of crack initiation and growth in viscoelastic media: 1. Theoretical development, *Int. J. Fracture*, **11**, 141.

Shu, L. S., and Onat, E. T. (1967). On anisotropic viscoelastic solids, in *Proc. Fourth Symp. Nav. Structural Mech.*, p. 203, Pergamon.

Smith, G. D. (1978). *Numerical Solution of Differential Equations: Finite Difference Methods*, 2nd edn, Clarendon Press.

Spiegel, M. R. (1965). *Theory and Problems of Laplace Transforms*, Schaum Publishing Co.

Tadmor, Z., and Gogos, C. G. (1979). *Principles of Polymer Processing*, Wiley.

Williams, J. G. (1980). *Stress Analysis of Polymers*, 2nd ed., Ellis Horwood, Chichester.

Williams, J. G., and Marshall, G. P. (1975). Environmental crack and craze growth phenomena in polymers, *Proc. R. Soc. Lond. (A)*, **342**, 55.
Williams, M. L. (1964). Structural analysis of viscoelastic materials, *AIAA J.*, **2**, 785.
Zak, A. R. (1967). Structural analysis of realistic solid propellant materials, *J. Spacecraft & Rockets*, **5**, 270.
Zienkiewicz, O. C. (1971). *The Finite Element Method in Engineering Science*, p. 395, McGraw-Hill.

Mathematical Methods in Engineering
Edited by G. A. O. Davies
© 1984, John Wiley & Sons, Ltd.

12

Nuclear Power

12.1 Introduction

The major use of nuclear power for peaceful purposes is in the generation of electricity, usually for base-load land power stations. A significant application, however, is to maritime or more particularly naval propulsion which is generally via direct mechanical drive. There are, indeed, other specialized applications but in commercial terms they are less significant and are not pursued specifically here. Nor do we deal explicitly with any other applications of nuclear energy to military ends, although it must be remarked that techniques developed under the pressure of weapons programmes have been reflected in civilian applications.

Nuclear engineering is not a primary engineering discipline, as electrical or mechanical engineering might be considered to be, but is rather a synthesis of many engineering disciplines and indeed of much else. Thus nuclear engineering is a high technology discipline involving advanced scientific concepts. It also demands, for successful application, an integration of engineering with many strands including economics, health physics, industrial psychology, etc., all of which will have their own mathematical needs and customary techniques. Because nuclear power has the two characteristics of high capital cost and potential risk with safety implications, there is indeed intense commercial or economic pressure to carry out refined calculations for both analysis and synthesis (design) purposes. The industry has a massive investment in man-years, for example, in the *software* of computer programs as well as recourse to some of the largest of mainframe digital computers.

The present commercial application of nuclear power takes the form of the fission reaction, the fissioning of heavy elements such as the 235 isotope of uranium. Consideration is now being given to the generation of electricity from a fusion reaction, such as that involving the deuterium and tritium isotopes of hydrogen. In such fusion reactions, the confinement of the charged particles is likely to be by means of magnetic fields with the consequent involvement of plasma physics and magnetohydrodynamics, which in turn rest heavily on *vector analysis* and *vector field theory* [see V, Chapter 13]. Consideration of space, however, and the fact that controlled fusion is a potential but not yet an

established technique with economic demonstration, precludes further consideration of the specialized mathematical problems fusion engineering undoubtedly
presents, and we shall speak only of fission engineering problems whilst noting
that much that is important in one is common to the other.

In our development, then, of the mathematics of nuclear engineering it will be
necessary to be selective and address only those points that are peculiar to
nuclear power and not deal explicitly with those aspects that are broadly
common to other fields of engineering application.

In the following section we deal with the representation of the transformation
of matter at the nuclear level by radioactive decay and neutron transmutation.
Subsequent sections discuss the problem of the distribution of neutrons and
other particles as a transport theory problem, followed by a section looking at
some of the approximation methods for this problem, particularly diffusion
theory, with particular emphasis on steady-state problems. In the following
section we turn to time-dependent reactor kinetics and control. These problems
involve relatively rapid transients and it is helpful to consider separately the
longer term changes involved with the nuclear fuel cycle, both within and
without the reactor itself, as a separate concluding section.

At the end of this chapter a few broad references are given for texts,
monographs, etc., that are mentioned in the text. These would extend this
introduction to the mathematic analysis of nuclear power and nuclear
engineering.

12.2 Radioactive decay and isotopic transformations

The radioactive decay of an isotope is characterized by a decay constant, λ,
leading to a different isotope where disintegrations per time is essentially
independent of the physical state of the material (temperature, pressure, etc.). To
this extent, therefore, the equations describing such transformations will have
the simplicity of constant coefficients. In many applications, there is a further
simplification that only the mean rate of transformation of a number of atoms
(or nuclei), $N(t)$, as a function of time is required. Taking the simplest case where
isotope A decays to a stable isotope B, we may write for the mean number of the
first isotope:

$$\frac{d\bar{N}_A}{dt} = -\lambda_A \bar{N}_A; \qquad \bar{N}_A(0) = N_{A0}, \qquad (12.1)$$

an *ordinary differential equation of first order with constant coefficients* [see IV,
§§7.2 and 7.4], subject to an initial boundary condition for the (given) initial
mean density $\bar{N}_A$, the overbar representing the *mean* or *expected* behaviour [see
II, §5.1] (Thomas and Abbey, 1973).

Associated with the loss of 'mother' isotope A is the gain of 'daughter' B and the change of this second isotope (assumed stable) may be represented via

$$\frac{d\bar{N}_B}{dt} = \lambda_A \bar{N}_A; \qquad \bar{N}_B(0) = N_{B0} = 0, \tag{12.2}$$

where the final term would be an appropriate initial condition if the decay process starts with pure separated mother.

The solution of equation (12.1) is elementary [see IV, §7.4.1] and takes the form

$$\bar{N}_{A(t)} = N_{a0} \, e^{-\lambda_A t}; \qquad t \geqslant 0. \tag{12.3}$$

The solution is undefined before the quoted *initial condition*. Correspondingly the build-up of stable daughter is

$$\bar{N}_B(t) = N_{B0} + [N_{A0} - \bar{N}_A(t)] \rightarrow N_{A0} - \bar{N}_A(t)$$

$$= N_{a0}(1 - e^{-\lambda_A t}). \tag{12.4}$$

The exponential solution [see IV, §2.11], shown in Figure 12.1 in non-dimensional form, illustrates the exponential fall-off of the original mother atoms and the exponential build-up of daughter atoms. In particular, it may be noted that the initial slop of the decaying exponential extrapolates to a t-value, $\tau_A = 1/\lambda_A$ [see III, §2.4], and it will be shown later from the *stochastic* equations for radioactive decay that τ_A may properly be interpreted as the *mean* life of the unstable isotope A. It should perhaps be noted that τ_A is related to the half-life $T_{A/2}$, and hence λ_A to $T_{A/2}$ by the expression

$$\lambda T_{1/2} = -\ln \tfrac{1}{2} = 0.693 \tag{12.5}$$

for the time, $T_{1/2}$, that leads to a reduction of the initial inventory by a factor of two. The solution shows that this time is independent of the amount of initial isotope.

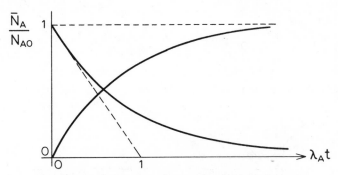

Figure 12.1: Non-dimensional decay of initially pure
mother and build-up of daughter.

Figure 12.2: Schematic simple radioactive decay chain.

Of course, the decay system may be much more complicated than shown so far. The daughter B may itself be radioactive, with characteristic decay constant λ_B, leading to a chain of radioactive isotopes (Figure 12.2).

The nature of the emitted radiations (alpha or beta particles, gamma rays or photons, neutrons, etc.) has not so far been mentioned but this may be significant not only intrinsically (for shielding purposes perhaps) but because they in turn may induce further radio-activity or transmutations.

A further complication is *branching*, where one isotope may decay in two (occasionally more) competing ways. This branching, as far as is known, is characterized by a fixed probability of dividing into branch 1, branch 2, etc., and so, for the purposes of calculating mean numbers, leads again to a constant coefficient of the relevant equations. In some cases the branches may rejoin as illustrated in Figure 12.3 for the uranium-238 radioactive series where all branches ultimately terminate in the stable lead isotope Pb-206.

So far the mathematical representation of the decay chain has been via a finite number or set of *first-order ordinary differential equations* with *constant*

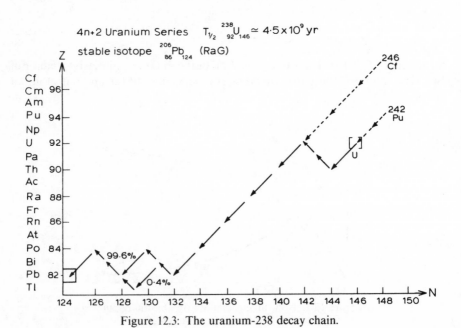

Figure 12.3: The uranium-238 decay chain.

coefficients and an appropriate number of *initial* conditions, equal to the number of first-order equations or the order of the differential set, provided by an assumed known initial inventory serving to make the problem well posed physically [see IV, §§7.4 and 7.9.2]. Given that such a system of equations has a unique solution then the applied mathematician, like the physicist or engineer, will be happy that the solution indeed represents the required answer on physical if not mathematical grounds. Because the equations are *linear* a number of solution techniques that can, if necessary, partition the required solution amongst a number of elementary solutions are possible; because the coefficients are constant, the solutions of the equations in principle are straightforward [see IV, §7.9.2]. Two methods suggesting themselves are (i) a *matrix operator notation* [see IV, §19.1.2] with a solution composed of *orthogonal vectors* of the defining homogenous matrix [see IV, §19.2.1] and (ii) the *Laplace transform method* [see IV, §13.4.6] where the time-dependent differential equations are transformed to a set of algebraic equations in which the initial conditions are explicitly represented (Bateman, 1910). After suitable algebraic manipulation, the system with constant coefficients can be expected to be in a form where inverse transforms are recognized in a purely mechanical manner without recourse to integral inverse transformations.

A feature of the Laplace transformation, in addition to the explicit form of initial conditions, is the availability of asymptotic properties for large time (small transform parameter).

When allowance is made for additional *source* terms (e.g. a cyclotron or a nuclear reactor *creating* isotopes) then the representation may become more complicated in that the coefficients of the equations may no longer be constant. A useful representation of the general problem, as it becomes more complex, is via *graph theory* [see V, Chapter 6] (or signal theory as it is known to some engineers) as illustrated by Figure 12.4 corresponding to the equations with full-range branching and production.

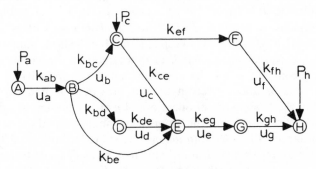

Figure 12.4: Graph theory representation of decay and transmutation.

For sufficiently simple (given) time-dependent coefficients, it may still be possible to employ *Laplace transform methods* at the expense of analytical *inversion* [see IV, §13.4.2], the *Bromwich integral*. If this fails or if a general method for routine work is required, then direct *numerical integration* [see III, Chapter 7] or solution of the equations is feasible, usually in these circumstances with the aid of a digital computer, using a number of standard techniques for *difference equations* that approximate the original differential equations, such as the *Runge–Kutta* methods of various orders [see III, §8.2.2]. One complication now is that in a complex chain it may well happen that two (or more) of the coefficients have similar magnitude (represented here by $\lambda_i \simeq \lambda_j$, say). When $\lambda_i = \lambda_j$, the analytical solution displays a *repeated root*. In principle this does not occur unless λ_i is exactly equal to λ_j, but in cases of approximate equality the *numerical* solution may be *ill-conditioned* [see III, §1.7] and difficult or at least expensive to achieve. It may be noted that the graph theory approach has recently been developed with promise in this area (Henley and Williams, 1973).

Of course more complicated problems still will involve a spatial variation of the concentrations and this brings us to transport problems and their methods of approximate solution, discussed in Section 12.3.

Stochastic formulation

·Given that the number of nuclei of each isotope considered is usually very large, say of the order of 10^{27} in one kilogram, the treatment of the *probability* of transmutation in terms of the *mean* or *expected number* is generally satisfactory [see II, §8.1]. There are occasions, however, when the likely spread around the mean is significant even in something like radioactive decay, and it raises the point that there is inherent information available from the study of such variability (or noise). This will lead ultimately to stochastic reactor or noise theory but it is convenient to introduce the concepts in simple form here.

Suppose N_a atoms of a radioactive species are present at $t = 0$ and we desire the probability P_n $(t \geqslant 0)$ of having exactly $n = 0, 1, 2, \ldots$ atoms present subsequently. This is a problem of *discrete probabilities* with a continuous parameter [see II, §18.0.4]; we may keep in mind that the large number of events may lead to a *continuous probability* approximation being of value. The disintegration constant λ has the meaning that the probability of one atom decaying in time δt about t, given that at present it is undecayed at t, is $\lambda t + o(\delta t)$, and the probability of one atom of n atoms present at t decaying is $(n\lambda \, \delta t + o(\delta t)$. We assume that the atoms decay independently at random and that, in accordance with *order notation* $o(\delta t)$ [see IV, Definition 2.3.3], the probability of two atoms decaying in δt tends to zero as δt tends to zero.

Then a balance for the probability $P_0(t)$ and $P_0(t + \delta t)$ can be constructed from the meaning of the conditional probabilities [see II, §6.5] by allowing for the two mutually exclusive ways in which the system can be in the state $n = 0$ at $t + \delta t$:

$$P_0(t + \delta t) = \qquad 1 \qquad \times \qquad P_0(t) \qquad + 1[\lambda \ \delta t + o(\delta t)]P_1(t) \qquad (12.6)$$

$$\begin{bmatrix} \text{conditional} \\ \text{probability} \\ \text{of not decaying} \\ \text{from state } n = 0 \end{bmatrix} \times \begin{bmatrix} \text{probability} \\ \text{of being in} \\ \text{state } n = 0 \end{bmatrix} + \begin{bmatrix} \text{conditional} \\ \text{probability} \\ \text{of decaying} \\ \text{from state} \\ n = 1 \end{bmatrix} \times \begin{bmatrix} \text{probability} \\ \text{of being in} \\ \text{state } n = 1 \end{bmatrix}$$

which rearranges to the *ordinary differential equation* [see IV, §7.2] in the limit of $\delta t \to 0$:

$$\frac{dP_0}{dt} = \lambda P_1(t), \qquad (12.7)$$

with which we associate the *initial condition* that $P_0 = 0$ for $N > 0$ and $P_0 = 1$ for $N = 0$ (the initial number N vanishing) [see II, §20.4.3].

For $n > 0$, similar probability arguments yield a relation between P_n and P_{n+1} in the form

$$P_n(t + \delta t) = (1 - n\lambda \ \delta t)P_n(t) + (n + 1)\lambda \ \delta t P_{n+1}(t) \qquad (12.8)$$

or

$$\frac{dP_n}{dt} = -n\lambda P_n + (n + 1)\lambda P_{n+1}; \qquad P_n(0) = \delta_{nN}, \qquad (12.9)$$

employing the *Kronecker delta* δ_{nN} [see I, (6.2.8)], and is seen to encompass the case $n = 0$ also.

We have an infinite set of first-order ordinary differential equations, linear with constant coefficients (in time, of course, not in n). A direct numerical solution would suffer from the necessity of truncating the equations and using some artifice (referred to often as 'closure') for the highest P_n in the last equation if $n_{max} < N$. Here, however, a general method of solution is available that converts the infinite set of ordinary differential equations to a single linear *partial differential equation* [see IV, §8.1], also of first order and therefore in principle soluble by the *method of Lagrange*. The conversion of the *set* $\{P_n\}$ to a single *generating function* $G(x, t)$ is done via the definition

$$G(x, t) = \sum_{n=0}^{\infty} x^n P_n(t), \qquad (12.10)$$

with $0 < x < 1$ a dummy variable [see II, §12.0]. $G(x, t)$ is well defined in this range since the P_n form a *proper stochastic variable*.

It is readily seen that:

(i) The original probabilities may be recovered from G by the device [see IV, §3.1]

$$\left. \frac{\partial^n G}{\partial x^n} \right|_{x=0} = n! \ P_n(t). \qquad (12.11)$$

(ii) Indeed, the P_n may be thought of as the *moments* of the generating function [see II, §12.2]. The *factorial moments* of the probability distribution about the origin [see II, §9.11] are conveniently given by

$$\left.\frac{\partial^j G}{\partial x^j}\right|_{x=1} = \sum_n n(n-1)\cdots(n-j+1)P_n(t) \tag{12.12}$$

(iii) $G(x, t)$ satisfies the first-order linear partial differential equation and associated boundary conditions

$$\frac{\partial G}{\partial t} = \lambda(1-x)\frac{\partial G}{\partial x}; \qquad G(x, 0) = x^N. \tag{12.13}$$

The general solution found by the method of Lagrange may be adapted to fit the boundary condition exactly and leads to the unique solution

$$G(x, t) = [(x-1)e^{-\lambda t} + 1]^N = (1 - e^{-\lambda t})\left(1 + \frac{x\,e^{-\lambda t}}{1 - e^{-\lambda t}}\right)^N$$

$$= (1 - e^{-\lambda t})^N\left[1 + \frac{x\,e^{-\lambda t}}{1 - e^{-\lambda t}} + \cdots + \binom{N}{n}\left(\frac{e^{-\lambda t}}{1 - e^{-\lambda t}}\right)^n x^n + \cdots\right] \tag{12.14}$$

and hence

$$P_n(t) = \binom{N}{n} e^{-n\lambda t}(1 - e^{-\lambda t})^{N-n}, \tag{12.15}$$

where $\binom{N}{n}$ is the *combinatorial coefficient* [see I, §3.8], a result which is perhaps obvious on physical grounds on observing that the decay of a single atom would be characterized by a probability set

$$P_0 = 1 - e^{-\lambda t}, \qquad P_1 = e^{-\lambda t} \qquad \text{and} \qquad P_{n>1} = 0.$$

Mention should be made that this development of the original equations takes the stance of given initial conditions and varies the subsequent time t. It is therefore characterized as the *forward stochastic equation*. An alternative view would be to specify the final state of the system and vary some earlier time, s say: the *backward Kolmogorov equation*. The resulting equations, at least in the present context, are equivalent because of the constancy of the coefficient and hence *translational invariance* of the solutions follows. However they do have a distinct origin and are both necessary, as equations, to a *variational* representation of the stochastic problem leading to variational and *perturbation* approximations.

The foregoing solution can be adapted to answer such problems as finding the probability of seeing exactly c atoms decay (counted) in an interval t or counted in an interval t starting when it was known that N atoms were present.

From equation (12.14) we may obtain moments which to second order are

$$\langle 1 \rangle = G(1, t) = 1 \qquad \text{(proper stochastic variable)},$$

$$\langle n \rangle = \frac{\partial G}{\partial x}\bigg|_{x=1} = N\,e^{-\lambda t} \qquad \text{i.e. the anticipated mean } \bar{N}(t)$$

and

$$\langle n(n-1) \rangle = \frac{\partial^2 G}{\partial x^2}\bigg|_{x=1} = N(N-1)\,e^{-2\lambda t},$$

the *variance* [see II, §9.1] from which we obtain

$$\sigma^2 = \langle (n - \langle n \rangle)^2 \rangle = N\,e^{-\lambda t}(1 - e^{-\lambda t}).$$

For one initial atom, we may readily utilize the solutions to obtain the mean life τ of one atom, known to exist at time zero, as

$$\tau = \frac{\displaystyle\int_0^\infty t P_1(t)\lambda\,dt}{\displaystyle\int_0^\infty P_1(t)\lambda\,dt} = \frac{1}{\lambda}\frac{\displaystyle\int_0^\infty x\,e^{-x}\,dx}{\displaystyle\int_0^\infty e^{-x}\,dx} = \frac{1!}{\lambda 0!} = \frac{1}{\lambda}, \tag{12.16}$$

employing the *integral definition of a factorial, $n!$* [see IV, §10.2.1]. If λ is so small that $\bar{N} = N\,e^{-\lambda t}$ is essentially constant over any intervals considered and N is large, the foregoing *binomial* type of solution degenerates [see II, §5.5] to the well-known *Poisson process* solution [see II, §20.1], since P_n is the probability of have exactly $N - n$ atoms decay in time t. Then the probability $Q_c(t)$ of having exactly c atoms decay in interval t is given by

$$Q_c(t) = \binom{N}{N-c} e^{-(N-c)\lambda t}(1 - e^{-\lambda t})^c.$$

But if $\lambda t \to 0$ whilst keeping $N\lambda t$ a constant, m, then

$$Q_c(t) \to \frac{1}{c!}e^{-m}N(N-1)\cdots(N-c+1)(\lambda t)^c,$$

and as $N \to \infty$,

$$Q_c(t) \to e^{-m}\frac{m^c}{c!}, \tag{12.17}$$

where m is the mean number of disintegrations expected in interval t (now assumed constant) and Q_c has degenerated to the Poisson form appropriate to alpha decay, i.e. long half-life.

The stochastic distribution for more complicated decay schemes may be formulated in the same way and in principle may be solved. Complications which inhibit exact solutions in the presence of arbitrary production terms, involved in a *multiplying stochastic chain process*, are considered in Section 12.5. It may not, however, be practical to carry out the analysis in a complex case with several terms in the decay scheme, but if numerical methods are resorted to, the problem of closure becomes significant unless the cut-off term is taken as of greater order than N, the initial number of atoms. A preferred method, then, is to take advantage of the linearity of the original equations and use *combinatorics* [see V, Chapter 15] with the elementary one initial atom solutions for these problems in the absence of production terms. It may finally be remarked that the backward stochastic equations are not generally mathematically linear in the generating function but have an alternative set of relations that express the combination of independent events inherent in the whole approach to this stochastic problem.

12.3 Particle transport theory

In the previous section, nuclear reactions were discussed in the context of time as the single independent variable. More generally, of course, reactions involve a distribution over space. In this section we consider particle transport theory in what is itself a limited framework but extending to the description of the particles as a function of their position and velocity as well as time. The major limitations we impose are (i) that the particles are neutral and suffer only nuclear reactions with the material of the region of space described, where force fields (e.g. electromagnetism, gravity, etc.) are excluded, and (ii) that the concentration of particles is so low (in comparison with the material) that particle–particle interactions are negligible. Thus this limited form of particle transport theory is appropriate for terrestrial considerations of neutrons and gamma photons, both of considerable significance of course in nuclear reactor theory.

The full description of a *stochastic* basis of such a model, involving say N neutrons, would involve each in three space and three velocity coordinates as well as time, i.e. a $7N$-dimensional *phase space*, and is evidently a challenging problem to which must be coupled the necessary equations to describe the effect of the neutrons on the material properties. The challenge cannot be met in any generality and the development of such stochastic problems can be regarded as a specialist interest. We will, however, refer later to an important branch of stochastic solution methods for ostensibly deterministic problems.

The deterministic form of the problem (more properly the reduction to a problem involving the behaviour in the mean) is even so in itself a formidable challenge that has produced many elegant developments in analysis and equally significant (if more of the nature of brute force) developments in digital computers, having higher process speeds and larger fast memories. Whether or not the resulting attitude is a rationalization, it can be said that the attitude to the full problem is that entire generality is not necessary—that problems naturally divide into specialized and local problems needing the full treatment of the transport fundamentals and into problems of wider scope but where a coarser approximation is acceptable. In the same spirit, we shall in this section develop the particle transport equations and those solutions which claim in some sense at least to be 'exact', while in Section 12.4 will be given a group of approximate solution methods of which the diffusion approximation is perhaps in practice the most important.

The restriction, well supported physically, to neutron–material reaction or photon–material reaction without particle–particle reaction has a major advantage in that the independence of the particles (at least in the absence of coupling phenomena affecting material properties) retains the linearity of the equations that generalize the discussions of the previous sections. We may therefore anticipate the use of a number of classical techniques of mathematical physics based on the linearity of the equations studied.

While it is perhaps pedagogically advisable to make our present distinction between the transport equation and the diffusion approximation, it must also be said that a number of the methods of attack in one case have their counterpart in the other and in that sense the distinction is artificial, even if common and practical. Figure 12.5 is meant to compensate for any deleterious effects of such a division by showing in a schematic form how various approaches to the formulation and solution of the particle transport problem are interrelated.

Integral formulation

There are two main approaches to formulating a transport equation: the *integro-differential* [see III, §10.6.2], or *Boltzmann equation* and the *integral equation*. Naturally the two forms in the final analysis are equivalent but they show different characteristics in respect of methods of solution and approximation. The integral form has a property of relating the behaviour at any point in space as well as velocity to every other; it therefore tends to smooth out errors introduced by approximations and is readily adapted (although not necessarily quickly solved) to numerical integration and therefore computer methods.

To describe the method we introduce the notation of the neutron flux and the macrocopic cross-sections, both appropriate to the neutron transport problem. A distinction must be made between the neutron flux ϕ which is defined as the density of neutrons (per unit of volume and per unit of velocity) at position $\mathbf{r}$ and velocity $\mathbf{v}$ multiplied by the neutron speed v:

Neutron flux: $\qquad\qquad\qquad \phi = vn(\mathbf{r}, \mathbf{v}, t),$

and the gamma or photon flux, defined as the comparable photon density times the photon velocity (a further concept in gamma ray problems is the energy flux where the photon flux is further multiplied by the energy of the photons:

Photon flux: $\qquad\qquad\qquad \boldsymbol{\phi} = \mathbf{v}n(\mathbf{r}, \mathbf{v}, t).$

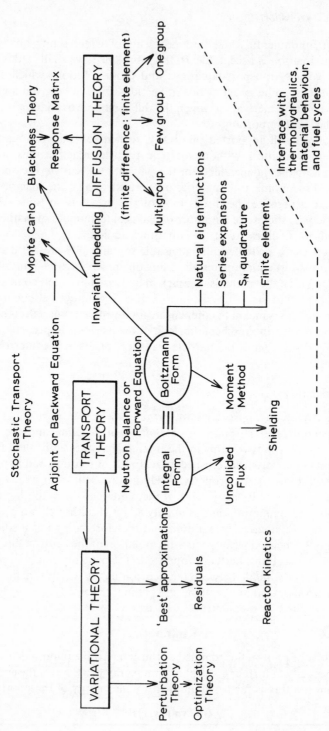

Figure 12.5: A transport theory schematic.

The important distinction to be made is that the photon flux is a true vector concept, photons per unit of time per unit of area *in a given direction*, while the neutron flux is a scalar concept and is distinct from the current density of neutrons, $j(r, v, t)$ which is the analogue of the photon flux. The use of the term 'flux' in neutron work is anomalous.

Second, the material properties describing the interaction potential of the material with neutrons are represented by a *macroscopic* cross-section Σ, itself a product of the nuclear volumetric densities of the material atoms and the *microscopic* cross-sections or target areas per atom for the reaction concerned. The product of Σ with the neutron flux ϕ is then an expression for the *reaction rate density* (RRD) of neutrons in a given field ϕ with a given material:

$$\text{RRD} = \Sigma\phi. \tag{12.18}$$

This is more exactly comparable with the customary gamma ray expression where μ plays the same role as Σ but is referred to as the (linear) *attenuation coefficient*.

Consider now the passage of a neutron without interaction across a distance r through a material of constant Σ. In analogy with the decay of a radioactive atom with time, the removal probability for such a particle to have (some) interaction in this distance is $1 - \exp(-\Sigma r)$. For a photon description, the analogous expression is $1 - \exp(-\mu r)$. Correspondingly, the mean uncollided flux, given a flux at the origin, is the streamed term $\phi e^{-\Sigma r}$ or $\phi e^{-\mu r}$. The integral form of the transport equation is then based on a summation of all the sources of particles which can stream towards a point weighted with their non-collision probability and contribute to the flux of interest. The extension to non-uniform properties by integrating along the path length is straightforward.

The complication and disadvantage of the integral form arises when the sources of particles are considered, particularly if these sources are not isotropic. Indeed, these complications are so severe that we shall not pursue the general form here. We remark, however, that the exponential removal term plays a fundamental role in many aspects of transport solutions. Second, with sufficiently simplified cases, the *kernal* of the *integral* may admit solution by the classic methods for *integral equations* and has had particular relevance in problems involving energy but not spatial dependence (where anistropy is less constricting). These solutions may again be analytic or numerically based [see III, Chapter 10].

An interesting alternative representation of transport theory that has links on the one hand with the stochastic problem, as a probability expression, and on the other hand with Bellman's *dynamic programming* (or the Hamilton–Jacobi formulation of a variational principle [see IV, §16.3]) is invariant embedding (Wing, 1962). This seeks to determine the reflection and transmission probabilities for particles and links them together in a closed equation. The major characteristic of this form is that it is non-linear, in strong contrast therefore to the integral and integro-differential descriptions of the same phenomenon. (In this, one is reminded of the adjoint or backward stochastic equation, and indeed the invariant embedding approach can be used in the stochastic formulation.) This apparent loss of simplicity is the cost of some gain in ease of solution which can best be described in terms of the differential equation analogue: that a second-order linear ordinary differential equation can always be reduced to a first-order equation at the cost of an introduction of non-linearity (*Riccati equation*) [see IV, §7.10.8].

Invarient embedding is not widely used in practice but, at a diffusion level, the concepts are utilized in practical form as response matrix theory (see 12.4).

Boltzmann formulation

The integro-differential or Boltzmann form of the equation will be developed and discussed in more specific terms as the more generally useful form. (Although it is customary to call our equation the Boltzmann transport equation, it is, strictly speaking, the special or linear form of the equation historically associated with a 'foreign' or contaminant gas at sufficiently low densities not to suffer self-scattering.) Consider, first, the form of the scattering function, $f_s(\mathbf{r}, \mathbf{v} \to \mathbf{v}', t)$, describing the probability density for neutrons (at time t, position $\mathbf{r}$, velocity $\mathbf{v}$) scattering and being distributed on exit from scattering with velocity $\mathbf{v}' = v'\mathbf{\Omega}$. As such it is normalized by an integration over all velocities (i.e. all speeds and solid angle 4π) as

$$\int f_s(\mathbf{v} \to \mathbf{v}') \, dv' \, d\Omega' = 1. \tag{12.19}$$

There are of course certain kinematic relations which constrain f_s, in particular the *conservation of kinetic energy* (for elastic scattering) and *conservation of momentum* (for all cases) between the scattered neutron (or photon) and the target atom (or electron).

One elementary model supposes that there is no speed nor energy change on scattering and indeed that the emitted neutrons are isotropically emitted from scattering, uniformly distributed over the 4π steradians. Then $f_s(\mathbf{v} \to \mathbf{v}') \to \delta(v' - v)/4\pi$ using the *Dirac delta distribution* notation. (This model may be extended in scope to neutrons in equilibrium under a Maxwell or thermal distribution and so is of some greater applicability.) It will be true in many cases that the material itself is isotropic, having no preferred orientation at a point $\mathbf{r}$, so that the scattering function depends only on the angle between the velocities and on their individual speeds. With μ also conventionally representing the cosine of the angle of scattering, this means $f_s \to f_s(v, v', \mu = \mathbf{\Omega} \cdot \mathbf{\Omega}')$ if $\mathbf{v} = v\mathbf{\Omega}$.

The total reactions that may be suffered by a particle at a point can be divided into the categories *scattering* and *absorption*. The former can be divided into elastic and inelastic scattering, the latter into capture and fission. Correspondingly we write

$$\Sigma = \Sigma^s + \Sigma^a = \Sigma^{es} + \Sigma^{is} + \Sigma^c + \Sigma^f. \tag{12.20}$$

Scattering and fission both contribute to the production of particles in a particle balance with terms proportional to the neutron density or photon density (or equivalent fluxes); these are therefore homogeneous terms in the balance equation. There may also be an independent and inhomogeneous source of particles, described by a source density $S(\mathbf{r}, \mathbf{v}, t)$ which will be a further coefficient of the equation.

Neutrons induced by fission will also have some distribution function, f_f, over $\mathbf{v}'$. In addition, the mean number of neutrons from fission is generally greater than one so that an equivalent $f_f(\mathbf{v} \to \mathbf{v}')$ is not normalized to unity. It is conventional to refer to the mean number of neutrons from fission as $\nu(\mathbf{v})$ and a further distribution function $\chi(\mathbf{v} \to \mathbf{v}')$ which is normalized to unity. The equivalent photon problem might be pair production.

Finally, we consider the streaming term which represents in the integro-differential equation that which was represented by the exponential related term in integral theory. Taking the *divergence operator* to act upon the spatial and not the velocity variables, this term is simply $-\nabla \cdot \mathbf{v} n = -\mathbf{\Omega} \cdot \nabla \phi$ (a loss term in the balance associated with losses from local reactions given by $-\Sigma \phi$) [see V, §13.4].

Combining these terms in a neutron balance (the photon balance equation is similar) gives

$$\frac{1}{v}\frac{\partial \phi}{\partial t} = \frac{\partial n}{\partial t} = \mathbf{\Omega} \cdot \nabla \phi - \Sigma \phi + \int \Sigma^s(\mathbf{v}') f_s(\mathbf{v}' \to \mathbf{v}) \phi(\mathbf{v}')\, dv'\, d\Omega' + S$$

$$+ \int \nu(\mathbf{v}') \Sigma^f(\mathbf{v}') \chi(\mathbf{v}' \to \mathbf{v}) \phi(\mathbf{v}')\, dv'\, d\Omega', \tag{12.21}$$

to which must be associated suitable initial conditions to describe some initial configuration. If a finite region of space is to be considered at any rate, there must be suitable boundary conditions to describe neutron sources from outside the domain of the equation if it is to be well posed. It may be remarked that despite its complexity, the transport equation is relatively easily formulated with few and well-understood assumptions. In comparison, other approximations, although much more easily solved, are based upon approximations, simplifications and assumptions whose consequences for accuracy of solution are much more difficult to understand. It is a major role of transport theory to provide a few specialized and benchmark solutions against which approximate methods can be unequivocally tested. In this spirit we pursue some major specializations of the transport equation.

Steady-state one-group transport theory

Even with the assumptions already made, the Boltzmann transport equation offers considerable difficulty. In the next section we study the diffusion approximation. Here we consider a number of further specializations and the techniques they admit in a treatment where the direction of particle travel is a major concern. It will indeed help 'to see the wood for the trees' if we omit the independent source, concentrate on a steady-state problem, assume isotropic scattering in our coordinate system and make the further assumptions that the material properties—and hence coefficients of the equations—are uniform in space as well as constant in time. Finally, accepting the one-speed model and a one-spatial-dimensional problem, the former leads to a reduction of the equation to

$$0 = -\mathbf{\Omega} \cdot \nabla \phi - \Sigma \phi + \frac{c\Sigma}{4\pi} \int_{4\pi} \phi(\mathbf{\Omega} \cdot \mathbf{\Omega}') \, d\Omega', \qquad (12.22)$$

where c, the multiplicity, represents both the neutron re-emitted from scattering and $v - 1$ additional neutrons from fission: $c\Sigma = \Sigma^s + v\Sigma^f$. The one-dimensional (in space) assumption permits an integration over azimuthal symmetry [see V, §2.1.5], introducing a factor of 2π and results in

$$0 = -\mu \frac{\partial \phi(x, \mu)}{\partial x} - \Sigma \phi + \frac{c}{2} \Sigma \int_{-1}^{1} \phi(x, \mu') \, d\mu'. \qquad (12.23)$$

Such an equation has to be supplemented by suitable spatial boundary conditions in the x direction. Although it is only of first order as a differential equation, boundary conditions are generally derived from the consideration of two faces, as shown in Figure 12.6. If the problem is one of the self-sustaining neutron flux in the slab ($c > 1$), then there are no neutrons incident from outside

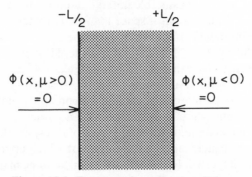

Figure 12.6: Transport boundary conditions.

the slab. This means that $\phi(\mu > 0) = 0$ for the right-directed neutrons at the left-hand face, and vice versa. Thus the apparently first-order equation has boundary conditions of which half are applied at one face and half at the other, and is therefore more of the nature of a conventional second-order equation leading to *elliptical* solutions [see III, §9.5, and IV, §8.5]. Our problem might be to find the critical size L as a function of the multiplicity available, c. This is generally referred to as a *critical-size problem*. For simplicity, measure L in units of the mean free path, $1/\Sigma$.

To solve such an equation one takes note of the linearity in ϕ and might suppose that an *eigenfunction–eigenvalue problem* is at hand or at any rate the opportunity to employ the *Fourier-type expansion function* techniques of mathematical physics and, in particular, *orthogonal series* [see IV, §7.8]. The integral term, however, is unusual and provides for the special nature of the transport eigenfunctions which will now be shown to lead to *singular integral equation theory* [see III, §10.2.3] (Case and Zweifel, 1967). That is to say, the *eigenvalues* will not only have discrete components but will also have a *continuous contribution* to the spectrum with correspondingly complicated *eigenfunctions*.

To demonstrate this, suppose that we may write the typical eigenfunction form via a *separation of variables* [see IV, §8.2], putting

$$\phi(x, \mu) = \phi_v(\mu) \, e^{-x/v}$$

(having taken note of the underlying exponential removal behaviour). Then

$$(\mu - v)\phi_v(\mu) = \frac{cv}{2} \int_{-1}^{1} \phi_v(\mu') \, d\mu' = \frac{cv}{2}, \qquad (12.24)$$

employing an admissible arbitrary normalization of the eigenfunction $\phi_v(\mu)$. Then

$$\phi_v(\mu) = \frac{cv}{2} \frac{1}{\mu - v}; \qquad v \neq \mu. \qquad (12.25)$$

However, consider also the possibility that $v = \mu$, which indeed admits a Dirac delta distribution component vanishing outside the range for which this is feasible:

$$\phi_v(\mu) = \frac{cv}{2} P \frac{1}{\mu - v} + \lambda(v) \, \delta(v - \mu), \qquad (12.26)$$

where P indicates the *principal value* of the integral [see IV, §9.10.2] when the integrand is singular (i.e. is defined to omit a contribution from the singularity) and λ is a so-far arbitrary normalization.

Reapplying the normalization, however, gives eigenvalues in the first case:

(i) $v \neq \mu$:

$$\Lambda(v) \equiv 1 - \frac{cv}{2} \int_{-1}^{1} \frac{d\mu}{v - \mu} = 1 - cv \tanh^{-1}\left(\frac{1}{v}\right) = 0, \qquad (12.27)$$

from which two discrete eigenvalues $\pm v_0$, $|R_e v_0| > 0$ emerge. In the second case:

(ii) $-1 \leqslant v \leqslant 1$, from which we obtain the so-far arbitrary function $\lambda(v)$ as

$$\lambda(v) = 1 - \frac{cv}{2} P \int_{-1}^{1} \frac{d\mu}{v - \mu} = 1 - cv \tanh^{-1} v. \qquad (12.28)$$

It may be shown that $\lambda(v)$ and $\Lambda(v)$ are related by *Plemelj formulae* (Davis, 1978; Kanwal, 1971); e.g.

$$\lambda(v) = \tfrac{1}{2}[\Lambda^{+}(v) + \Lambda^{-}(v)]. \qquad (12.29)$$

This set of eigenfunctions is complete and sufficient to satisfy the boundary conditions such as we have put forward, at least within the arbitrary overall normalization corresponding to the arbitrary power level in this linear, critical problem. The solutions clearly show the exponential dominant discrete components that will hold within a wide slab and which relate to diffusion theory solutions, as well as the continuous spectrum which provides for a local decaying solution to match the boundary or other interface conditions. Amongst other things, it indicates that diffusion theory should be adequate for large systems, well away (a few mean free paths) from localized strong absorbers or sources. However, this idea must be tempered as being appropriate to the criticality case from which it has been derived and should not suggest that in inhomogeneous cases (especially shielding problems) diffusion theory is valid.

The eigenfunction set may be shown to be not only complete but to have *orthogonal* properties over the full range of μ (-1 to $+1$) and the half-range (0 to ± 1) so that the *expansion coefficients* may be determined [see IV, §20.4]. This determination leads to an *integral equation* of straightforward form for the expansion coefficients which may readily be evaluated by quadrature [see III, Chapter 7], usually at small cost on a digital computer. It is seen that the use of the natural eigenfunctions and subsequent analysis have eliminated the differential equation and led to a replacement of the singular integral equation with this ordinary integral equation. This represents a primary characteristic of the method compared to the alternatives discussed next.

The resulting equations for the expansion coefficients turn out to be very similar in form to the classical astrophysics methods of solving analogous particle transport problems—the methods of *Weiner–Hopf* (Davis, 1978;

Hochstadt, 1973; Williams, 1973) [see §11.4] and *Chandrasekar*. Here an integral formulation is made and, in the classical grey-body problem, a solution constructed with appropriate *analytical properties*. There is a view that the Weiner–Hopf method is just as rewarding as the singular eigenfunction method. We take the view, however, that an eigenfunction expansion is certainly more in the mainstream of mathematical physics and serves to elucidate the fundamental properties of the transport problem in a far more transparent way (McCormick and Kuscer, 1973).

Angular expansion series methods

The singular integral method is difficult to apply to more realistic problems, involving multiregions, multienergy and anisotropic scattering. A more common treatment of the transport equation is to use an expansion function technique employing suitable and often orthogonal series [see IV, Chapter 20], with a truncation that is dictated by a compromise between desired accuracy and resulting tedium. The following table lists some of these approaches:

Singular natural eigenfunctions
Spherical harmonics P_N
Double P_N
Fourier transformed B_L
Moment expansion
Discrete ordinates S_N
Finite element expansions

In this table, the first entry of the singular eigenfunction expansion method has been singled out for its theoretical elegance as the natural mode method for the problem. Remaining entries are the methods of attacking the angular dependence which, if not potentially better, are certainly more commonly used in practice. The first of these, the *spherical harmonics method*, is widely used and deserves some detailed discussion (see Henry, 1975; Meghreblian and Holmes, 1960).

Spherical harmonics makes use of the full set of eigenfunctions arising from the *solution of Laplace's equation in spherical coordinates* [see IV, §7.8.2 and Example 8.2.5]. The orthogonality properties of such series are utilized and, as they are used in comparable geometries, many simplifications arise. The expansion is carried out both for the principal series to represent the angular flux (with the lowest term representing the important scalar flux integrated over all angles) and, as far as it is necessary, for the angular dependence of the scattering distribution function $f_s(\mu)$. The principal components are represented in the spherical harmonics as typified for the one-group flux:

$$\phi(x, \mu) = \sum_{n=0}^{\infty} P_n(\mu)g_n(x), \tag{12.30}$$

which we now apply to our one-group isotropic problem for comparison with the singular eigenfunction expansion method.

The usual aspects of a series expansion apply. That is, the series may in principle be complete (although the *Gibbs phenomenon* [see IV, §20.6.4] may be present) but the expansion coefficients $g_n(x)$ may need to be extended to many terms to give adequate *convergence*. If practical necessity dictates the use of a finite number of terms, the problem of closure arises, i.e. what to do about the term in g_{N+1} occurring in the highest equation, for g_N.

The spherical harmonics method is attractive because its close relation with the geometry of the problem at hand makes the lowest terms, P_0 and P_1, relate to the fundamental behaviour of the true flux, at least at sufficient distances from sources and sinks, and indeed these components are readily identified with diffusion theory. (In a time-dependent expansion, however, we obtain the more physical *Telegrapher's* equation rather than time-dependent diffusion theory from a P_1 expansion; see Meghreblian and Holmes, 1960.)

The assumed fundamental solution is substituted into the equation together with the same $\exp(-x/\nu)$ assumption of the eigenfunction approach and use is made of the orthogonal properties of the P_n or *Legendre polynomials* [see IV, §10.3] to obtain a series of ordinary differential equations for the $g_n(x)$. By further use of the recurrence relations for these polynomials it is possible to reduce these to a *three-banded* matrix form in the present instance or a *sparse-banded* [see III, §4.12] matrix in general. That is, we now obtain

$$\frac{n+1}{2n+1} \frac{dg_{n+1}}{dx} + \frac{n}{2n+1} \frac{dg_{n-1}}{dx} = (c\delta_{n0} - 1)g_n; \qquad n \geqslant 0, \tag{12.31}$$

and, associated with this, an equation to determine the corresponding ν_n, giving $N + 1$ solutions after truncation to order N, that can be shown to approximate the two discrete and the infinite set of continuous eigenvalues of the singular or natural harmonics. Truncation at $N = 1$ indeed leads as anticipated to a diffusion-like system of equations.

For higher order problems, especially involving anisotropic scattering as well as regionwise property changes, numerical methods for the solution of these equations must be turned to and convergence is found not to be rapid. Boundary conditions also have to be represented and in the general case we do not have the spherically symmetric boundary and interface conditions which would lend themselves to a spherical harmonics treatment, again a disadvantage of the method.

Variants of the spherical harmonics method are (i) the double spherical harmonics method, where separate expansions over hemispheres are attempted suitable for plane interfaces, and (ii) the B_L method which utilizes a *complex Fourier series* for the spatial dependency [see IV, §20.5.6].

The *method of moments* (Goldstein, 1959), i.e. integrals of the flux weighted with μ^n, is a variant of spherical harmonics suitable for an infinite system (and hence shielding) theory in which the attraction is the elimination of the spatial differential equations in favour of a set of finite integral equations which are moreover *closed*, i.e. no terminating approximation is necessary in a solution to finite order. Its major disadvantage is the limitation to the infinite system.

The major alternative to a series expansion method for dealing with the angular dependence of transport theory is a direct attack on the integral term of the Boltzmann equation. The most well-established method is Carlson's S_N method (Engle, 1963) which is essentially a *quadrature* [see III, §7.3] scheme of classical form for the integral over N points, usually chosen as 2^n, i.e. $N = 2, 4, 8,$ 16, etc., leading to a corresponding number of simultaneous and now algebraic equations for the flux at discrete angles. The angular flux has to be interpolated over these discrete solutions [see III, §2.3] and the total flux obtained by integration over them. This approach leads to a technical difficulty of the method that has plagued its applications to shielding theory, i.e. the ray effect where the lumping leads to anomalous discrete fluxes of a mathematical and non-physical nature.

A more recent comer to the field of transport theory which may yet replace S_N theory is the *method of finite elements* (Chapter 15), developed in structural mechanics in particular (see Chapter 10) and having application also to the solution in the diffusion approximation. The method has the attraction of a formulation that is variationally based and, at least for one-group energy theory, admits to precise statements on the sign of the error involved (Lewis, 1981). The general philosophy of the finite element method will be referred to again in Section 12.4.

Even when we limit consideration to direct methods capable, in principle, of yielding an exact answer or at least an answer as accurate as desired, there are seen to be a variety of methods proposed to solve the neutron transport equation; new ones are continuously being advanced (or hope springs eternal in the supervisor's breast). In practice, however, the realistic application to realistic problems is bound to involve extensive digital computing and the investment in the program package of established methods makes it exceedingly difficult for an alternative to be developed to the stage where it is tried and economically competitive. The S_N method (in the form of ANISN perhaps) has come to a dominant position in the field of transport methods capable of determining angular fluxes with required precision (Engle, 1963).

Experimental analysis

Before turning to the approximate methods of a lower order for solving the transport equation, it is opportune to make a few remarks on the analytical and mathematical methods appropriate to the problem of measuring the interpreting experimental values of such things as cross-sections. Of course, many of the problems of experimental analysis are of wider standing than nuclear engineering; they call for *probability and statistics*. It has become appreciated that the question of the *correlation* of data [see II, §9.8] and the influence of the inherent errors on the derived information—what is reported is rarely a primary measurement—has considerable influence on the accuracy to be attributed to either a measurement or a subsequent calculation. Interpretations of a *correlation matrix* are frequently desirable. A problem that again is not peculiar to nuclear engineering, but shows up clearly in the field, is the inversion of a few experimental results to yield a 'best-fitted' representation of the data involved. This may be referred to as 'stripping' and in mathematical terms is often a matter of *inverting a matrix* [see III, §4.5]—perhaps a *matrix operator*—to obtain a best representation of the desired matrix from a limited number of known vectors. There is scope for the design of experiments to avoid unnecessary difficulties associated wtih *ill-conditioned equations* [see III, §§1.7 and 6.1.5] militating against such invertion.

12.4 The diffusion and other approximations

The principal approximation theory used for the widest range of applications of nuclear reactor theory (with the exception of shielding theory) is the diffusion theory approximation, a second-order differential equation form [see IV, §8.1] which is capable of reproducing the mean behaviour of neutrons or other particles at sufficient distance from discontinuities in the geometry and properties of the system. The exception of shielding theory arises precisely because the concern of shielding theory is not with the average particle but rather with the exceptional particle that, despite the imposition of the shielding material, has still managed to penetrate to regions where it is not wanted. Before turning to diffusion theory as such, it will be helpful to give a taxonomy of the types of problem, a classification that guides the appropriate choice of method of solution. This is shown in Figure 12.7.

The first division into homogeneous and inhomogeneous problems is a mathematical one which in part corresponds to the critical, source-free self-sustaining system or a system (complete with shielding) with a known source and a requirement to determine a result in a specific region. The intermediate class of problems is related to the determination of some secondary characteristic in a critical system; the latter implies that the power level is arbitrary and that any secondary characteristic of physical interest must be a normalization-

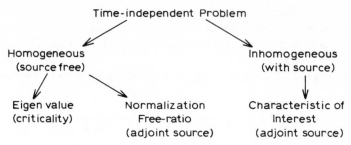

Figure 12.7: A taxonomy of reactor problems.

free ratio (a physical example is a breeding ratio in a critical reactor). It will be found that both of these problems (i.e. excluding criticality as such) introduce an adjoint source into the equation adjoint to the neutron balance or forward equation.

The diffusion approximation

A direct approach to the diffusion approximation for a one-group model can be given as follows. Consider the vector equation of conservation representing the balance of neutron density $n(\mathbf{r}, t)$ in the presence of a neutron current $\mathbf{j}(\mathbf{r}, t)$ [see IV, §5.3] (a *vector*) streaming through the medium. We may write directly

$$\frac{\partial n}{\partial t} = -\mathbf{V} \cdot \mathbf{j} + S - \Sigma^a v n + v \Sigma^f v n, \tag{12.32}$$

where the final two terms give the absorption and fission production rate of neutrons (scattering is no longer explicitly represented since no change of either direction of speed is involved in this model) and $S(\mathbf{r}, t)$ is the independent neutron source density rate. If the model is to be extended to multienergy neutron groups, this may be done in two ways, both utilizing a term analogous to the fission term to represent the scattering between groups as well as fission induced in one group but providing neutrons into another. These two methods will be the continuous approach where the additional source is an integral term over all the energies concerned or the discrete case where the additional source is a sum over a number (finite or infinite) of energy groups.

This equation for the conservation of neutrons (exact within the limitations of the model) is coupled with some expression for the neutron current $\mathbf{j}$, and it is this that introduces the major nature of the diffusion approximation. The simplest statement is to accept the approximation of Fick's 'law' which may be written for $\mathbf{j}$ in terms of the gradient of the neutron density or

$$\mathbf{j} = -D_0 \nabla n = -D \nabla \phi, \tag{12.33}$$

where D_0, D are alternative forms of the diffusion coefficient. It is the role of transport theory to provide expressions for the diffusion coefficient D and other corrections in order to make, it is to be hoped, the diffusion approximation into a worthwhile device.

Substituting (12.33) into (12.32) gives the one-energy group diffusion theory equation:

$$\frac{1}{v}\frac{\partial \phi}{\partial t} = \mathbf{V}\cdot D\mathbf{V}\phi + S - \Sigma^a\phi + \nu\Sigma^f\phi. \tag{12.34}$$

To this must be added appropriate boundary conditions and interface conditions (as well as initial conditions), again fabricated from solutions to specialized transport equations.

Evidently the new equation is a *partial differential equation of linear form of order two* in view of the term $\mathbf{V}\cdot D\mathbf{V}$. In special circumstances it reduces to well-known forms of mathematical physics [see IV, Chapter 13], particularly the *diffusion equation*, the heat conduction (Fourier) equation, the *Poisson equation* and conceivably the *Laplace equation* [see IV, §8.1]. In general, it has additional terms however. As an example, suppose (i) D is uniform, (ii) there is no independent source S and (iii) write $\nu\Sigma^f = \eta\Sigma^a$, $D/\Sigma^a = L^2$ (thermal diffusion area). Then we have

$$\frac{1}{vD}\frac{\partial \phi}{\partial t} = \nabla^2\phi + \frac{\eta - 1}{L^2}\phi \tag{12.35}$$

and a relationship with both the *heat conduction* equation and the *Helmholtz equation* becomes obvious.

One method of solution, then, for relatively simple geometries is the *separation of variables* [see IV, §8.2] followed by *Fourier series expansions* [see IV, Chapter 20] as necessary to match the boundary conditions. Where these are the 'natural' boundary conditions arising in criticality problems, then generally the lowest mode only of the Fourier expansion is needed in simple geometries. If more complicated geometries are considered (the fully reflected cube, for example) then complete separation is not possible but a Fourier series (followed by truncation) solution can be developed. The equation can be separated in some twenty-one variant geometries but, alas, for practical reasons, these generally do not correspond with the way reactors are built.

In respect of the boundary conditions, the most immediately physical conditions are those of the third or *mixed type*, i.e. $\lambda\mathbf{V}\phi + \phi = 0$ at the boundary while at planes of symmetry Dirichlet conditions ($\mathbf{V}\phi = 0$) will be appropriate. The mixed conditions are more difficult to apply; fortunately the size of $1/\lambda$ in relation to reactor size is often so small that Neumann ($\phi = 0$) conditions can be used instead—use of the so-called *extrapolation distance*.

Finite difference versus finite element

The classical methods are suitable only for simple geometries, e.g. a single uniform core region surrounded by a single uniform reflector region. These at least serve to demonstrate the general nature of the diffusion equation and its solutions, including the relationship to the discrete eigenvalue components of transport theory. The method of the *response matrix* (Lindhal and Weiss, 1981) seeks to exploit classical diffusion theory solutions by establishing geometric relationships once and for all for a given geometry, the method becoming efficient and attractive if this geometric cell is repeated many times, as may well happen in a reactor with thousands of fuel pins, say.

In the absence of such patterns and when all aspects of *invariance* of the equations and their solutions have been exploited (*rotational, translational invariance*, etc.), one is forced back into numerical solution procedures for partial differential equations which here are of dominantly *elliptical* [see III, §9.5], i.e. two-point, form. That is, a marching solution from a well-defined 'initial' boundary is not feasible. Two major techniques are available: *finite difference* and finite element, the latter being a relative newcomer to reactor theory. The use of one or the other seems to be a necessity in any practical problem of detail where the diffusion of neutrons is coupled with a thermo-hydraulic description of the reactor. Indeed, detailed three-dimensional solutions to such a full problem are probably still outside the capacity of normal engineering studies.

Finite differences of various orders, developed from the original *relaxation methods* (Allen and Southwell, 1955; Southwell, 1940) [see §7.2], are widely used and have the well-known advantages of *narrow banding* and *sparse matrices* (for $\nabla \cdot D\nabla$, although the matrices representing other terms in the model may not be sparse) [see III, §4.12]. *Acceleration techniques* [see III, §3.43] improve performance and convergence is not generally a problem for the time-independent problems (but see the next section for time dependence).

Finite element methods (Hansen and Kang, 1975) are being adopted into reactor theory having been developed in disciplines such as civil and structural engineering (Chapter 10). They, too, may lead to *sparse-banded matrices* but they have fundamental properties that make them additionally attractive, particularly in the present linear problem.

(i) The finite element methods of Chapter 15 have a *variational* basis [see IV, Chapter 12] that may admit conclusions to be drawn about the sign (and magnitude) of residual errors. Variational methods in general are discussed later but we have to say this property is at its most powerful only in a one-energy group model.

(ii) Finite element methods have a flexibility of geometry that is attractive. The construction of their *meshes* can be automated (*computer-aided design methods*).

Once again, however, the invested capital of existing multigroup finite difference computer codes makes it possible that these advantages will not be realized in practice despite current development interest.

Source-sink theory

A major defect of diffusion theory is its failure (compared to an exact or relatively accurate transport solution) around material discontinuities, interfaces and external boundaries and especially therefore in practice close to highly absorbing fuel elements in thermal neutron reactors. An attractive approach, due to the Russian school of Galanin and others, is the source-sink method where the absorption (and corresponding fission source) is approximated as being concentrated at a point (point-sink) or, in view of the usual axial symmetry of a reactor, as a line-sink. The method is really a specialized aspect of integral transport theory and is particularly useful for control-rod studies since these closely approximate strong absorbers in a line-sink (Marchuk, 1959).

Variational and perturbation methods

In one sense, *variational theory* is not an approximation but rather a concise description of the governing *Euler–Langrange equations* [see IV, Chapter 12] of a problem; it may well be a goal of a theoretician to provide a variational principle from which all consequences of the model can be derived (Lewins, 1965). It is a peculiarity of the neutron problem where energy variations are taken into account that there is an inherent physical asymmetry; the process of slowing down neutrons by scattering is physically distinct from the process of 'speeding up' neutrons by producing them in fission. It follows that the inherently non-symmetric problem is not *self-adjoint* and that the second or adjoint stationary condition, or equation, of a variational method is distinct from the first (forward or state equation). This may be seen mathematically in the non-self-adjointness of the scattering distribution function by comparing the two terms

$$\int f(\mathbf{v}' \to \mathbf{v}) v(\mathbf{v}') \Sigma^{\mathrm{f}}(\mathbf{v}') \phi(\mathbf{v}') \, dv' \, d\Omega', \tag{12.36}$$

which is not essentially the same as

$$\int f(\mathbf{v} \to \mathbf{v}') v(\mathbf{v}) \Sigma^{\mathrm{f}}(\mathbf{v}) \phi^{+}(\mathbf{v}') \, dv' \, d\Omega' = v\Sigma^{\mathrm{f}}(\mathbf{v}) \int f(\mathbf{v} \to \mathbf{v}') \phi^{+}(\mathbf{v}') \, dv' \, d\Omega'. \tag{12.37}$$

(The kernal of the one-group integral equation is not *self-adjoint*, but can be symmetrized by elementary means, an anticipated consequence of the existence of a Maxwellian equilibrium. In either form, the one-group equations in the steady state can be taken as self-adjoint—but see the desirable distinction to be drawn.)

This essential difference allowed, indeed forced, the adjoint equation of nuclear reactor theory to be given an independent physical interpretation, referred to as the 'importance' concept, in which the variational method employs not a quadratic functional in the density but rather a bi-linear functional in the density and its associated adjoint function. (Confusingly, when the flux formulation of the neutron equations is used, the associated adjoint function is actually adjoint to the neutron density but is called the adjoint flux.) The *importance*, n^+ say, satisfies the equation adjoint to (12.21), which has the form

$$-\frac{1}{v}\frac{\partial n^+}{\partial t} = \mathbf{\nabla}\cdot\mathbf{\Omega} n^+ - \Sigma n^+ + \Sigma^s \int f_s(\mathbf{v} \to \mathbf{v}')n^+(\mathbf{v}')\, dv'\, d\Omega'$$

$$+ v\Sigma^f \int \chi(\mathbf{v} \to \mathbf{v}')n^+(\mathbf{v}')\, dv'\, d\Omega' + H^+. \tag{12.38}$$

Here $H^+(\mathbf{r}, \mathbf{v}, t)$ is an *adjoint source* which corresponds physically to some weighting (e.g. a detector distribution), responding to neutrons in the system to give a global characterization or characteristic of interest $\bar{\bar{\mathbf{R}}}$:

$$\bar{\bar{\mathbf{R}}} = \int_{\text{time}} \int_{\text{velocity}} \int_{\text{space}} H^+ n \, dx\, dy\, dz\, dv\, d\Omega\, dt = \langle H^+ n \rangle, \tag{12.39}$$

where the notation $\langle\,\rangle$ indicates the full integration [see IV, §6.4]. Obviously different cases can be obtained by specializing the choice of H^+.

The commuting property of the adjoint operator then results in the two forms

$$\bar{\bar{\mathbf{R}}} = \langle H^+ n \rangle = \langle n^+ S \rangle, \tag{12.40}$$

and from the second it is evident that n^+ is the importance or contribution of every (source) neutron to the measurable characteristic.

It may be noted that if a problem consists of examining the effects of different sources S for a fixed detector distribution H^+, then it is more efficient to solve the adjoint equation once for n^+ and determine $\bar{\bar{\mathbf{R}}}$ by quadratures over each $\langle n^+ S \rangle$ [see III, Chapter 7] than to have to solve the neutron equation for an n corresponding to each choice of S.

It may also be pointed out that the density equation and its adjoint are the deterministic equivalents (for the mean behaviour) of the *forward* and *backward* Kolmogorov equations of the *stochastic process*; hence the equation for n may indeed be referred to as the forward equation and for n^+ as the backward equation.

The variational method can be more succinctly expressed in a matrix and vector notation over a generalized variable $\mathbf{x}$ and where we omit the explicit indication of *transposed vectors* [see I, §6.5] by using the convention that all integrated expressions are to be *inner products* [see IV, (20.6.9)]. Then we may write the forward and backward equations respectively as

$$\frac{\partial N}{\partial t} = MN + S; \qquad -\frac{\partial N^+}{\partial t} = M^*N^+ + H^+, \qquad (12.41)$$

with

$$\bar{\mathbf{R}} = \int\!\!\int N^+ S\, d\mathbf{x}\, dt = \int\!\!\int H^+ N\, d\mathbf{x}\, dt, \qquad (12.42)$$

where $M(\mathbf{x}, t)$ and $M^*(\mathbf{x}, t)$ are commuting operators under integration over all 'space' and time. This shows incidentally that the first-order time dependence introduces an inherent non self-adjointness even if $M = M^*$.

We now employ a *Lagrangian* $\bar{\mathbf{L}}$ (integral of a *Lagrange density*):

$$\bar{\mathbf{L}} = \int\!\!\int H^+ N\, d\mathbf{x}\, dt + \int\!\!\int N^+ \left(-\frac{\partial N}{\partial t} + MN + S \right) d\mathbf{x}\, dt$$

$$= \int\!\!\int N^+ S\, d\mathbf{x}\, dt + \int\!\!\int N \left(\frac{\partial N^+}{\partial t} + M^*N^+ + H^+ \right) d\mathbf{x}\, dt \rightarrow \bar{\mathbf{R}}. \quad (12.43)$$

Evidently

$$\delta\bar{\mathbf{L}}(\delta N^+),\ \delta\bar{\mathbf{L}}(\delta N) \rightarrow 0$$

for first-order variations. We therefore have a *stationary variational principle* with our forward and backward equations as *stationary conditions* and N, like N^+, is seen to be a generalized *Lagrange multiplier* [see IV, §15.1.4].

Suitable boundary conditions to justify commutation must be available but any non-natural boundary conditions can formally be absorbed into the source S or adjoint source H^* using a *Dirac delta* notation, so this is no real obstacle.

Only, however, if the following conditions are met will the problem be a true maximum (or minimum) principle, since this depends on the existence of a quadratic form in $\int\!\!\int N^2\, d\mathbf{x}\, dt$:

(i) The problem must be time independent.
(ii) $M^* = M$; i.e. a self-adjoint operator.
(iii) $H^+ = S$; a suitable problem $\bar{\mathbf{R}}$.

Such a variational formulation has two distinct major applications (Stacey, 1974):

(i) As an approximation method for computing a single character of interest $\bar{\mathbf{R}}$ (which may be adapted to eigenvalues also). The method is at its most powerful when it can be formulated as a quadratic and therefore becomes a maximum/minimum principle. In such applications, N and N^+ are coupled with such devices as the *Raleigh–Ritz procedures* to exploit the stationary conditions [see III, §4.6].

(ii) As a scheme to obtain approximate equations themselves (whose solutions can be regarded as a class of *trial functions*). Thus diffusion equations can be obtained by a variational reduction of the transport equation; few-group energy constants obtained as a variational (adjoint weighting) collapse of a multigroup problem; and finite element methods obtained as variational equations for the accepted class (polynomial, etc.) of trial functions (see Chapter 15).

In another viewpoint, the variational method is the father of *perturbation methods* in reactor physics and grandfather of *optimization methods* (including *Pontryagin's optimal control*; see Section 14.11). Since it is difficult to solve the equations for other than specialized cases, it is natural to seek to represent a problem as a perturbation from a solvable specialized case. To first order, for a specified characteristic of interest (i.e. specified H^+) the perturbation in $\bar{\mathbf{R}}$ can be formulated by the quadrature,

$$\delta\bar{\mathbf{R}} = \int\int (N\,\delta H^+ + N^+\,\delta M\,N + N^+\,\delta S)\,d\mathbf{x}\,dt, \qquad (12.44)$$

where N^+, N are taken from the unperturbed solution and where δH^+, δM, δS are the perturbations of the desired problem from the solvable problem, giving $\delta\bar{\mathbf{R}}$ correct to first order.

Higher order variational and perturbation methods may be developed and at least partially extended into non-linear problems by a *recurrence relationship* (Greenspan, 1976). Specialized methods employing additional generalized forward and backward state equations enable the problem of the homogeneous ratio in a critical system to be dealt with variationally. A particular use of the adjoint weighting to reduce the complexity of equations is to derive the form of the ordinary differential equations for 'lumped' reactor kinetics and control from the transport model (see Section 12.5).

Probability methods

If we cannot solve the deterministic neutron transport equations (for a behaviour in the mean) in any generality then we certainly face an impossible task in proposing a stochastic transport theory in general terms. Thus only some very specialized results (including the derivation of the appropriate backward equations and, with more difficulty, the forward equations that justify the equations in the mean) are possible and they are too specialized to pursue here. There is a sense, however, in which probability methods are of considerable significance as further methods of approximately solving the (forward in the mean) transport equation. These range from simple removal theory where the probability of a particle suffering removal in crossing the chord of a certain geometry is evaluated and used in conjunction with other calculations through the related ideas of *response matrix theory* (Lindhal and Weiss, 1981). In this latter, the effect on one boundary of a region is determined for a variation in the 'driving' term at another boundary. If these regions are repeated often in the reactor—as may be true with some thousands of fuel pins, for example—the effort once performed may prove a sound investment in chaining up the effects throughout the whole system.

The most sophisticated use of probability theory, however, is in *Monte Carlo methods* [see III, §7.9] of determining the behaviour in the mean by probablistic simulation of the behaviour as a series of discrete particles (Moore, 1976). This makes use of *random numbers* (more probably *pseudo-random numbers* available in a computer) to judge at each moment of the particle's history what is the next event by choice, according to reaction cross-sections, etc., from the available consequences. When enough particles have been tracked (they may be considered to have 'died' on capture and a new source particle must be initiated) some estimate is available of their distribution and hence an estimate of related effects (e.g. a computation of $\bar{\mathbf{R}}$) is possible.

The major issue is how to obtain a precise estimate of the behaviour in the mean, i.e. to reduce the *variance* [see II, §9.1] of the Monte Carlo sampling technique. The method is of considerable value where three-dimensional and detailed geometries (i.e. variation of properties throughout the reactor) are involved because, once the task of obtaining enough fast core storage to provide this representation has been solved, the actual random number generation, tracking and recording ('scoring') is relatively trivial and not affected much by the geometric complexity. The Monto Carlo method is indeed the leading contender in this class of problem, gaining little from a reduction to two- or one-dimensional problems.

There are technical difficulties. To save having long computer CPU (central processor unit) running times and costs, it is desirable to *accelerate the convergence* of the computation. This may be done with the use of an adjoint weighting appropriate to the problem at hand, probably not computed by the Monte Carlo method but from some simpler approximation, even 'guessed' by an experienced analyst; here the method is called 'importance sampling'.

In criticality problems ($S = H^+ = 0$), there is a mathematical difficulty that the expected value of the desired eigenvalue as estimated from the mean behaviour will have a large, indeed unbounded, variance. This is a purely mathematical problem since a real physical problem is never exactly linear. Nevertheless, the mathematical consequences require renormalization to diminish the effect in applications to criticality studies and related homogeneous ratios in critical systems.

In inhomogeneous problems where the characteristic $\bar{\mathbf{R}}$ (or rather H^+) is localized (e.g. a shielding problem where the concern is for the rare neutron penetrating the shield rather than the behaviour in the mean) the problem of economizing in computer time is extreme and calls for such devices as splitting, Russian roulette, etc., as well as 'importance sampling'.

Conclusion

In this discussion we have concentrated on the transport of neutrons or photons and the approximate methods of solving the transport equation or its simplifications, chiefly as a steady-state problem. The technical demands (and indeed the economic and social demands) of nuclear power engineering have.put great pressure on analysts and computer manufacturers to meet this challenge; it may indeed be claimed that a major impetus for *mainframe computer* development in the 'fifties came from such challenges.

It must also be stressed that there are many other facets to the problem of describing the behaviour of a nuclear reactor, its associated equipment and environs. The distribution of neutrons and photons is a minority concern overall, although a vital one, and considerations of thermohydraulics (with its inherent non-linearities), stress behaviour of components and materials changes induced must also be treated.

The field has borrowed techniques from classical mathematical physics and analysis, and, more recently, finite element theory. In return it has given to other disciplines its development of an understanding of particle transport theory— so evidently a discrete and continuous problem at the same time—that has in turn made a range of techniques available in other disciplines as well as providing a major source of pressure to improve digital computer capabilities.

12.5 Reactor kinetics and control

Stochastic description and reactor noise studies

The description of the multiplication chain as a *stochastic process* [see II, Chapter 18] has a limited role to play in reactor dynamics; any detailed modelling including spatial and energy dependence quickly succumbs to the difficulties of solution. The thought is that measurements of the reactor, or more generally system behaviour, will have an *information content* inherent not only in the *average* value of the parameters observed but in their *fluctuations*. Such fluctuations, expressed via the *variance* as well as the *mean* [see II, §§8.1 and 9.1], may serve, for example, to presage a development of an operating abnormality. In that the measurement required may be taken continuously and on-line, i.e. without having to divert the operation of the system for experimental purposes, there are valuable potential advantages in such methods of what is commonly called 'noise' analysis, although it has to be said that the promise is not fully realized in practice as yet (Williams, 1974).

The low-power stochastic equations with time dependence in the presence of a fission multiplying chain are readily derived and, in the generating function notation of Section 12.2, may be written as

$$\frac{\partial G}{\partial t} = \left[\frac{g(x) - x}{\tau_f} + \frac{1 - x}{\tau_c} \right] \frac{\partial G}{\partial x} + (1 - x)SG, \qquad (12.45)$$

$$G(x, 0) = x^N,$$

$$g(x) = \sum_{x=0}^{\infty} x^\nu p_\nu.$$

Here $g(x, t) \to g(x)$ is an auxilliary generating function representing the distribution of ν neutrons from fission and τ_f, τ_c are the mean times before fission and capture. A connection may be made for the properties in the mean which will govern the deterministic behaviour studied in the next sub-section:

$$\left. \frac{\partial g}{\partial x} \right|_{x=1} = \langle \nu \rangle; \qquad \Lambda = \frac{1}{\langle \nu \rangle \tau_f}; \qquad \rho = 1 - \frac{\langle \nu \rangle \tau_f}{\tau_c}. \qquad (12.46)$$

An extension of the model is readily made to cover (i) delayed neutron precursors (see later) and (ii) the observable fraction of neutrons that are captured in a detector. These extensions bring no substantial increase in mathematical difficulty and will be ignored here.

Equation (12.45) is a *first-order and linear partial differential equation* [see IV, §8.1] and can in principle therefore be solved by the *method of Lagrange*, yielding as solutions an arbitrary function. The problem arises that the boundary conditions associated with a given number of initial neutrons does not necessarily serve to match the form of this solution function.

It is found, however, that if it may be supposed that the maximum number of neutrons produced in fission has a maximum at $\nu = 2$, then $p_{\nu > 2} = 0$ and the solution function is sufficiently simple that the initial boundary condition can be determined explicitly in terms of this function; then the problem is in principle solved (Pacilio *et al.*, 1980). Although this simplification is not generally true, it is possible to represent approximately a realistic fission distribution by a *polynomial expansion* of $g(x)$ in $(1 - x)$ to a *quadratic* term which satisfies the simple case and gives a very reasonable representation to the model. This quadratic expansion is the normally used form of the stochastic equations (although a series expansion of the initial boundary condition by the *method of Lagrange* would also in principle be feasible although perhaps not producing a sufficient improvement in accuracy to be worth while). The major alternative to this treatment is to reduce the full stochastic equation to a set of equations for the behaviour of the *mean* and of the *variance* if this is thought to be a sufficient description. It would not, however, serve to determine any specific probability, e.g. the extinction probability P_0, the probability that there will be no neutrons in what might be a supercritical reactor with a (stochastic) source—a point of some interest for specialized safety studies.

In a more general problem for power, there will be other equations that describe the dependence of the properties—τ_f, τ_c particularly—on the power, i.e. the neutron history. This means that the fluctuating processes are no longer independent; the forward equations lose their *linearity* and the backward equations their corresponding *combinatorial* properties. The *Langevin method* (Williams, 1974) is now perhaps the only satisfactory way of constructing a useful model that appears to encapsulate sufficient of the physics to be realistic as well as being amenable to solution.

Dynamic equations

The full behaviour of a nuclear reactor cannot in practice be described in detail. We have seen in earlier sections the pressure to make simplifications such as constant properties, uniform properties, time independence, etc. When, however, the nature of the problem is to describe the time-dependent behaviour for the reactor, its internal components and the peripheral components associated with it may extend perhaps as far as the electricity user at the end of a distant supply line. One is then correspondingly forced to drop the detail in other directions, particularly of course being content with a description of the behaviour in the *mean* (deterministic model).

Useful results can indeed be obtained from the extreme device of reducing all spatial and velocity dependence (at least of the neutrons and photons) to an *ordinary differential equation* description in which the reactor properties may be said to have been 'lumped'—sometimes referred to as a 'point model'. Such equations may be referred to as the *neutronic dynamics equations* and are then to be compared to descriptions of, say, how power affects temperature affects reactor properties affects in turn the number of neutrons on a lumped, ordinary differential basis.

To make such a reduction from the general to the particular will usually be done using a suitable weighting function (obtained ideally in the framework of a *variational principle* [see IV, Chapter 12] to minimize the inherent error, as discussed in the previous section) and an integration over the reactor extent, space and velocity. There are, of course, 'half-way houses' in this scheme, to divide the reactor into some (small) number of regions—whether in space or energy—but the principle remains the same although the number of residual equations increases.

The result of such a procedure may be expressed as the reactor (neutronic) dynamic equations, which must now incorporate the phenomenon of delayed neutrons, and their precursors, that play a significant role in reactor kinetics although negligible in the steady state (and hence have been ignored in the discussion to this point). We obtain (see Lewins, 1978)

$$\frac{dn}{dt} = \frac{\rho - \beta}{\Lambda} n + \sum_{i=1}^{I} \lambda_i c_i + s; \qquad n(0) = n_0 \qquad (12.47)$$

and

$$\frac{dc_i}{dt} = \frac{\beta_i}{\Lambda} n - \lambda_i c_i; \qquad i = 1, 2, \ldots, I; \ c_i(0) = c_{i0}. \tag{12.48}$$

In this set of ordinary linear differential equations we have the single independent variable t (time) and $I + 1$ dependent variables, $n(t)$ (now the neutron number or population) and $c_i(t)$, the populations of delayed neutron precursors of the ith type. The model supposes that precursors (products of the fissioning process) decay with characteristic constants to release a neutron after some (probable) decay period; such delayed neutrons once released are (in this model) the equivalent of the prompt neutrons immediately released in fission. (Differences are properly taken into account by the adjoint or 'importance weighting' and included in the definitions of the β_i.)

Remaining coefficients of these equations are

$\rho(t)$ the reactivity,

β_i the delayed neutron fraction in the ith species:

$$\beta = \sum_i \beta_i,$$

λ_i the decay constant,

Λ the neutron reproduction time

and s the independent source rate.

Although β is small, say 0.7 per cent of neutrons via the delay mechanism, the mean decay times $1/\lambda_i$ are generally many thousands of times longer than the neutron reproduction time, essentially the same as its lifetime, between 10^{-3} and 10^{-6} s for Λ. The particular formulation of the equations above is because, for nearly all cases, all coefficients save ρ and s may be taken as constant with corresponding simplification of the solution method.

The reactivity ρ is a measure of the difference between the rate of production of neutrons by neutrons and the rate of removal of neutrons in the whole reactor. For a critical (i.e. just self-sustaining) reactor in the steady state to be feasible it is easily seen that ρ and s must be zero, although a steady state is feasible with $\rho < 0$ if there is a source rate s.

Solution methods without feedback

If the coefficients are *given*, in particular if we know the reactivity at all times and a set of initial conditions, we have the simpler case, dealt with in this section, since the equations are linear. This corresponds to a low-power reactor or a reactor at start-up. Operation at power, however, leads to the physical effects where, say, the neutron population change affects power affects temperature affects reactivity. To our neutronics equations must be added further coupled equations which render the enlarged set non-linear, to be treated subsequently.

In the present case we therefore have a single time-dependent coefficient ρ if we may ignore source variation. When ρ is itself constant then the solution is easily accomplished by a number of techniques, including particularly *matrix and vector eigenvalue methods* [see IV, §7.8] and the *Laplace transform* [see IV, §13.4]. For the former, we may write a vector $\mathbf{N} = [n, c_1, c_2, \ldots, c_I]$ and construct a matrix equation of the form [see IV, §7.9]

$$\frac{d\mathbf{N}}{dt} = \omega\mathbf{N} = \mathbf{MN} + \mathbf{S}; \qquad \mathbf{N}(0) = \mathbf{N}_0, \tag{12.49}$$

seeking eigenvalues ω and eigenfunctions to satisfy $\mathbf{N}(0)$. When the Laplace transform method in *transform variable p* is used, the *characteristic equation* for p [see IV, §7.4] is found to be

$$\Lambda p + \sum_i \frac{p\beta_i}{p + \lambda_i} = \rho. \tag{12.50}$$

Indeed, on clearing fractions this is seen to be a polynomial in p of degree $(I + 1)$ with, therefore, according to the *fundamental theorem of algebra*, $I + 1$ roots [see I, §14.5]. In this case it may be shown that the roots are all real and distinct (if they were not distinct we could have coalesced groups with the same value of λ_i without loss of physical detail). Indeed, the roots lie between ordered pairs of the Λ and λ_i. When note is taken of the magnitudes of the coefficients, it is found that one root (p_0 say) is substantially different in magnitude and (for a super-critical reactor, $\rho > 0$) in sign from the rest and is such as to dominate the *asymptotic* behaviour. That is, p_0 determines the ultimate neutron behaviour and the remaining I roots $p_1, \ldots, p_I$ determine the *transient* behaviour, to satisfy in total the $I + 1$ *initial conditions* (see also Figure 12.8).

Regarded as a *matrix equation*, the disparity in the roots makes the matrix *stiff* (Gear, 1971), and in generalizations to time-dependent reactivity where a direct numerical solution is sought, this will be a considerable disadvantage, the ratio of the roots being perhaps as much as 10^5. To achieve numerical stability from an *elementary marching solution* [see III, §9.3] the *time step* must be of the order of the shortest inverse root, but the solution is required over times at least as long as the largest of the inverse roots, calling for many time steps if a simple technique of solution is used. Instead, one must use high-order *predictor-corrector methods* [see III, §8.3.1] for adequate performance.

For a limited number of time-varying reactivities, e.g. sinusoidal ρ or a linear or 'ramp' reactivity, exact solutions may still be possible analytically from the *Laplace transform* method [see IV, §13.4] at the expense of having to undertake an *analytic inversion* [see IV, §13.4.2] (the *Bromwich integral*). For extreme but useful cases, approximation methods equivalent to *singular perturbation theory* are possible, putting dn/dt or dc/dt to zero. However, in general, a numerical solution is necessary with the disadvantages already referred to of a stiff matrix [see III, §8.3.6].

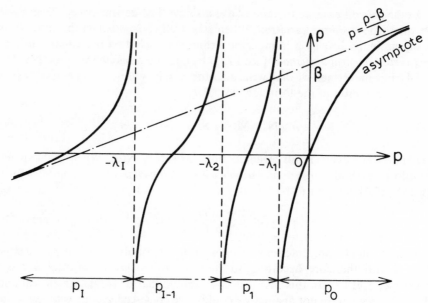

Figure 12.8: Schematic solution of the characteristic dynamics equation.

Power dynamics

Given that the only time-dependent coefficient significant at power in the neutronics equation is ρ, it is seen that the only term to give computing problems is ρn. A general method for dealing with the equations at power, coupled to the remaining description providing non-linearity overall, is to suppose that in the normal operation of the reactor, $n(t)$ lies close to some constant value n_0, while $\rho(t)$ is—for good operating practice—retained close to $\rho_0 = 0$.

Thus a *linearization* technique in which one writes $n = n_0 + \delta n$, $\rho = \rho_0 + \delta \rho$ consists of neglecting the term $\delta \rho \, \delta n$ in the expression $(n_0 + \delta n)(\rho_0 + \delta \rho)$. The resulting linearized equations will be found to have essentially the same form in δN as they did in the original N, except that the retained term $\delta \rho n_0$ plays the role of the neutron source rate s which no longer appears as significant in a power reactor. So linearized, the variation of δn with $\delta \rho$ and the coupling equations (linearized if necessary) then make the problem suitable for conversion to *Laplace transform* [see IV, §13.4] and associated *Fourier transform* [see IV, §13.2] techniques of an elementary kind again. Such a procedure enables the conventional methods of the *frequency–space* analysis of *control engineering* (see also Chapter 14) to be utilized to the full (Kerlin, 1976).

It would seem that the method depends essentially on the smallness of the departure of n from n_0, raising the question of how well the solution can be trusted. This question is relieved to some extent by the observation that if the stability studies in the frequency space indeed predict *instability*, the prediction is chiefly valuable for the opportunity to prevent such behaviour. It has been found that this conversion to classical control engineering in the frequency space has worked well for its purpose—the normal operation of the reactor as an element in a system.

However, this relief does not extend to the problem of actually solving the coupled equations at power for large departures from the nominal operating level, a procedure of importance in safety studies if not in daily practice. Some progress can be made by returning to the state space, i.e. a time-dependent not frequency-dependent formulation with the opportunity to employ *topological methods*, particularly *Liapunov functions* (see Chapter 14) [see also IV, §7.11.5]. Whilst these are relatively straight-forward with only two dependent variables, they become very difficult to utilize successfully in any higher dimensionality. This observation is related to the fact that the two-variable problem presents itself as a first-order ordinary differential equation for which methods exist (*Piccard*, for example) [see also IV, §7.11.1], guaranteeing a solution; no such guarantee is available in higher dimensionality and higher order.

More generally, then, such problems are driven back to a direct numerical solution of the differential equations. This may be done for specific cases but at a cost that would seem to preclude generic or parametric results. Unfortunately this is often what is wanted in safety studies which must demonstrate satisfactory behaviour in an excursion under *any* combination of initiating conditions.

The state-space representation is also the starting point for modern control theory, e.g. *Pontryagin's optimization* and *Bellman's dynamic programming* (see Chapter 14) [see also IV, Chapter 16]. The development of cheap, rapid microcomputers that may be dedicated on-line suggests that there will be major developments in the nuclear reactor (as any other) system control in the near future, based on a *direct optimization* of the system.

Safety and reliability

It is hoped that matters of safety and reliability will relate to an ubiquitous philosophy permeating the whole of an engineer's approach to nuclear power. It may already be clear, however, that since failure is at least in some sense an inherent dynamic concept, then safety and reliability studies are often closely associated with reactor kinetics and control.

Reliability engineering is a well-developed topic with applications far wider than just to nuclear power; it combines probability theory, on the one hand often conveniently manipulated as a *Boolean algebra* [see I, Chapter 16], with *statistics* to determine suitable values of failure probabilities, etc. Reliability and risk analysis are applied specifically to nuclear power by McCormick (1981).

Risk studies will again be a concept applicable to a wide range of technology. It may involve all the concepts of reliability theory, with extensions for more specific methods of evaluating the sequences leading to failure, such as *fault trees* and *event trees* (Kaufmann, Grouchko and Cruon, 1977). Risk studies have a particular difficulty, involving psychology and the social sciences, in relating mathematical definitions of risks—i.e. the probability of an occurrence weighted with its consequences—to a publicly *perceived risk*. In this nuclear power may well epitomize the problem in matters of its public acceptability. There is a case for viewing probability in this context not as an axiomatic concept, nor as the simple 'equally likely' concept, and not even as a limiting concept, but rather as the fourth of the alternative foundations of probability—a measure of the subjective willingness to 'bet'. The matter is not unrelated to the low observed frequency of reactor accidents that have large consequences. The methods of *Bayesian probability theory* [see II, §16.4] can be valuable in this area (Apostolakis, 1980).

12.6 Nuclear fuel cycles

Nuclear fuel cycles include the entire processing of nuclear fuel from prospecting and mining of uranium (and thorium), through refining enrichment (if desired) in the desired isotope U-235, to fabrication of fuel elements followed by utilization in a reactor—the in-core fuel cycle. The cycle is then continued with removal, temporary storage, reprocessing and further storage with ultimate disposal of the waste products. If in reprocessing, unused fissile material (uranium or the bred plutonium, etc.) is returned to the cycle it may be described as a recycle.

The cycle may be made more complex if the waste uranium from the enrichment process is utilized in a fast reactor to provide further plutonium as well as power. Figure 12.9 illustrates typical flows in the fuel cycle.

While it is well known that a nuclear reactor is capital expensive (i.e. the labour and equivalent costs of manufacture are incurred before any return is available) it is perhaps not appreciated that over the, say, thirty-year life of a nuclear power plant, the nominal value of the fuel used is of the same order as the initial capital costs. There is therefore substantial incentive to make the optimum use of the fuel, not only by breeding and recycle but by optimizing in-core fuel usage.

Much of the out-of-core fuel cycle is conventional mining and chemical engineering, complicated by the radiation and possibly criticality hazards that are already touched upon in discussing shielding and critical systems. Specialized approaches to criticality problems have been developed, however, since the requirement is on the safety of assuring non-criticality, and hence bounded methods including variational techniques may be cheap and acceptable (Moore, 1976; Thomas and Abbey, 1973).

Enrichment of isotope is also required for associated nuclear purposes (boron for control rods, deuterium for heavy water reactors and in future fusion reactors etc.) but will be illustrated here in terms of the uranium enrichment plant, although this in turn has features of conventional chemical engineering distillation and other separation plants (see Section 9.4). A peculiar feature of isotope enrichment is the small physicochemical difference between isotopes in current commercial systems (gaseous effusion—Benedict and Pigford, 1981 and centrifuge—Olander, 1972), leading to small separation factors per stage and extremely large plants with many hundreds of stages and large internal flow rates. The need to optimize the size of such a plant at each enrichment level, on grounds of both capital cost and running cost, is more pressing than in conventional distillation, say, and accounts for a more complex theory (Benedict and Pigford, 1981).

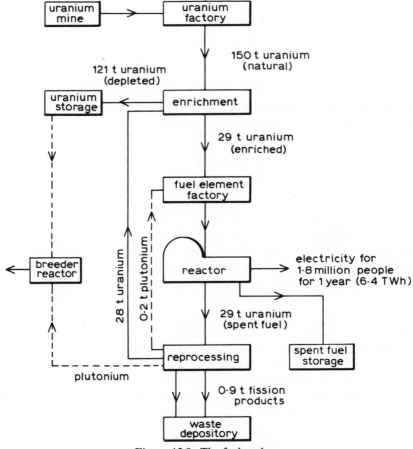

Figure 12.9: The fuel cycle.

Enrichment of isotopes

A 'stage' in an isotope separation plant may be understood to have a feed stream and two exiting streams, one at an increased enrichment of the desired isotope (product) and one decreased (waste), as shown in Figure 12.10. Some

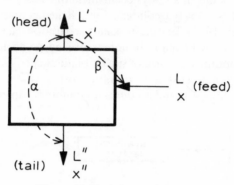

Figure 12.10: A separation stage.

physical process is used to vary the desired isotope concentration x with its flow rate L of the mixture between the product side, x' and L', and the waste side, x'' and L''. The stage separation factor α is defined as

$$\alpha = \frac{\zeta'}{\zeta''} = \frac{x'/(1 + x')}{x''/(1 + x'')}, \tag{12.51}$$

where ζ is the abundance ratio. This definition is used because for a wide range of ζ, the same separation factor α is valid for all stages, which would not be true when expressed for the ratio of isotopic concentrations. If x is low, however, say up to 3 per cent for a commercial nuclear fuel requirement from a natural enrichment of 0.715 per cent U-235, then we may approximate $\zeta \cong x$ and our equations become considerably simpler. We shall follow this approximation and indicate finally a more accurate result which may be obtained from the literature.

For the separation of U-235 from U-238, the classic effusion process (wrongly called gaseous diffusion) has a separation factor of only some 1.004 and even the more recent ultracentrifuge process has $\alpha \cong 1.05$. Thus a further good approximation is to take $|\alpha - 1| \ll 1$; it also follows that with the large number of stages, a continuous approximation will be acceptable.

The head separation factor β is defined analogously to α but in respect of the enrichment from stage input to head

$$\beta = \frac{\zeta'}{\zeta''} = \frac{x'/(1 + x')}{x/(1 + x)} \simeq \frac{x'}{x}. \tag{12.52}$$

The value β may be varied at choice, and between stages, but will be limited to $\beta \leqslant \alpha$, of course.

Enrichment units may be arranged in sequence to form a cascade so that the small enrichment per stage can be repeated to give the desired overall increase from the available feed material. We suppose that there is one feed at x_f and one product at x_p. A simple cascade will reject the tail fraction at each stage to waste; this is exceedingly uneconomical although it gives a theoretical lower bound for the minimum number of stages, which will be obtained if the waste at each stage is rejected at essentially the same concentration as the feed, implying that there is a vanishing head flow at each stage and vanishing product flow P:

$$\frac{x_p}{x_f} \to \frac{x_p}{x_w} = (\alpha)^{N_{min}}; \qquad N_{min} \simeq \frac{\ln x_p/x_w}{\alpha - 1}. \tag{12.53}$$

Invariably for uranium, a countercurrent cascade is used where the tails from each stage save the last are returned for further processing (Figure 12.11) and only the bottom stage rejects waste at some chosen x_w. Sufficiently many units are installed at each stage to accumulate the desired flow rates L_n.

By varying the proportion of head flow to tail flow, β can be made to approach α; this again leads to a minimum number of stages in the countercurrent cascade combined with a vanishing product flow P and hence an unbounded reflux ratio $R = L''/P$ at each stage. It may analogously be shown that the minimum reflux ratio implies an infinite number of stages. Thus both the minimum number of stages and the minimum reflux ratio are theoretical bounds and not practical values. The so-called 'ideal' cascade to be described is in fact a practicable design which both minimizes the total internal flow rates over the whole cascade (for a given duty) and leads to a number of stages which

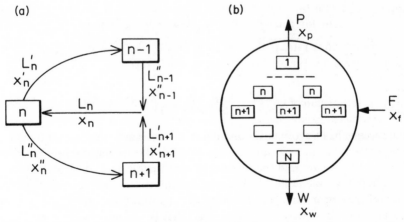

Figure 12.11: The countercurrent cascade and its flows: (a) stage interconnection and (b) total cascade.

is double the minimum number (without incurring infinite reflux) and a reflux ratio at any stage which is double the minimum (without incurring an infinite number of stages).

A mass balance may be drawn for the total cascade down from the first or product stage to the nth stage, both for the total mass and the mass of the desired isotope:

$$F = P + W; \qquad x_f F = x_p P + x_w W, \qquad (12.54)$$

leading to the relations

$$\frac{F}{P} = \frac{x_p - x_w}{x_f - x_w}; \qquad \frac{W}{P} = \frac{x_p - x_f}{x_f - x_w}. \qquad (12.55)$$

These six external variables are thus related by two algebraic equations which serve to eliminate two of them; we may, for example, determined F/P and W/P in terms of selected x_p, x_f and x_w. Typical values for the latter are 3, 0.715 and 0.2 per cent respectively, from which it may be determined that the external feed per kilogram of 3 per cent enriched product is ~ 4.5 kg, and N_{min}, the minimum number of stages, is some 350, with realistic N, say 700.

A similar mass balance and isotope balance may be drawn for the nth stage only to give, in the enrichment section:

Enrichment: $$R_n = \frac{L_n''}{P} = \frac{x_p - x_{n+1}'}{x_{n+1}' - x_n'}, \qquad (12.56)$$

yielding an expression for the reflux ratio at that stage.

Discard of the waste from the feed stage would again be wasteful and one uses a stripping section below the feed for which:

Stripping: $$R_n = \frac{x_p - x_f}{x_f - x_w} \frac{x_{n+1}' - x_w}{x_{n+1}' - x_n''}. \qquad (12.57)$$

The minimum reflux ratio in the feed stage is obtained from $x_n' \to x_n'' \to x_f$ and is shown for the same assumed operating external variables:

$$R_f|_{min} = \left. \frac{x_p - x_{n+1}'}{x_{n+1}' - x_n'} \right|_f \cong \frac{x_p - x_f}{x_f(\alpha - 1)} \simeq 743 \text{ kg/kg}, \qquad (12.58)$$

so the cascade has large refluxes and many stages as anticipated, justifying both the necessity for analysis as an ideal cascade and the use of a continuous model.

The 'ideal' cascade may be defined as one operating with a proportion of head-to-tail flows such that the two contributions to the feed at each stage, from the tail of the upper stage and the head of the lower stage, are equal, and thus suffer no entropy increase due to mixing. That is:

Ideal cascade: $$x_{n-1}'' = x_{n+1}' = x_n \qquad (12.59)$$

It follows that $\alpha = \beta^2$ or, since $\alpha \to 1$, $\beta - 1 = \frac{1}{2}(\alpha - 1)$. We then have:

Enrichment:
$$R_n \to \frac{x_p/x_n - 1}{\beta - 1}$$
twice the minimum reflux ratio

Stripping:
$$R_n \to \frac{x_p - x_f}{x_f - x_w} \frac{1 - x_w/w_n}{\beta - 1}$$

Number of stages:

$$x'_n = x_{n-1} = \beta x_n \to N = \frac{\ln x_p/x_w}{\ln \beta} \simeq \frac{\ln x_p/x_w}{\beta - 1}. \tag{12.60}$$

Figure 12.12 shows the proportions of reflux ratio at any stage (x axis) against the stage number (y axis) in a continuous model based on these equations. The area so shown is at the same time a measure of the capital cost of the plant and of the running cost. It may be obtained in the form

$$A = \int_{n(x_p)=0}^{n(x_w)=N} R(x)\, dn(x) = \left(\frac{2}{\alpha - 1}\right)^2 [PV(x_p) + WV(x_w) - FV(x_f)], \tag{12.61}$$

where $V(x) = -\ln x$ (approximate, low enrichment). For our example, the total reflux over all stages A is approximately 250 Mg/kg product. However, when the exact expressions in terms of abundance ratios ζ are employed (necessary for higher enrichment plant as $x \to 0.5$) then the corrected expression replaces $-\ln x$ with a value function $V(x)$ defined as

$$V(x) = (2x - 1) \ln (x/(1 - x)). \tag{12.62}$$

It will be seen from this result that the problem (approximate or more exactly solved) is a *conservative* or *potential problem* in which a value can be ascribed to the product at x_p in terms only of the feed at x_f and the chosen waste at x_w

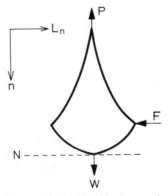

Figure 12.12: The ideal cascade.

(recollecting that these also serve to define F/P and W/P), together with the stage separation factor α. Of course, there is a further normalization to either the capital cost per stage or the running cost per stage, as the case may be. The value function or *potential* $V(x)$ is also known as the separative work and is measured in units (SWU) of kilograms per kilogram of product. When the relative cost of capital and running is known, the results may be used to optimize the choice of tails as, say, x_w, against any commercial value put upon depleted uranium. It is easily seen that $V(x)$ is symmetrical about $x = \frac{1}{2}$ and unbounded as $x \rightarrow 1, \rightarrow 0$ (indicating the impossibility in this model of obtaining pure isotope).

In-core fuel cycles

The rate of use or burn-up of fuel and production of fission products and breeding of further fissile material is to slow that the time scale admits the neglect of delayed neutrons and other short-term kinetic effects. Unfortunately, this is about the only accurate simplification admissible; realistic treatment of fuel (Benedict and Pigford, 1981) requires a detailed description of the power history (hence the distribution of neutrons), the temperature history (hence the thermohydraulics) and of the isotopic changes (depletion equations) linked together as illustrated in Figure 12.13.

The depletion equations are developments of the transformation equations in Section 12.2 and may be illustrated as follows:

U-238:
$$\frac{dN^{28}}{dt} = -\sigma_a^{28} \phi N^{28},$$

U-239:
$$\frac{dN^{29}}{dt} = \sigma_c^{29} \phi N^{29} - \lambda^{29} N^{29},$$

Np-239:
$$\frac{dN^{39}}{dt} = \lambda^{39} N^{39} - \lambda^{39} N^{39}, \tag{12.63}$$

Pu-239:
$$\frac{dN^{49}}{dt} = \lambda^{39} N^{39} - \sigma_a^{49} \phi N^{49},$$

Pu-240:
$$\frac{dN^{40}}{dt} = \sigma_c^{49} \phi N^{49} - \sigma_a^{40} \phi N^{40},$$

etc.

Therefore, these are locally *ordinary differential equations* coupled with the *partial differential* (or *integro-differential*) [see III, §10.6.2] description of the other modules of Figure 12.13. Equations (12.63) have *variable coefficients* [see IV, §7.3].

One small further simplification may be made in which the effects of

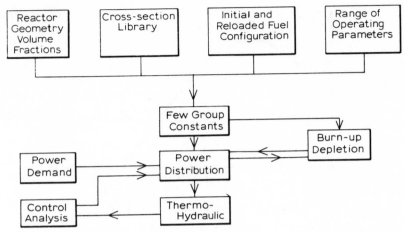

Figure 12.13: Schematic for in-core fuel computations.

radioactive decay are ignored, either because they happen on an appreciably shorter time scale or on a noticeably longer time scale. (The approximation introduces a few per cent error and may serve for survey purposes.)

It is impossible to describe here the numerical analysis involved in realizing Figure 12.13. We make a drastic simplication for illustrative purposes of uncoupling the depletion equations and treating the flux ϕ as an independent variable, together with time. These two may be combined as a single variable θ, the fluence or flux-time, defined as

$$\theta = \int_0^t \phi(t') \, dt'. \tag{12.64}$$

Note that the original equations had time-dependent coefficients in $\phi(t)$: even if the reactor is run at constant total power (and, as in the uncoupling, constant local power) the burn-up of fuel would require increasing flux to correspond with constant thermal power. The typical equations now, however, have *constant coefficients* [see IV, §7.9.2] and may be readily integrated:

$$\frac{dN^{28}}{d\theta} = -\sigma_a^{28} N^{28},$$

$$\frac{dN^{49}}{d\theta} = \sigma_c^{28} N^{28} - \sigma_a^{49} N^{49}, \tag{12.65}$$

$$\frac{dN^{40}}{d\theta} = \sigma_c^{49} N^{49} - \sigma_a^{40} N^{40}, \text{ etc.}$$

As a result of these processes and material changes (and ignoring the temperature history) the reactivity ρ will vary. Figure 12.14 illustrates a typical

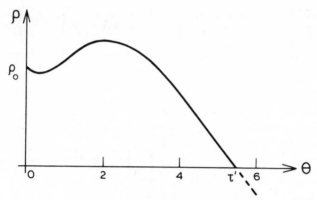

Figure 12.14: Reactivity with burn-up.

change with fluence—which is conventionally expressed in thermal energy per mass of fuel as shown. In both light water reactors (LWR) and advanced gas-cooled reactors (AGR), the development of reactivity against burn-up is quite well approximated by a *linear relation*

$$\rho_1(t) = \rho_{10}\left(1 - \frac{\theta}{\tau'}\right), \tag{12.66}$$

where ρ_0 is the reactivity of the core on the first loading at the beginning of a fuel cycle (BOC) and τ' is therefore the cycle 'time' or burn-up.

If the reactor is operated in a simple way, loaded and operated until reactivity is depleted to zero (at which the chain process must stop), it will be noted that before the end of the cycle (EOC) the operation of the reactor at constant power will essentially demand a zero net reactivity. To secure this despite the excess reactivity of the fuel will require control rods to competitively absorb neutrons.

An alternative mode of operation may be called n-batch as opposed to the foregoing one-batch operation. Here $(1/n)$th of the fuel is taken out at the EOC and replaced with fresh fuel; the reactor is now at a positive reactivity and may be operated until at the next EOC its reactivity due to the fuel has fallen again to zero. It will be seen that n-batch operation permits a greater utilization of fuel, to considerable economic advantage, inherently because the neutrons previously wasted in the control rods may now be used in the fuel to produce further fission, even though there is a negative contribution of some fuel to the net fuel reactivity.

To see these effects we continue the decoupling by assuming (i) that each $1/n$ fraction of the fuel contributes a partial fuel reactivity for which the exposure of burn-up of that particular batch determines the reactivity contribution as $1/n$ of the reactivity of a one-batch core exposed to the same burn-up and (ii) that the burn-up curve may be extended into negative regions, i.e. linearly extrapolated in our model past $\theta = \tau'$, the one-batch cycle time. The cycle times in n-batch operation, assuming similar fresh fuel to be loaded at each EOC, are by no means the same, cycle to cycle, and will be denoted as τ_j for the jth cycle in n-batch operation. It will also be apparent that a particular fuel batch would be expected to reside for n cycles in all (with the dwell time longer than the cycle time), although when first operations commence, early batches must be removed to accommodate fresh fuel before residing so long, an effect leading to the variable cycle time. After analyzing this, we later consider the problem of making the n-batch cycle times more uniform.

In the present linear model we have, after sufficient cycles,

$$\rho(\theta) = \frac{1}{n}\left[\rho_1\left(\theta - \sum_{i=j-n}^{j-1}\tau_1\right) + \rho_1\left(\theta - \sum_{i=j-n}^{j-2}\tau_2\right)\right.$$
$$\left. + \cdots + \rho_1(\theta - \tau_{j-2} - \tau_{j-1}) + \rho_1(\theta - \tau_{j-1})\right]$$

fresh fuel <--> old fuel

$$\text{for } \sum_{i=j-n}^{j-1}\tau_i \leqslant \theta \leqslant \sum_{i=j-n}^{j}\tau_i.$$

Therefore, at the EOC, when $\rho = 0$,

$$0 = n - n\frac{\tau_j}{\tau'} - (n-1)\frac{\tau_{j-1}}{\tau'} - (n-2)\frac{\tau_{j-2}}{\tau'} - \cdots - \frac{\tau_{j-n}}{\tau'}. \tag{12.67}$$

On the other hand, $n - 1$ initial cycles starting with $j = 1$ terminate with the conditions

$$0 = n\left(\frac{1 - \tau_1}{\tau'}\right) \qquad\qquad\qquad ; \qquad j = 1,$$

$$0 = 1 - \frac{\tau_2}{\tau'} + (n-1)\left(1 - \frac{\tau_2 + \tau_1}{\tau'}\right) \qquad ; \qquad j = 2, \tag{12.68}$$

$$0 = 1 - \frac{\tau_3}{\tau'} + 1 - \frac{\tau_3 + \tau_2}{\tau'} + (n-2)\left(1 - \frac{\tau_3 + \tau_2 + \tau_1}{\tau'}\right); \qquad j = 3,$$

etc. $\qquad\qquad\qquad\qquad\qquad\qquad\qquad\qquad ; \qquad j < n.$

Equation (12.67) constitutes a *recurrence relation* for τ_j given $n - 1$ preceding values τ_{j-1} to τ_{j-n} [see I, §14.12]. The $n - 1$ relations peculiar to the first operation of the reactor, equations (12.68), constitute relations by means of which the recurrence relation can be initiated; once started it yields all subsequent cycles by recurrence. As a recurrence relation, this may be readily solved either numerically or graphically [see I, §14.13], illustrated in Figure 12.15 for a three-batch operation.

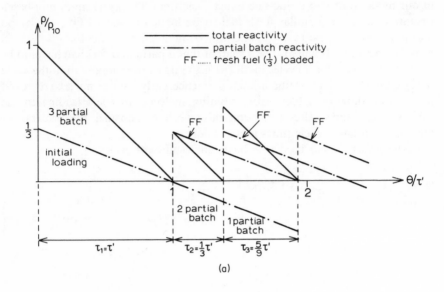

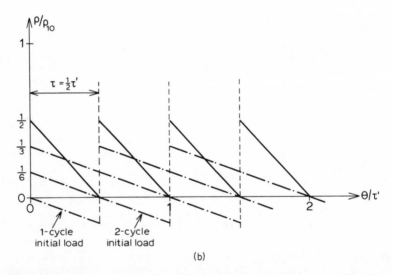

Figure 12.15: Graphical solution of the recurrence relations: (a) uniform batch loading, variable cycle time, for three-batch operation, (b) non-uniform initial loading, constant cycle time, for three-batch operation. (Note that the desired initial loading has one nominal batch, to dwell three cycles, one half-reactivity batch to dwell two cycles and one zero reactivity batch to dwell one cycle only before removal.)

It may be noted in particular that:

(i) $j = 1$: $\tau_1/\tau' = 1$,
   i.e. the first cycle time is the same as the one-batch cycle time.
(ii) $j = 2$; $\tau_2/\tau' = 1/n$,
   i.e. the second cycle is $(1/n)$th of the first cycle. This may be an un-
   acceptable distortion.

It may be supposed that after sufficient cycles, the system settles to a constant cycle time. Assuming convergence, we have $\tau_j = \tau_{j-1} = \cdots = \tau_{j-n} = \tau$, whence

$$n = [n + (n - 1) + (n - 2) + \cdots + 2 + 1]\frac{\tau}{\tau'} = \frac{n(n + 1)}{2}\frac{\tau}{\tau'}, \quad (12.69)$$

whence $\tau/\tau' = 2/(n + 1)$. Thus the dwell time in n-batch operation is $2n/(n + 1)$ times the dwell time in one-batch operation. It is seen that as $n \to \infty$, which may be considered as a continuous reloading operation, the fuel utilization is doubled; it has reached 75 per cent of its optimum in a three-batch mode of operation.

An alternative solution method is to regard equation (12.67) as a *difference equation of order* $(n - 1)$, which is seen to be again *linear* and with *constant coefficients* [see I, §14.12]. The $n - 1$ peculiar initial equations will serve to provide the necessary $n - 1$ *initial boundary conditions*. The method of solution, analogous to *linear ordinary differential equations*, is here to express the solution as the sum of a *complementary function* and a *particular integral* [see IV, §7.3.2]; for the latter, the steady-state asymptotic solution of equation (12.69) serves.

The complementary function is obtained assuming $\tau_j = p\tau_{j-1} = p^2\tau_{j-2} = \cdots = p^{n-1}\tau_{j-n+1}$ in the equivalent homogeneous equation, whence

$$\sum_{i=1}^{n} i\frac{\tau_{j-i}}{\tau'} = 0 = \left(\sum_{i=1}^{n} ip^{i-1}\right)\tau_{j-n+1}. \quad (12.70)$$

Since τ_j does not in general vanish, p must satisfy the characteristic equation

$$\sum_{i=1}^{n} ip^{i-1} = 0. \quad (12.71)$$

According to the *fundamental theorem of algebra* [see I, §14.5] there are $n - 1$ roots, p_i, not necessarily distinct and possibly complex, serving to provide $n - 1$ solutions of the form $A_i p^j$ for τ_j, enough to match, therefore, the $n - 1$ initial conditions and the inhomogeneous equation.

For convergence, it is necessary that all roots p_i lie within the unit circle, i.e. have magnitude less than one. This may be proved with the aid of *Rouche's theorem* (Wood, 1982). If the reactivity as a function of burn-up is not linear but nevertheless monotonic, it permits a transformation of the more general problem to the same form and thus guarantees convergence again.

Given large fluctuations in the initial cycles, however, an alternative problem may well be to find the appropriate values of partial batch initial reactivities that will lead to a constant cycle time from the start. Essentially most of the initial fuel loading can be of batches with a lower reactivity that in turn can be provided with a lower-than-usual enrichment, thus achieving a further economic advantage.

In obtaining these results it must again be emphasized that the device of decoupling the material balance from the remainder of the in-core fuel cycle computation, leading to a partial reactivity, is extreme, although it illustrates the potential advantages available in a detailed model of fuel management (Graves, 1979). A related question may be to determine that distribution of fuel within the core that will in some specified sense optimize the operation. Considerable work has been done on this more general problem using *dynamic programming* [see IV, Chapter 16] (Turney and Wade, 1977) and some work with (*Pontryagin*) *optimization theory* has also been done for control management (see Chapter 14) [see also IV, Chapter 15]. An elegant variational optimization result, *Haling's* (1963) *principle*, has the additional advantage of decoupling the control-rod problem within a fuel cycle from the remaining fuel cycle calculations.

J.D.L.

References

Allen, D. N. De G., and Southwell, R. V. (1955). Motion of a viscous fluid past a cylinder, *Quart. J. Mech. Appl. Math.*, VIII, 129–145.

Apostolakis, G. (1980). Bayesian methods in risk analysis, in *Advances in Nuclear Science and Technology*, Vol. 13, Plenum.

Bateman, H. (1910). The solution of a system of differential equations occurring in the theory of radioactive transformations, *Proc. Camb. Phil. Soc.*, **15**, 423.

Benedict, M., and Pigford, T. H. (1981). *Nuclear Chemical Engineering*, McGraw-Hill.

Case, K. M., and Zweifel, P. F. (1967). *Linear Transport Theory*, Addison-Wesley.

Davis, B. (1978). *Integral Transforms and their Applications*, Applied Mathematical Sciences, Vol. 25, Springer-Verlag.

Engle, W. W. Jr. (1963). *A User Manual for ANISN*, K1693, U.S. AEC.

Gear, C. W. (1971). *Numerical Initial Value Problems in Ordinary Differential Equations*, Prentice-Hall.

Goldstein, H. (1959). *Fundamental Aspects of Reactor Shielding*, Addison-Wesley.

Graves, H. W. (1979). *Nuclear Fuel Management*, Wiley.

Greenspan, E. (1976). Developments in perturbation theory, in *Advances in Nuclear Sciences and Technology*, Vol. 9, Academic Press.

Haling, R. K. (1963). *Optimal Strategy for Maintaining an Optimal Power Distribution*, TID 7672.

Hansen, K. F., and Kang, C. M. (1975). Finite element methods in reactor physics analysis, *Advances in Nuclear Science and Technology*, Vol. 8, Academic Press.

Henley, E. J., and Williams, R. (Eds), (1973). *Graph Theory in Modern Engineering*, Academic Press.

Henry, A. F. (1975). *Nuclear Reactor Analysis*, MIT Press.

Hochstadt, H. (1973). *Integral Equations*, Wiley.

Kaufmann, A., Grouchko, D., and Cruon, R. (1977). *Mathematical Models for the Study of the Reliability Systems*, Mathematics in Science and Engineering, Vol. 124, Academic Press.

Kerlin, T. W. (1976). *Frequency Response Testing in Nuclear Reactors*, Academic Press.

Lewins, J. (1965). *Importance; the Adjoint Function*, Pergamon.

Lewins, J. (1978). *Nuclear Reactor Kinetics and Control*, Pergamon.

Lewis, E. E. (1981). Finite element approximations, *Advances in Nuclear Science and Technology*, Vol. 13, Plenum.

Lindhal, S.-O., and Weiss, Z. (1981). The response matrix method, *Advances in Nuclear Science and Technology*, Vol. 13, Plenum.

McCormick, N. J. (1981). *Reliability and Risk Analysis*; Methods and Nuclear Power Applications, Academic Press.

McCormick, N. J., and Kuščer, I. (1973). Singular eigenfunction expansions in neutron transport theory, *Advances in Nuclear Science Technology*, Vol. 7, Academic Press.

Marchuk, G. I. (1959). *Numerical Methods for Nuclear Reactor Calculations*, Russian trans. by Consultants Bureau, New York.

Meghreblian, R. V., and Holmes, D. K. (1960). *Reactor Analysis*, McGraw-Hill.

Moore, J. G. (1976). The solution of criticality problems by Monte Carlo methods, *Advances in Nuclear Science and Technology*, Vol. 9, Academic Press.

Olander, D. R. (1972). Technical basis of the gas centrifuge, *Advances in Nuclear Science and Technology*, Vol. 6, Academic Press.

Pacilio, N., et al. (1980). Analysis of reactor noise, *Advances in Nuclear Science and Technology*, Vol. 12, Plenum.

Southwell, R. V. (1940). *Relaxation Methods in Engineering Science: A Treatise on Approximate Computation*, Oxford University Press.

Stacey, W. H. Jr. (1974). *Variational Methods in Nuclear Reactor Physics*, Academic Press.

Thomas, A. F., and Abbey, F. (1973). *Calculation Methods for Interacting Arrays of Fissile Materials*, Pergamon.

Turney, W. B., and Wade, D. C. (1977). Optimal control applications in nuclear reactor design and operation, *Advances in Nuclear Science and Technology*, Vol. 10, Plenum.

Williams, M. M. R. (1973). The Weiner–Hopf technique, *Advances in Nuclear Science and Technology*, Vol. 7, Academic Press.

Williams, M. M. R. (1974). *Random Process in Nuclear Reactors*, Pergamon.

Wing, G. M. (1962). *An Introduction to Transport Theory*, Wiley.

Wood, J. (1982). *Computational Methods in Reactor Shielding*, Pergamon.

Mathematical Methods in Engineering
Edited by G. A. O. Davies
© 1984, John Wiley & Sons, Ltd.

13

Systems Engineering

13.1 What is systems engineering?

Systems engineering is a relatively new field—it is not really a single subject but is rather a collection of techniques and methods which assist engineers to make optimal decisions in the face of competition, uncertainty and imprecision. It has its basis in the earlier development of 'operations research' (or 'operational research') in the United Kingdom during World War II. The optimal allocation of limited defence resources was a major contribution to the winning of the Battle of Britain. After the War the methods which had been successfully applied to defence operations were adapted to the needs of large industrial corporations such as those associated with petroleum and chemical engineering. The centre of these activities shifted from the United Kingdom to the United States. The increased availability of large high-speed computers greatly facilitated this trend and the emphasis on corporate decision-making led to the feeling that 'operations research' was an unsuitable title for this field and many considered 'management science' to be a more descriptive title. New concepts were absorbed from other developing technologies such as communications, cybernetics, control systems and econometrics. The managerial applications tended to treat the involved technology as a 'black box'. Thus a management consultant would consider the operations of a corporation concerned with the production of pre-cast concrete units in much the same way as he might tackle the operations of a car assembly plant or a soap factory. He would not, for example, be involved in the structural engineering or automobile engineering aspects of the project.

However, engineers who became skilled in management techniques soon found applications to their own technology. For example, a structural engineer might find mathematical programming had been developed for managerial applications but could be adapted to the identification of new structural theorems. Thus the new field of 'systems engineering' has been created by using management techniques to advance engineering itself.

13.2 Mathematical models

All engineers are familiar with the need to verify proposed theories by the creation and testing of physical models which represent, usually to some reduced scale, the actual properties of the real system. Such iconic models are usually costly to construct and test and are of limited application. Sometimes the systems being tested can be represented through some mechanical and/or electrical analogy, but such analogic models have very limited applications. The most common form of model is one in which the various aspects of the considered system are represented mathematically (Figure 13.1). Such mathematical models can be grouped in various categories [see IV, Chapter 14]. The most common mathematical model assumes complete certainty in that the input is specified completely and the parameters of the system are also known exactly. The model displays deterministic regularity in that identical mathematical experiments can be repeated with identical responses. However, uncertainty generally enters into real systems and emphasis will be placed herein on two differing types of uncertainty. First, experiments will not be identically repeatable because of the intrinsic randomness of the system. Such models may display statistical regularity in that consistent statements may be made with regard to the statistics of their responses rather than consistent statements about their responses to individual inputs. Such statistical considerations are of importance with respect to studies of traffic conditions, material properties, hydrology, wind loading and in many other important engineering problems. Frequently all sources of uncertainty are considered to be statistical, but a distinction should be drawn between the statistical uncertainty associated with random experiments and that uncertainty which stems from imprecision. It is usually assumed that imprecision is totally undesirable. However, in many problems (such as those associated with major planning decisions like the Third London Airport, the Channel Tunnel and the Trans-Alaskan pipeline) the

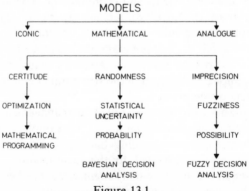

Figure 13.1

quantity of precise information available to the decision-makers is a source of embarrassment. It is necessary to carry the main weight of the detailed information in an imprecise but logical manner so that the various components of the decision analysis can be seen clearly and the decision-making facilitated. Such an integrated process is only possible with the proper use of imprecision. This process is particularly important when professional opinions are obtained from authoritative sources and when these opinions are incorporated in the decision-making.

The mathematical theory involved in handling statistical uncertainty is that of *probability* and the formal application of this theory to decision-making is generally termed *Bayesian decision-making*. The mathematical treatment of imprecision (or *fuzziness*) has led to a new theory of *possibility* and applications of fuzzy decision analysis are being developed currently [see the Guidebook for Economists, Chapter 17].

13.3 Optimization

The simplest mathematical models are based on deterministic regularity. The characteristics of the engineering systems are considered to be known with complete certainty and consequently the problem becomes one of *optimization* subject to known *constraints* [see IV, §15.1.3]. However, the simplest class of optimization is where there are no constraints. From calculus, the necessary conditions for optimality can be established readily (Figure 13.2). The first class of constrained optimization to be studied in detail was where the constraints are all equations and the variables are unrestricted in sense. This is classical (or *Lagrangian*) *optimization* (Figure 13.3), in which the constrained optimization of the function z is converted into unconstrained optimization of the *Lagrangian* (L) [see IV, §15.1.4].

However, most problems of engineering optimization involve *inequality constraints* and such a problem is termed a *mathematical program* (MP). The

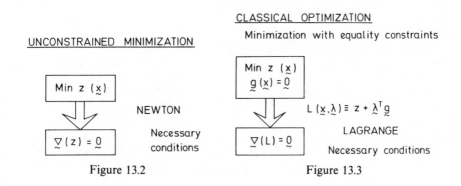

CLASSICAL OPTIMIZATION
Minimization with equality constraints

UNCONSTRAINED MINIMIZATION

Min $z\,(\underset{\sim}{x})$

$\nabla(z) = \underset{\sim}{0}$

NEWTON

Necessary conditions

Min $z\,(\underset{\sim}{x})$
$\underset{\sim}{g}\,(\underset{\sim}{x}) = \underset{\sim}{0}$

$L\,(\underset{\sim}{x},\underset{\sim}{\lambda}) \equiv z + \underset{\sim}{\lambda}^T \underset{\sim}{g}$

LAGRANGE

$\nabla(L) = \underset{\sim}{0}$

Necessary conditions

Figure 13.2   Figure 13.3

<u>MATHEMATICAL PROGRAMMING</u>
Minimization with inequality constraints

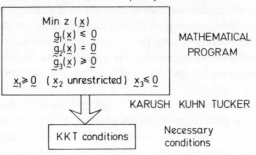

$$\text{Min } z\,(\underline{x})$$
$$g_1(\underline{x}) \leq \underline{0}$$
$$g_2(\underline{x}) = \underline{0}$$
$$g_3(\underline{x}) \geq \underline{0}$$
$$\underline{x}_1 \geq \underline{0} \quad (\underline{x}_2 \text{ unrestricted}) \quad \underline{x}_3 \leq \underline{0}$$

MATHEMATICAL
PROGRAM

KARUSH KUHN TUCKER

KKT conditions

Necessary
conditions

Figure 13.4

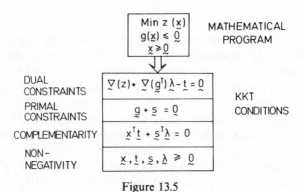

$$\text{Min } z\,(\underline{x})$$
$$g(\underline{x}) \leq \underline{0}$$
$$\underline{x} \geq \underline{0}$$

MATHEMATICAL
PROGRAM

DUAL
CONSTRAINTS $\nabla(z) + \nabla(g^T)\,\underline{\lambda} - \underline{t} = \underline{0}$

PRIMAL
CONSTRAINTS $\underline{g} + \underline{s} = \underline{0}$

COMPLEMENTARITY $\underline{x}^T\underline{t} + \underline{s}^T\underline{\lambda} = 0$

NON-
NEGATIVITY $\underline{x}, \underline{t}, \underline{s}, \underline{\lambda} \geq \underline{0}$

KKT
CONDITIONS

Figure 13.5

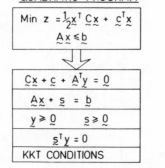

<u>QUADRATIC PROGRAM</u>

$$\text{Min } z = \tfrac{1}{2}\underline{x}^T\,\underline{C}\,\underline{x} + \underline{c}^T\underline{x}$$
$$\underline{A}\,\underline{x} \leq \underline{b}$$

$\underline{C}\underline{x} + \underline{c} + \underline{A}^T\underline{y} = \underline{0}$

$\underline{A}\underline{x} + \underline{s} = \underline{b}$

$\underline{y} \geq \underline{0} \qquad \underline{s} \geq \underline{0}$

$\underline{s}^T\underline{y} = 0$

KKT CONDITIONS

LINEAR
COMPLEMENTARITY
PROBLEM
(LCP)

Figure 13.6

necessary conditions for the solution of an MP are termed the *Karush–Kuhn–Tucker* (KKT) conditions [see IV, §15.4.1] (Figure 13.4) and an illustration of these conditions is given for a broad class of MPs in Figure 13.5 and the special case of a *quadratic program* (QP) [see IV, §15.6.3] is given in Figure 13.6.

Frequently engineering problems occur more naturally as systems of equations and inequalities rather than as MPs. It can be shown that if such a system can be organized in the form of a set of KKT conditions then (under certain conditions of *convexity* [see IV, §15.2]) an equivalent MP can be inferred (Figure 13.7.) In Figure 13.8, the KKT conditions for a *linear program* (LP) are stated and then a second (or *dual*) LP is inferred from the above equivalence theorem [see I, Chapter 11]. The two (primal–dual) LPs are alternative representations of the same problem. The program which involves the lesser amount of computation will be selected and using *the simplex algorithm* the solution to the other program will be obtained simultaneously. Thus both the primal and dual variables will be evaluated at optimality [see I, §11.4].

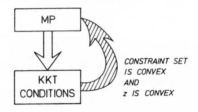

KUHN-TUCKER EQUIVALENCE

Figure 13.7

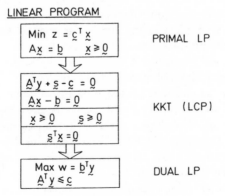

Figure 13.8

If a problem occurs naturally as an MP then its KKT conditions may be evaluated as a preliminary step to evaluating its solution. However, if a problem occurs naturally as a system of equations and inequalities then from the equivalence theorem a primal–dual pair of MPs may be generated and these programs encode the extremal theorems of the considered class of problem. The former application is a straightforward use of mathematical programming as a managerial tool, but the latter application is an example of the power of systems engineering to utilize such a managerial aid to generate new results within the appropriate technology. Perhaps this distinction can be made clearer by an illustration concerning structural engineering applications.

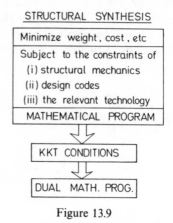

Figure 13.9

For example, the process of automated optimal structural design is termed 'structural synthesis'; this process is outlined in Figure 13.9. The design objective of minimizing weight or cost (for example) is sought subject to the constraints imposed by structural mechanics (equilibrium, compatibility), by the design codes (limiting stress, displacements) and by the relevant technology (standard rolled steel sections, pre-cast concrete members). Such a problem is clearly represented by an MP whose KKT conditions may be considered and a dual MP derived. A special case (Munro, 1979a) is the minimization of a linearized weight function for the plastic limit design of a steel frame of known geometry and topology. The variables of the problem may be the design variables ($\mathbf{d}$) and the indeterminacies($\mathbf{p}$). The design variables for a flexural frame will be the plastic moments of resistance which fix the steel sections. Using the safe theorem of plastic limit analysis, any solution which is statically admissible will correspond to a safe solution, and the problem is to find amongst these safe solutions that one which minimizes the weight. This problem can be stated in the form of the primal LP of Figure 13.10.

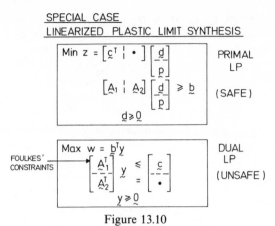

Figure 13.10

Dualizing this program leads to another LP and it would be of interest to give a physical interpretation to the dual variables. The second set of dual constraints become the mechanism compatibility conditions if the dual variables (**y**) are the non-negative components of the mechanism deformation rates. The first set of dual constraints are additional properties of the optimal solution and are termed the *Foulkes' constraints*. Thus, even in this managerial use of linear programming, new results are obtained which give further insight into the studied problem. It can be shown (Munro, 1979b) that the dual LP presents less computational effort when using the standard form of the Simplex algorithm.

However, a much more convincing demonstration of the advantages of a systems approach to structural problems is revealed when the equivalence theorem is adapted to the generation of extremal principles in discretized structural mechanics. Here the fundamental structural relations are organized in the form of KKT conditions (Figure 13.11) before inferring the equivalent pair of primal–dual MPs. For example, if a mesh description (Munro and Smith,

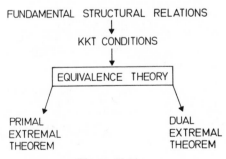

Figure 13.11

MESH DESCRIPTION OF FUNDAMENTAL
STRUCTURAL RELATIONS AT PLASTIC COLLAPSE

$A\underset{\sim}{x} = \underset{\sim}{b}$	KINEMATIC ADMISSIBILITY =MECHANISM COMPATIBILITY + SCALING
$A^T\underset{\sim}{y} + \underset{\sim}{s} = \underset{\sim}{c}$	STATIC ADMISSIBILITY =EQUILIBRIUM + YIELD
$\underset{\sim}{x}, \underset{\sim}{s} \geq \underset{\sim}{0}$	NON−NEGATIVITY
$\underset{\sim}{s}^T\underset{\sim}{x} = 0$	PARITY

LCP

Figure 13.12

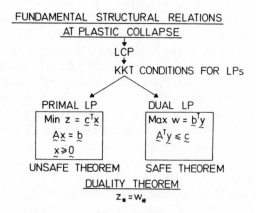

FUNDAMENTAL STRUCTURAL RELATIONS
AT PLASTIC COLLAPSE

LCP

KKT CONDITIONS FOR LPs

PRIMAL LP

$$\text{Min } z = \underset{\sim}{c}^T\underset{\sim}{x}$$
$$A\underset{\sim}{x} = \underset{\sim}{b}$$
$$\underset{\sim}{x} \geq \underset{\sim}{0}$$

UNSAFE THEOREM

DUAL LP

$$\text{Max } w = \underset{\sim}{b}^T\underset{\sim}{y}$$
$$A^T\underset{\sim}{y} \leq \underset{\sim}{c}$$

SAFE THEOREM

DUALITY THEOREM
$z_* = w_*$

Figure 13.13

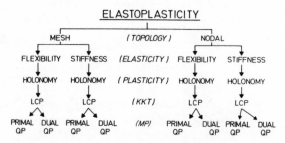

Figure 13.14

1972) of the fundamental structural relations at plastic collapse of a structure are stated using a *piecewise linearized* (PWL) yield criterion (Figure 13.12) then they constitute a *linear complementarity problem* (LCP) of precisely the type which has been shown to correspond to LPs. The application of the equivalence theory is outlined in Figure 13.13 and the two LPs will be seen to encode the two limit theorems. These theorems are, of course, well known, but this same formalism can be applied to successively more complex problems and the corresponding extremal theorems can be generated.

This process is outlined in Figure 13.14 for the holonomic elastoplasticity of structures with PWL yield criteria and using a *finite element* (FE) model (Maier, 1968; Smith, 1974; Smith and Munro, 1978). Choices are available with respect to (i) a mesh or nodal description of the FE connectivity and (ii) a flexibility or stiffness matrix representation of the elastic properties of the elements. Thus four alternative LCPs can be derived and these are equivalent to four pairs of primal–dual QPs. These QPs in turn encode the extensions of the dual variational principles of elasticity to this class of problem.

13.4 Randomness

The basic concept in the statistical modelling of engineering systems is the *decision tree* (Figure 13.15) [see VI, §19.4.2]. The various actions (a_i) available to the decision-maker are listed and the appropriate states of nature (s_j) are identified for each action. The subject of *utility theory* is concerned with the allocation of a utility value (u_{ij}) to each action-state pair (a_i, u_j). This is a measure of the 'goodness' or 'worth' of the eventuality (e_{ij}) identified by that action-state pair. The prescription of statistical (or *Bayesian*) *decision analysis* is to take that action which maximizes the expected utility [VI, §19.1]. Alternatively, it may be more convenient to express the 'badness' of the eventuality

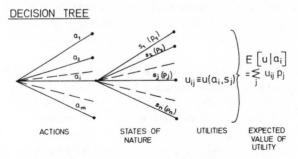

DECISION TREE

$$u_{ij} \equiv u(a_i, s_j)$$

$$E\left[u \mid a_i\right] = \sum_j u_{ij}\, p_j$$

ACTIONS   STATES OF NATURE   UTILITIES   EXPECTED VALUE OF UTILITY

<u>PRESCRIPTION</u> of Bayesian Decision Analysis
"Take that action which <u>MAXIMIZES</u> the expected utility"

Figure 13.15

through a loss function and, in this case, the action which minimizes the loss will be sought.

In general, engineers will be compelled to make decisions when faced with sparse prior knowledge and, in the process of operating the system, new knowledge will be gained. New statistical knowledge can be incorporated in the decision-making by means of *Bayes' rule* [see II, (16.4.9)], which will update the probabilities (Figure 13.16). This evolutionary process will ensure that at any stage the optimal decision will be determined by utilizing all the statistical knowledge available to the engineer.

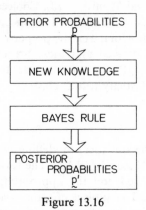

Figure 13.16

The fundamental problem is to evaluate the 'best' estimate of the prior probabilities. The prior knowledge will be assumed to be in the form of known expected values $(\bar{g}_i)$ of functions (g_i) of a random variable (X) (Figure 13.17). The probabilities (p_j) must satisfy the constraints imposed by the prior knowledge as well as the known properties (normality, non-negativity) of all probability distributions (Figure 13.18). However, in general, there will be a multiplicity of probability distributions which satisfy these constraints. Which distribution is best?

The Shannon (1948) measure of statistical uncertainty is the entropy (H) defined by the strictly concave function [see IV, §15.2.6] given in Figure 13.9. Suppose two feasible probability distributions have different entropies; then the distribution with the smaller entropy implies less uncertainty (and therefore greater certainty) than the distribution of greater entropy. However, this greater certainty is not justified by the prior statistical knowledge and hence if the distribution of smaller entropy were preferred then it would be a biased choice. Thus the least-biased distribution (Jaynes, 1968; Munro, 1979b) is that one which maximizes the entropy subject to the constraints of prior statistical knowledge (Figure 13.19).

PRIOR PROBABILITIES

RANDOM VARIABLE X

STATES $\quad s_1, - -, s_j, - -, s_n$

PROBABILITIES $\quad p_1, - -, p_j, - -, p_n$

FUNCTION OF A RANDOM VARIABLE

$$g_i (X)$$

EXPECTED VALUE

$$\sum_j g_i (X = s_j) \; p_j \equiv \bar{g}_i$$

Figure 13.17

PRIOR KNOWLEDGE

$$\bar{g}_1, \bar{g}_2, - - -, \bar{g}_i, - - -, \bar{g}_m$$

$\{p_1, - -, p_j, - -, p_n\}$ must satisfy the constraints

$$\sum_j g_i (X = s_j) \; p_j \equiv \bar{g}_i \qquad i = 1, 2, - -, m$$
$$p_j \geqslant 0 \qquad j = 1, 2, - -, n$$
$$\sum_j p_j = 1$$

WHAT IS THE "BEST" ESTIMATE

OF $\{p_1, - -, p_j, - -, p_n\}$?

Figure 13.18

MEASURE OF STATISTICAL UNCERTAINTY

$$H = - K \sum_{j=1}^{n} p_j \ln p_j \qquad \text{ENTROPY}$$

EVALUATION OF PRIOR PROBABILITIES

$$\text{Max} \left(\frac{H}{K}\right) = - \sum_j p_j \ln p_j$$
$$\sum_j p_j = 1$$
$$\sum_j g_i (X = s_j) \; p_j = \bar{g}_i$$
$$p_j \geqslant 0$$

PRINCIPLE OF
MAXIMUM
ENTROPY

Figure 13.19

It will be seen that the determination of prior probabilities has been reduced to the solution of a mathematical program of the type discussed earlier for deterministic optimization. Because of the special nature of this problem, the non-negative constraints can be dropped and the MP is reduced to a classical optimization problem. This process is a purely objective way of handling the statistics and removes any personalistic consideration from the evaluation of the prior probabilities.

A simple illustration of the application of this method to a managerial problem is concerned with the ready-mixed concrete industry (Munro and Jowitt, 1978). In this industry many relatively simple decisions have to be made in the face of rapidly changing requirements. These decisions relate to the production and transport of concrete in answer to orders which are frequently received with little or no prior warning. Normally the decisions for a group of mixing plants are made at a single control centre by someone who receives the orders by telephone, makes all decisions regarding the operation of plant and allocation of mixer trucks and then communicates the decisions by telephone to the plants and the transport depots. The decisions are subject to continual reappraisal, partly because of operational exigencies within the group (e.g. the breakdown of a truck) or outside the group (e.g. the malfunctioning of a pump on a site to which concrete is being delivered) but mainly because of unpredictable changes in the state of orders. The efficient operation of a group is clearly dependent on the competence of the central controller, who is frequently under great stress. The ability to accept incoming calls and to process orders into good decisions represents a potential constriction on the channels of control. Simplistic attempts to remove this blockage are frequently of no avail. For example, the introduction of more telephones or more controllers may aggravate the problem by making decision-making more difficult or by removing the facility for making a single overall decision for the entire group. This situation is worsened by the facts than an important sales incentive is the ready availability of a wide range of

SINGLE MIXING PLANT

Possible mixes	$a_1, a_2, \ldots, a_i, \ldots, a_m$
unfilled orders at start of period	$U_1, U_2, \ldots, U_i, \ldots, U_m$
orders received during period	$X_1, X_2, \ldots, X_i, \ldots, X_m$
Historical average orders received during period	$\mu_1, \mu_2, \ldots, \mu_i, \ldots, \mu_m$
Capacity	V

Figure 13.20

CONSIDER ONE MIX (i^{TH}mix)

$$p_{x_i} (s_q^j) \equiv p_q \qquad q = 0,1,---,\infty$$

$$\text{Max } H = -\sum_q p_q \ln p_q$$
$$\sum_q p_q = 1$$
$$\sum_q p_q q = \mu$$
$$p_q \geqslant 0$$

Solution $p_q = \left[\dfrac{1}{1+\mu}\right]\left[\dfrac{\mu}{1+\mu}\right]^q$

Figure 13.21

mixes and that the quick acceptance and processing of all orders is vital to the commercial success of a company.

The significant parameters for a simplified model of a single mixing plant are shown in Figure 13.20. The controller must decide on the optimal mix to be produced during a considered period. Prior knowledge is needed of the unfilled orders at the start of the period and also of the average orders received in the past during corresponding periods. The total order to be received for any mix during the considered period is, of course, unknown and must be regarded as a random variable (X_i) [see II, §4.1]. It would assist the decision-making if the probabilities of the order for any mix attaining any particular magnitude could be evaluated. Using the maximum entropy formalism an explicit solution can be obtained as shown in Figure 13.21. This permits the expected value of the orders to be calculated for each mix and hence the expected values of unfilled orders and of surplus concrete for any action, i.e. for any mix to be produced. The loss function can be constructed for any time period and a selection can be made of the mix whose production will minimize the loss. Fuller details of this have been presented elsewhere, but the main point is that the maximum entropy formalism has provided the key to this managerial problem.

However, in the spirit of the systems approach to engineering, a further example will be given of this technique with a non-managerial application to a problem in engineering science. The selected problem is concerned with granular materials (Jowitt and Munro, 1975). Such materials are usually represented either as a continuum or as possessing a regular packing microstructure. Consider an array of uniform spherical particles. A study has been made of a number of packing arrangements and twenty-eight distinct microstates have been identified (Figure 13.22). Any macrosample (of volume V) is assumed to be divided into n units of volume, where n is large. It is required to estimate the number (n_j) of units of volume in the jth microstate. Alternatively, if one could focus on one particle of the macrosample one might speculate as to the

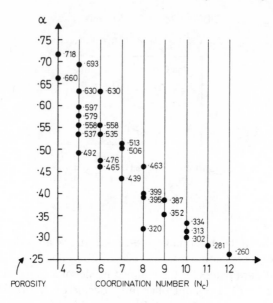

11	BASE FIGURES
42	PACKING ARRANGEMENTS
28	DISTINCT MICRO-STATES

Figure 13.22

probability ($p_j = n_j/n$) that the neighbourhood of the randomly selected particle will be in the jth microstate. The porosity (α_j) of the jth microstate is known and suppose that the porosity ($\bar{\alpha}$) of the macrosample is obtained by experiment. The least-biased probabilities (p_j) can now be evaluated by the program of Figure 13.23. The results for the full range of values of porosity ($\bar{\alpha}$) are sketched in Figure 13.24 and the supremal value of porosity will be termed the critical porosity ($\bar{\alpha}_c$). It will be seen that, in keeping with the statistical mechanics viewpoint, it can be concluded that if any sample of lower porosity is disturbed then it will tend to adopt a new configuration of higher porosity whilst a macrosample whose porosity is higher than this critical value will tend to adopt a new configuration of lower porosity when disturbed. This purely theoretical result is broadly confirmed by the actual behaviour of granular materials.

From the calculated probabilities and the coordination numbers of the microstates the statistical mean coordination number can be evaluated for any macrosample porosity. These results are plotted as a broken line in Figure 13.25, the dots representing experimental results.

$$\text{Max}\left(\frac{H}{k}\right) = -\sum_j p_j \ln p_j$$

$$\sum_j p_j = 1$$

$$\sum_j \alpha_j p_j = \bar{\alpha}$$

$$p \geqslant 0$$

$$j = 1, 2, \ldots, 28$$

Figure 13.23

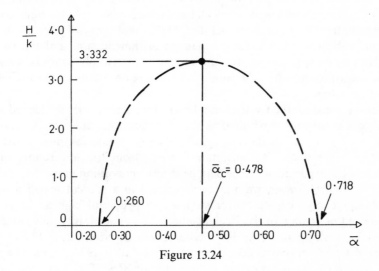

Figure 13.24

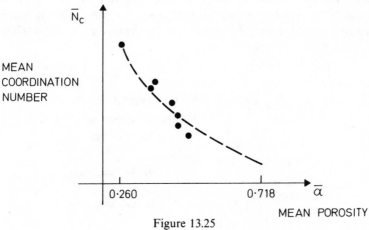

Figure 13.25

Thus the same techniques which have been found convenient in the solution of strictly managerial problems have been of assistance in treating a fundamental problem of statistical geomechanics.

13.5 Imprecision

The distinction between randomness and imprecision has been drawn earlier. This distinction can perhaps be made clearer by a consideration of *sets* [see I, Chapter 1]. The universal set (U) may be the student population of a college and the set A may be the engineering students whilst set B may be the undergraduates. The union is the set of all the students who are engineers or undergraduates or both. The intersection is the set of all the students who are both engineers and undergraduates. The *Venn diagram* [see I, §1.2.2] is shown in Figure 13.26. The plus sign indicates a union rather than an arithmetic sum and the various candidates for entry to the set are shown to the right of a solidus whilst a zero or unity is shown to the left of the solidus, depending on whether the candidate is without or within the set.

In conventional set theory, the boundaries of the sets are crisply defined and a candidate is clearly either within or without the set. An event is a sub-set of the sample space and probability is a measure of an event normally associated with a random experiment. The uncertainty arises from the uncertainty of the outcome to the experiment rather than from any imprecision in the definition of the event itself. However, we may be interested in a student group which is defined imprecisely—e.g. the 'clever' students. In structural engineering we may be interested in 'strong' or 'flexible' members. Thus uncertainty may enter into our decision-making due to the imprecision of our descriptions. In Figure 13.27 the idea of a crisp set is extended (Blockley, 1980; Zadeh, 1973) to that of a *fuzzy set* in which the boundaries are defined imprecisely. The supports for the candidates are now allowed to take any value from zero to unity, depending on the compatibility of the candidate with respect to the concept being represented fuzzily. Just as the plus sign indicated union in the discrete case, so the integral

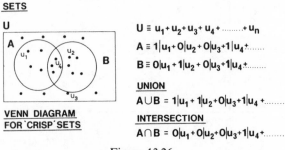

SETS

$U \equiv u_1 + u_2 + u_3 + u_4 + \dots\dots + u_n$

$A \equiv 1|u_1 + 0|u_2 + 0|u_3 + 1|u_4 + \dots\dots$

$B \equiv 0|u_1 + 1|u_2 + 0|u_3 + 1|u_4 + \dots\dots$

UNION

$A \cup B = 1|u_1 + 1|u_2 + 0|u_3 + 1|u_4 + \dots\dots$

INTERSECTION

$A \cap B = 0|u_1 + 0|u_2 + 0|u_3 + 1|u_4 + \dots\dots$

VENN DIAGRAM
FOR 'CRISP' SETS

Figure 13.26

FUZZY SETS

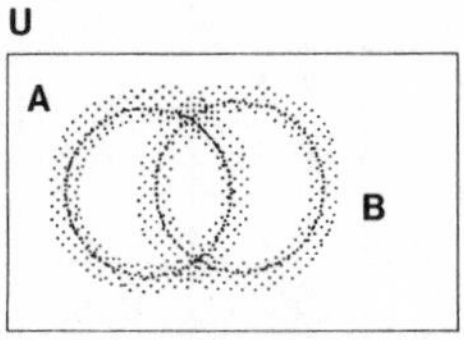

U

A ... B

VENN DIAGRAM
FOR 'FUZZY' SETS

$$U = u_1 + u_2 + \ldots + u_n$$

$$A \equiv \mu_1 | u_1 + \mu_2 | u_2 + \ldots + \mu_n | u_n$$

$$A \equiv \sum_{i=1}^{n} \mu_i | u_i \quad \text{(DISCRETE)}$$

$$A \equiv \int_U \mu_A(u) | u \quad \text{(CONTINUOUS)}$$

Figure 13.27

sign is adapted to continuous union. Some operations (*union, intersection, concentration, dilation* and *complement*) on fuzzy sets are indicated on Figure 13.28, in which the symbol ∨ represents the greater of the pair of quantities linked by the symbol. Similarly the symbol ∧ incidates the lesser of the pair of quantities linked by the symbol. A considerable advance was made in the utilization of fuzzy set theory by its association with linguistic variables. Labels (e.g. strong, weak, flexible, tall, old), hedges (e.g. very, quite, fairly, extremely), the negation (not) and connectives (e.g. and, but, or) can be assembled into relatively complex statements and their fuzzy representations can be compounded from the previous operations. Some illustrative examples are given in Figure 13.29 for strength, where the four candidates represent stress grades of timber joists, for example.

The problem of decision-making in the ready-mixed concrete industry has been considered previously from a statistical viewpoint. However, it has been found that weather and traffic conditions vary markedly from day to day and exert a major influence on the incoming orders. The controller is concerned

$$A \cup B = \int_U \left[\mu_A(u) \vee \mu_B(u) \right] \Big/ u$$

$$A \cap B = \int_U \left[\mu_A(u) \wedge \mu_B(u) \right] \Big/ u$$

$$\text{CON}(A) \equiv A^2 = \int_U \left[\mu_A(u) \right]^2 \Big/ u$$

$$\text{DIL}(A) = A^{0.5} = \int_U \left[\mu_A(u) \right]^{0.5} \Big/ u$$

$$\neg A = \int_U \left[1 - \mu_A(u) \right] \Big/ u$$

Figure 13.28

SET	LINGUISTIC VARIABLE	u			
		1	2	3	4
A	WEAK	0	0·1	0·7	1
B	STRONG	1	0·7	0·1	0
$\neg A$	NOT WEAK	1	0·9	0·3	0
$\neg B$	NOT STRONG	0	0·3	0·9	1
A^2	VERY WEAK	0	0·01	0·49	1
A^3	HIGHLY WEAK	0	0·001	0·343	1
A^4	VERY VERY WEAK	0	0·0001	0·2401	1
$\neg(A^2)$	NOT VERY WEAK	1	0·99	0·51	0
$\neg(B^2)$	NOT VERY STRONG	0	0·51	0·99	1
$[\neg(A^2)] \cap [\neg(B^2)]$	NOT VERY WEAK & NOT VERY STRONG	0	0·51	0·51	0
$[A] \cap [\neg(A^2)]$	WEAK BUT NOT VERY WEAK	0	0·1	0·51	0
$[A^4] \cup [B^4]$	VERY VERY WEAK OR VERY VERY STRONG	1	0·24	0·24	1

Figure 13.29

with making proper allowance for these effects based on a subjective assessment of the prevailing conditions. The orders for any mix will be reduced to five states (0, 1, 2, 3, 4) and some fuzzy sets for different levels of demand are shown in Figure 13.30. The linguistic label **large** is shown as a fuzzy set but if the statement is made that 'the demand is large' then the corresponding fuzzy supports become the possibility distribution for demand.

The controller could describe the traffic conditions with respect to the time that would be taken to cover a particular distance by a mixer truck. Thus the candidates are $\{10, 15, 20, 25, 30\}$ where the numbers are the durations of this standard journey in minutes. In a similar way weather conditions might be

DEMAND

$$\text{large} = 0|0 + 0\cdot2|1 + 0\cdot5|2 + 0\cdot9|3 + 1|4$$

$$\text{small} = 1|0 + 0\cdot9|1 + 0\cdot5|2 + 0\cdot2|3 + 0|4$$

$$\text{very small} = 1|0 + 0\cdot81|1 + 0\cdot25|2 + 0\cdot04|3 + 0|4$$

$$\text{very large} = 0|0 + 0\cdot04|1 + 0\cdot25|2 + 0\cdot81|3 + 1|4$$

$$\text{not very large} = 1|0 + 0\cdot96|1 + 0\cdot75|2 + 0\cdot19|3 + 0|4$$

Figure 13.30

TRAFFIC

$$\underset{\sim}{\text{good}} = 1|10 + 0.8|15 + 0.4|20 + 0.2|25 + 0.1|30$$

$$\underset{\sim}{\text{bad}} = 0.1|10 + 0.2|15 + 0.7|20 + 0.8|25 + 1|30$$

$$\underset{\sim}{\text{not good}} = 0|10 + 0.2|15 + 0.6|20 + 0.8|25 + 0.9|30$$

$$\underset{\sim}{\text{very good}} = 1|10 + 0.64|15 + 0.16|20 + 0.04|25 + 0.01|30$$

WEATHER

$$\underset{\sim}{\text{clement}} = 1|1 + 1|0.8 + 0.4|0.6 + 0.2|0.4 + 0.1|0.2 + 0|0$$

$$\underset{\sim}{\text{inclement}} = 0|1 + 0|0.8 + 0.6|0.6 + 0.8|0.4 + 0.9|0.2 + 1|0$$

$$\underset{\sim}{\text{very clement}} = 1|1 + 1|0.8 + 0.16|0.6 + 0.04|0.4 + 0.01|0.2 + 0|0$$

Figure 13.31

considered with respect to sunshine and precipitation and for the purposes of the present illustration the candidates are $\{1, 0.8, 0.6, 0.4, 0.2, 0\}$ where the higher numbers represent the better weather states. Some sample fuzzy sets (or possibility distributions) are shown in Figure 13.31.

The next stage is to construct the fuzzy relationship between traffic conditions and demand. In Figure 13.32 a moderately complex statement is made by the controller expressing a subjective view of this relationship, and this is converted to a fuzzy representation. The fuzzy product (**good** × **large**) is obtained in matrix form by considering any pair of candidates and inserting the lesser support. Similarly, in Figure 13.33 the product (**not good** × **not very large**) is obtained and finally the two matrices are combined by taking the larger support to derive the fuzzy relationship ($\mathbf{R_{TD}}$) between traffic and demand.

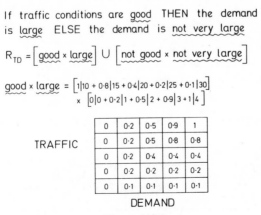

If traffic conditions are $\underset{\sim}{\text{good}}$ THEN the demand is $\underset{\sim}{\text{large}}$ ELSE the demand is $\underset{\sim}{\text{not very large}}$

$$R_{TD} = \left[\underset{\sim}{\text{good}} \times \underset{\sim}{\text{large}} \right] \cup \left[\underset{\sim}{\text{not good}} \times \underset{\sim}{\text{not very large}} \right]$$

$$\underset{\sim}{\text{good}} \times \underset{\sim}{\text{large}} = \left[1|10 + 0.8|15 + 0.4|20 + 0.2|25 + 0.1|30 \right] \\ \times \left[0|0 + 0.2|1 + 0.5|2 + 0.9|3 + 1|4 \right]$$

0	0·2	0·5	0·9	1
0	0·2	0·5	0·8	0·8
0	0·2	0·4	0·4	0·4
0	0·2	0·2	0·2	0·2
0	0·1	0·1	0·1	0·1

TRAFFIC (left of table)

DEMAND

Figure 13.32

Similarly,

not good × not very large =

0	0	0	0	0
0·2	0·2	0·2	0·19	0
0·6	0·6	0·6	0·19	0
0·8	0·8	0·75	0·19	0
0·9	0·9	0·75	0·19	0

R_{TD} =

0	0·2	0·5	0·9	1
0·2	0·2	0·5	0·8	0·8
0·6	0·6	0·6	0·4	0·4
0·8	0·8	0·75	0·2	0·2
0·9	0·9	0·75	0·19	0

Figure 13.33

If weather conditions are clement THEN the demand is large ELSE the demand is very small

$R_{WD} = \left[\text{clement} \times \text{large} \right] \cup \left[\text{inclement} \times \text{very small} \right]$

$$R_{WD} = \begin{bmatrix} 0 & 0·2 & 0·5 & 0·9 & 1 \\ 0 & 0·2 & 0·5 & 0·9 & 1 \\ 0·6 & 0·6 & 0·4 & 0·4 & 0·4 \\ 0·8 & 0·8 & 0·25 & 0·2 & 0·2 \\ 0·9 & 0·81 & 0·25 & 0·1 & 0·1 \\ 1 & 0·81 & 0·25 & 0·04 & 0 \end{bmatrix}$$

Figure 13.34

Controller decides :—

"transport conditions are very good"
and "weather is very clement"

$$D = (T \circ R_{TD}) \cup (W \circ R_{WD})$$

$$D = \left[0·2|0 + 0·2|1 + 0·5|2 + 0·9|3 + 1|4 \right]$$

Figure 13.35

In Figure 13.34 the fuzzy relationship ($\mathbf{R_{WD}}$) between weather and demand is constructed from a subjective view of the controller. Finally the subjective views of the controller are obtained for an actual day and written as possibility distributions for traffic conditions ($\mathbf{T}$) and weather conditions ($\mathbf{W}$). The corresponding possibility distribution for demand ($\mathbf{D}$) is obtained as shown in Figure 13.35 by first taking $\mathbf{T}$ as a row vector and multiplying into the matrix $\mathbf{R_{TD}}$ [see I, §6.2], but the normal multiplication is replaced by ($\wedge$) and the addition by ($\vee$). Secondly, the $\mathbf{W}$ vector is multiplied into the $\mathbf{R_{WD}}$ matrix in a similar maximin fashion, and finally the union of the two fuzzy sets leads to the possibility distribution for demand, as shown in Figure 13.35.

Thus a fuzzy support (or possibility) for each order state has been obtained, and this can be used to determine the optimal decision. The key problem is to combine the probabilities with the possibilities for those problems which have random and fuzzy components. This is the subject of some current research (Munro, 1979b).

13.6 Closure

The foregoing is a brief and very personal selection of some aspects of engineering decision-making tackled in a systems way. A broader viewpoint can be achieved from a study of the increasing literature on this subject (Neufville and Stafford, 1971; Stark and Nicholls, 1972).

J.M.

References

Blockley, D. I. (1980). *The Nature of Structural Design and Safety*, McGraw-Hill.
Jaynes, E. T. (1968). Prior probabilities, *IEEE Trans. Systs. Sci. Cybern.* SSC-4, **3**, 227.
Jowitt, P. W., and Munro, J. (1975). The influence of void distribution and entropy on the engineering properties of granular media, *Proc. ICASPZ*, Aachen.
Maier, G. (1968). Quadratic programming and theory of elastic-perfectly-plastic structures, *Mecannica*, **3**, Dec. 1968, 265–273.
Munro, J. (1979a). Optimal plastic design, *Engineering Plasticity Math. Programming* (Eds Cohn, M. Z., and Maier, G.), Chap. 7, Pergamon.
Munro, J. (1979b). Uncertainty and fuzziness in engineering decision making, *Proc. First Canadian Seminar 'Systems Theory for the Civil Engineer'*, Calgary, pp. 113–133.
Munro, J., and Jowitt, P. W. (1978). Decision analysis in the ready-mixed concrete industry, *Proc. Inst. Civ. Engrs.*, **65**, March, 41–52.
Munro, J., and Smith, D. L. (1972). Linear programming in plastic analysis and synthesis. *Proc. Int. Symp. Computer-aided structural design*, Warwick.
Neufville, R. de, and Stafford, J. H. (1971). *Systems Analysis for Engineers and Managers*, McGraw-Hill.
Shannon, C. E. (1948). The mathematical theory of communication, *Bell Systems Tech. J.*, **27**, 279–428, 623–656.
Smith, D. L. (1974). *Plastic Limit Analysis and Synthesis of Structures by Linear Programming*, Ph.D. thesis, University of London.

Smith, D. L., and Munro, J. (1978). On uniqueness in the elasto-plastic analysis of frames, *J. Struct. Mech.*, **6**, 85.

Stark, R. M., and Nicholls, R. L. (1972). *Mathematical Foundations for Design, Civil Engineering Systems*, McGraw-Hill.

Zadeh, L. A. (1973). Outline of a new approach to the analysis of complex systems and decision processes, *IEEE Trans. Systems: Man and Cybernetics*, **SMC-3**, 28–44.

Mathematical Methods in Engineering
Edited by G. A. O. Davies
© 1984, John Wiley & Sons, Ltd.

14

Control Engineering

14.1 Introduction

Dynamical engineering systems can be sub-divided into those concerned with the conversion and transfer of power and those concerned with the conversion and transfer of information. A gas turbine is an example of the former class and a television camera of the latter. In practice, however, all systems contain aspects of both power flow and information flow and the connection between the two is usually expressed in *control engineering* terms. Control engineering can be said to be the use of information to control the flow of power, and is largely centred on the concept of *feedback*.

Feedback is the use of some measure of the outputs of a system to adjust the system inputs so as to make those outputs change in a desired way. It can be seen that this concept concerns a loop and immediately questions of speed of response and stability arise. The control engineer's major preoccupations are with these two problems.

The simplest form of control is *open loop* control. In this case the system is assumed to be sufficiently predictable to allow assumptions to be made regarding the outputs so that there is no need to actually measure them. This form of control is common with digital systems, e.g. when the final drive element is a stepping motor.

Closed loop control is used where disturbances or the basic system complexity make the responses unpredictable. Closed loop systems usually involve feedback; however, sometimes *feedforward* is also included to provide a further element of control. An example of feedback would be if the head of water in a reservoir were used as the source of information to control the supply of water to the reservoir. Feedforward would be if the head of water were used to control the outflow of water from the reservoir.

In all the forms of control mentioned it is necessary to have means of describing the dynamics of systems and methods of measuring these dynamics where they cannot reasonably be deduced from basic scientific laws. It is reasonable, then, to start with a statement of the mathematical methods used to describe systems and then give some consideration to methods of finding approximations to these *mathematical models* when real systems are being tested.

From then on it will be assumed that a mathematical model of the system and the information (signals) flowing within it are known. The remaining sections are concerned with design and analysis techniques.

14.2 Mathematical models of systems

It is assumed throughout this section that the systems in question are stable in the sense used in linear systems theory. That is, any transient input which dies away to zero will produce a transient response (output) which also dies away to zero.

All the methods of linear systems modelling discussed in this section are described in standard books on control engineering. Further information is available, for example, from McGillem and Cooper (1974) or Saucedo and Schiring (1968). Models for non-linear systems can be studied further via Atherton (1975) or Smith (1966).

Continuous linear systems

(a) *The impulse response function.* A *pulse* can be thought of as a signal which is zero for all time except for the time period $0 < t < \delta t$, say, when it has constant value. If this constant value is equal to $1/\delta t$ and δt is allowed to become very small, then this pulse becomes the *unit impulse*. The response of a system to a unit impulse is known as the *impulse response function* and is characteristic of the system.

Related to the impulse response, in fact the integral of it [see IV, §4.1], is the *step response*. Engineering specifications of systems are often given in terms of the step response: particularly, the *rise time, overshoot, settling time,* etc.

(b) *The transfer function.* If the impulse response function of a system is $h(t)$, then its *Laplace transform* [see IV, §13.4]

$$H(s) = \int_0^\infty h(t)\, e^{-st}\, dt$$

is known as the *transfer function* of the system.

If the system input is $x(t)$ and its output is $y(t)$, then

$$H(s) = \frac{Y(s)}{X(s)}, \tag{14.1}$$

where $X(s)$ and $Y(s)$ are the Laplace transforms of $x(t)$ and $y(t)$ respectively.

The signals x and y can also be shown to be related by the *convolution integral*

$$y(t) = \int_a^b x(\tau)h(t - \tau)\, d\tau$$

or

$$y(t) = \int_c^d h(\tau)x(t - \tau)\, d\tau,$$

where the limits a, b, c and d depend on the signal x and on the nature of the system. A common situation in which x and h both equal zero for $t < 0$ gives $a = c = 0$ and $b = d = t$.

For multiple input–multiple output systems the Laplace transforms of the input and output vectors are related via a *transfer function matrix* which will be discussed further in Section 14.10.

(c) *The frequency response function.* If $x(t)$ is a sinusoid, say $Ae^{j\omega t}$, for $-\infty < t < \infty$, then the convolution integral can be used to show that $y = Ae^{j\omega t}H(j\omega)$, where

$$H(j\omega) = \int_{-\infty}^{\infty} h(t)\, e^{-j\omega t}\, dt,$$

which is the *Fourier transform* of $h(t)$ [see IV, §13.2]. $H(j\omega)$ is called the *frequency response function*. Engineering specifications of systems are often stated implicitly in terms of the frequency response function. This is done by using concepts such as *bandwidth* and *resonant frequency*, for example. These concepts will be elaborated in Section 14.6.

(d) *Differential equations.* The input, x, and output, y, of a lumped, time-invariant system can be represented by a differential equation of the form

$$P(D)x = Q(D)y,$$

where P and Q are polynomials in the operator $D = d/dt$ [see IV, §7.1]. In this situation the transfer function is a *rational function* [see I, §14.9] of the complex variable s.

It can be shown that

$$\frac{Y(s)}{X(s)} = \frac{P(s)}{Q(s)}$$

and so, by comparison with (14.1), $P(s)/Q(s)$ is therefore the transfer function. The impulse response function can be recovered from this by using *partial fractions* [see I, §14.10] and inverse Laplace transforms of standard first- and second-order rational functions [see IV, §13.4.2].

(e) *State-space models.* An alternative to the use of an nth-order differential equation is to use n first-order equations to relate a vector of state variables $\mathbf{x}$ to their derivatives [see IV, §7.9]. For the linear systems being considered an input–output relationship of the form $\dot{\mathbf{x}} = \mathbf{A}\mathbf{x}$ arises when this *state-space* description is used, where $\mathbf{A}$ is a square matrix [see I, §6.2]. If the system is non-linear [see IV, §7.11] we have $\dot{\mathbf{x}} = \mathbf{f}(\mathbf{x})$, $\mathbf{f}$ being a non-linear vector function [see IV, §5.3]. Control system design using the state-space approach will be elaborated in Section 14.10.

Sampled linear systems

In this section it is assumed that information regarding the signals at various points in the system is only available at discrete, equal-spaced points in time. The models are described very briefly as they are directly analogous to those described in the previous section for continuous systems.

(a) *Pulse response.* The pulse response is the response of a sampled system to a series of input samples which are zero for all values of time, except the sample at zero time which has unit value.

(b) *Transfer function.* The series of samples which is the pulse response of a system can be represented by a series of impulse functions delayed by integral numbers of sample periods and weighted by the value of the impulse response at those times. This can be written as

$$h^*(t) = \sum_{n=0}^{\infty} h(nT)\, \delta(t - nT),$$

where T is the sample period.

Taking the Laplace transform of this [see IV, §13.4] gives

$$H^*(s) = \sum_{n=0}^{\infty} h(nT)\, e^{-nTs}$$

or, writing e^{Ts} as z, then

$$H(z) = \sum_{n=0}^{\infty} h(nT)\, z^{-n},$$

which is the *Z-transform* of $h(t)$ [see IV, §13.5] and is the *transfer function* for a sampled system.

The input and output sequences x_s and y_s of a sampled system can be related by a *convolution sum* of the form

$$y_s(nT) = \sum_{l=0}^{\infty} x_s(lT) h(nT - lT)$$

and the Z-transform of this results in

$$Y(z) = X(z)H(z)$$

or the alternative statement of the Z-transfer function

$$H(z) = \frac{Y(z)}{X(z)}$$

(c) *Frequency response function.* The frequency response function of a sampled system, as for a continuous system, can be obtained by substituting $s = j\omega$ in the transfer function. This can be shown to be (Saucedo and Schiring, 1968)

$$H*(j\omega) = \frac{1}{T} \sum_{n=-\infty}^{\infty} H[j(\omega + n\omega_s)]$$

where $\omega_s = 2\pi/T$.

It can be seen that this function is periodic with period ω_s.

(d) *Difference equations.* Samples x and y from the input and output sequences of most sampled systems can be related by difference equations of the form

$$P(z^{-1})x = Q(z^{-1})y, \tag{14.2}$$

where P and Q are polynomials in the difference operator $z^{-1} \cdot [z^{-1}x(nT) = x((n-1)T)]$ [see I, §14.12].

Alternatively, written in terms of the Z-transform,

$$\frac{Y(z)}{X(z)} = \frac{P(z^{-1})}{Q(z^{-1})},$$

which is a rational function of z.

(e) *State-space models.* As with continuous systems it is possible to use a state variable description in which n first-order difference equations are used instead of one nth-order equation.

Non-linear systems

This sub-section describes methods of modelling continuous non-linear systems. Each method described has its counterpart in sampled systems theory, but these will not be described separately.

It should also be mentioned that in many engineering situations non-linear elements are approximated by linear models by using the *Taylor–Maclaurin* theorem [see IV, Theorem 5.8.1], specifically when the signal fluctuations about a fixed level are small. The section covers only models of non-linear systems or elements for which the signal variation is large compared with any range of approximate linearity.

Many non-linear systems are analysed for the purposes of control by directly incorporating the known non-linearity into a numerical solution or simulation of the response. Such procedures are special to each system and are not therefore considered here.

(a) *The Volterra series.* A generalization of the impulse response function–convolution integral model of a linear system is the Volterra series describing a non-linear system. The input $x(t)$ and the output $y(t)$ of a continuous dynamical system can be connected by the general equation [see IV, §4.7]

$$
y(t) = \int_{-\infty}^{\infty} h_1(\tau)x(t-\tau)\,d\tau
$$

$$
+ \int_{-\infty}^{\infty}\int_{-\infty}^{\infty} h_2(\tau_1,\tau_2)x(t-\tau_1)x(t-\tau_2)\,d\tau_1\,d\tau_2 + \int\!\!\int\!\!\int_{-\infty}^{\infty}\cdots. \qquad (14.3)
$$

It can be seen that the class of linear systems involves the first term of the series only which is the normal convolution integral.

(b) *Orthogonal expansions.* The Volterra series (14.3) has been expanded by Wiener (1942), Bose (1956) and others as sets of *orthogonal functions* [see IV, §20.4]. Many non-linear systems can be described by the general form $y \simeq f(a_1(t),\ldots,a_n(t))$, where f is a non-linear functional of the set of spectral coefficients a_j describing the past history of $x(t)$. Expanding f as an orthogonal series gives

$$
y \simeq \sum_{i_1=1}^{m}\sum_{i_2=1}^{m}\cdots\sum_{i_n=1}^{m} C_{i_1 i_2 i_3 \cdots i_n}\Phi_{i_1}(a_1)\Phi_{i_2}(a_2)\cdots\Phi_{i_n}(a_n),
$$

$$
\text{where} \qquad C_{i_1\cdots i_n} = \frac{\overline{y\Phi_{i_1}(a_1)\cdots\Phi_{i_n}(a_n)\omega}}{V_n},
$$

the bar denoting time averaging, ω being a weighting function and

$$
V_n = \overline{\omega(\Phi_{i_1}(a_1)\cdots\Phi_{i_n}(a_n))^2}.
$$

The finite number of spectral coefficients a_j restricts the model to systems with finite memory time. The finite length of the series (m) requires certain convergence properties of the non-linear functions Φ within the range of variation of the coefficients a_j and the use of time averages restricts the model to systems with *ergodic* input signals [see II, §19.6] and implies some restrictions to the statistical properties of x. Apart from these restrictions, which are not very severe, the model is quite general.

The expansion suggested by Wiener was in terms of a *Laguerre spectrum* [see IV, §10.5.1] for the functions a_j defining the dynamics and *Hermite functions* [see IV, §10.5.2] for the non-linear processing due to the function Φ_j.

This requires $x(t)$ to be *Gaussian white noise* [see IV, §18.1.3] so that the spectral coefficients are independent and Gaussian with a joint probability

$$p_n = \frac{1}{(2\pi)^{n/2}} \exp\left(-\frac{a_1^2 + a_2^2 + \cdots + a_n^2}{2}\right).$$

[see II, §13.4.1]. Choosing the weighting function ω to be

$$\omega = \exp\left(\frac{a_1^2 + a_2^2 + \cdots + a_n^2}{2}\right)$$

gives $V = \sqrt{2\pi}$ and the functions Φ_j are the Hermite functions $H_j(a_i)\,e^{-a_i^2/2}$.

(c) *Quasi-linearization*. This technique applies to systems in which the non-linear element can be separated out as a single-valued (memory-less) non-linear function or in special cases a double-valued non-linear function.

For the single-valued non-linear function $y = f(x)$ it is usual to define two *equivalent linear gains*

$$K_0 = \frac{\displaystyle\int_{-\infty}^{\infty} f(x)p(x)\,dx}{x_0}$$

and

$$K_1 = \frac{\displaystyle\int_{-\infty}^{\infty} (x - x_0)f(x)p(x)\,dx}{\displaystyle\int_{\infty}^{\infty} (x - x_0)^2 p(x)\,dx}$$

where x_0 is the mean value of x and $p(x)$ is the *amplitude probability–density distribution* of $x(t)$ [see II, §10.1]. K_0 is the gain to the mean value and K_1 to the fluctuating component of the input. The gains are given these values in order to minimize the mean square error of the quasi-linear model. Clearly the gains are functions of the signal parameters.

In feedback control engineering two situations are normally encountered, one in which the signal is Gaussian and $p(x) = (1/\sigma\sqrt{2\pi}) \exp[-(x - x_0)/2\sigma^2]$ and the other in which $x(t)$ is a sinusoid with the d.c. level x_0. In this latter case the gains can be shown to reduce to $K_0 = y_0/x_0$, where y_0 is the d.c. component of the *Fourier series* of y [see IV, §20.5] and $K_1 = Y_1/X_1$, where Y_1 is the fundamental of the Fourier series of y, and X_1 the fundamental of the input Fourier series. The gain K_1 or the function $-1/K_1$ is sometimes given the name the *describing function*.

It is possible to formulate equivalent linear gain models of non-linearities subjected to a combination of inputs. In all cases each gain is a function of the parameters of all the components and so signal interaction through the non-linearity is accounted for.

If the driving signal $x(t)$ is a sinusoid and the non-linearity is double valued, it is still possible to use this type of model for analysis. In this situation the gain to the sine wave becomes complex of the form $K_1 = K_R + jK_I$, where $j = \sqrt{-1}$ again and

$$K_R = \frac{Y_R}{X} \quad \text{and} \quad K_I = -\frac{Y_I}{X}$$

where Y_R is the in-phase component of the fundamental of $y(t)$ and Y_I is the quadrature component.

Descriptions of the applications of these methods of quasi-linearization to feedback control problems are given in Section 14.13, and are described in detail by Atherton (1975).

14.3 Estimating the dynamics of systems

When designing a control system it is necessary to have some form of mathematical model of the system [see IV, Chapter 14]. The engineer will choose whichever model is most convenient for the design method to be used and the measurement data which is available. Most of the commonly used models are defined in Section 14.2. The process of estimating models is often called *identification*.

Model estimation methods can be classified as either parametric or non-parametric depending on whether system parameters, such as transfer function coefficients or time constants, are to be estimated or whether one or other of the functions describing the system is to be estimated as a set of data points.

Parametric models

(a) *Use of the basic laws of science.* This very basic method assumes that the dynamic mechanisms of the system are understood, and thus the structure of a model is known to be correct in principle, and that experiments can be performed to find any unknown parameters.

This type of analysis usually leads in the first place to a set of differential or difference equations and can be applied to electrical, mechanical and electro-mechanical systems. It is possible, for example, to obtain a satisfactory model of an electromechanical servo-mechanism in this way by the use of Newton's laws of motion together with fundamental laws of electric circuit theory and electro-magnetism. Parameters, such as the moment of inertia, torque per unit input current, etc., are either available from manufacturers' data sheets or can be directly measured by experiment.

(b) *Frequency response or transient response testing.* In this method the structure of the system transfer function is assumed and unknown parameters estimated by measurement of discrete points on the response—e.g., the resonant frequency.

It is theoretically possible to obtain a sufficient number of *simultaneous equations* to enable a solution to be obtained for each unknown parameter [see I, §5.10]. Alternatively, an *overdetermined set* of equations can be obtained and the method of *least squares* used for estimation [see VI, §8.1]. Another way is to utilize gross features of the response (say the rate of decay of oscillations in the step response) to estimate parameters (such as the damping factor in a second-order system).

(c) *Random signal methods.* Several methods exist in which a system model is assumed in parametric form, the output of the model and of the real system are compared and the model parameters systematically adjusted to reduce the difference between the two outputs. The characteristics of the input signal used, the nature of the noise, the procedure for adjusting the parameters and the criterion for computing the error provides the large variety of methods in the class often referred to as *hill climbing* problems, although in this case a global *minimum* in the error surface is being sought. A common technique for seeking a minimum is to use the method of *steepest descents* [see IV, §9.14], in which the parameters are adjusted so as to proceed down the steepest gradient to the minimum.

The most powerful general identification procedures using these methods are the methods of *least squares, generalized least squares* and *maximum likelihood* [see VI, Chapters 6 and 12]. In all these cases the model structure is based on *a priori* physical knowledge and on the engineer's experience, although iterative versions of these procedures exist in which the order and structure of the model is automatically adjusted. The procedures are best summarized by referring to sampled data models, as it is invariably necessary to use digital computers to perform the required calculations.

The system models can be generalized (in the single-input–single-output case) to $A(z^{-1})y(i) = B(z^{-1})x(i) + e(i)$, where $x(i)$, $y(i)$ and $e(i)$ are respectively the input, output and noise sequences and $A(z^{-1})$ and $B(z^{-1})$ are polynomials in the difference, delay or shift operator as defined in equation (14.2). A can be considered to have a unit constant term and B to have a zero constant term; the other coefficients of A and B are the parameters θ being sought. For an nth-order model there are $2n + 1$ parameters.

Each time an estimate $\hat{\theta}$ of these parameters is made it will be different because of the finite averaging times involved. In order to assess the usefulness of methods attention is generally focused on the *mean vector* $\boldsymbol{\theta}_2 = \mathbf{E}[\hat{\theta}]$ and the *covariance matrix* $\mathbf{E}[(\hat{\theta} - \theta_2)(\hat{\theta} - \theta_2)']$ [see II, §13.3.1]. An estimator is said to be *asymptotically unbiased* if $\theta_2 \to \theta$ as the record length increases.

If no other factors are considered the choice of estimation algorithm is mainly dictated by the properties of the noise sequence $e(i)$, and it is made primarily to ensure unbiasedness.

If the sequence of differences (residuals) $r(i) = A(z^{-1})y(i) - B(z^{-1})x(i)$ is formed from the set of measurements x and y, the *least squares* estimator seeks to minimize $\Sigma[r(i)]^2$. It can be shown that the resulting estimates $\hat{\theta}$ are asymptotically unbiased provided the noise sequence is uncorrelated (i.e. 'white noise') and independent of x and y [see VI, §8.2.3]. If, however, the noise sequence is correlated other techniques are required if asymptotic unbiasedness is to be guaranteed.

Assuming the noise sequence can be modelled by an *autoregressive sequence* of the form $\varepsilon(i) = H(z^{-1})e(i)$, where H is a polynomial and $\varepsilon(i)$ is white noise, then new sequences $\tilde{x}(i) = H(z^{-1})x(i)$ and $\tilde{y}(i) = H(z^{-1})y(i)$ can be computed which do satisfy the requirement that the assumed noise sequence is uncorrelated. The least squares estimator can now be applied to the set of residuals $A(z^{-1})\tilde{y}(i) - B(z^{-1})\tilde{x}(i)$ and unbiased estimates of the parameters result. This procedure is known as the method of *generalized least squares*. $H(z^{-1})$ is generally not known in advance and an interactive procedure is required to build up an appropriate autoregressive model for the noise.

A further generalization of this situation arises when the noise sequence is modelled by the generalized linear filtering of the white noise sequence $\varepsilon(i)$ of the form

$$e(i) = A(z^{-1}) \frac{C(z^{-1})}{D(z^{-1})} \varepsilon(i),$$

where C and D are also polynomials and $\varepsilon(i)$ is a sequence of normal, uncorrelated zero mean samples, giving a method which is equivalent to the *maximum likelihood estimation*.

With each of these methods, particularly maximum likelihood, the minimization procedure is difficult to perform analytically and numerical techniques such as the *Newton–Raphson* iteration are required [see III, §5.4]. All the techniques can be programmed to provide estimates of the covariance matrix. In each of the techniques mentioned the noise is assumed to be additive. If, however, this is not the case, considerable difficulty arises. However, it is sometimes possible to use alternative methods, particularly in the case, say, of multiplicative noise by applying the technique of *instrumental variables* (Eykhoff, 1974).

Non-parametric models

(a) *Frequency response estimation.* A point on the frequency response function of a system can be obtained by measuring the gain and the phase or the real and imaginary parts of the output when a sinusoid is applied to the input. Machines for generating these waves and providing the measurement data are known as *transfer function analysers*. The measurement time for each estimated point and the variance associated with that estimate are inversely proportional. This method is a special case of the method of cross-correlation leading to equation (14.5).

An alternative method to point-by-point estimation using sinusoidal test signals is the use of a random test signal. If a test signal $x(t)$ can be considered to have arisen from an ergodic random process (McGillem and Cooper, 1974) [see II, §19.6], then its distribution of power with respect to frequency can be described by the *power–density spectrum* $\Phi_{xx}(\omega)$ [see VI, §18.10]. The frequency response function can then be identified as

$$H(j\omega) = \frac{\Phi_{xy}(j\omega)}{\Phi_{xx}(\omega)}, \tag{14.4}$$

where $\Phi_{xy}(j\omega)$ is the *cross power–density spectrum* of the system input $x(t)$ and its output $y(t)$.

If the power–density spectrum of the test signal $x(t)$ is a constant over all frequencies (i.e. $x(t)$ is white noise) then the frequency response function is proportional to the cross-spectral density. Devices called *spectral analysers* based on narrow band filters, multipliers and averagers exist for measuring these functions.

Parametric models can be established from the set of data points describing the frequency response function by assuming a model form and using curve-fitting techniques such as *least squares*.

The asymptotic approximations of Bodé(Section 14.6) are often useful in the early stages of this procedure.

(b) *Impulse response estimation.* The system impulse response can be estimated by repeatedly measuring and averaging the system step response and graphically differentiating it. However, a much more accurate technique is to use a random test signal as described in the previous section. This results in an equation of the form

$$\phi_{xy}(\tau) = \int_{-\infty}^{\infty} \phi_{xx}(\tau_1)h(\tau - \tau_1) \, d\tau_1, \tag{14.5}$$

where $\phi_{xx}(\tau)$ is the *autocorrelation function* of $x(t)$ and $\phi_{xy}(\tau)$ is the *cross-correlation function* of the input and output signals x and y [see II, §22.2]. The impulse response function $h(t)$ is computed by deconvolving (usually numerically) equation (14.5). Equation (14.4) is the Fourier transform [see IV, §13.2] of equation (14.5). Again a simplification results if $x(t)$ is white noise, viz. $\phi_{xy}(\tau) \propto h(\tau)$.

There are many practical considerations which need to be considered in the estimation of impulse or frequency response functions, e.g. what record length to use. Readers are referred to texts such as Davies (1970) for further information on these subjects.

A finite length of a random signal cannot be assumed to have the statistical properties, e.g. the autocorrelation function, of the infinite record. Furthermore, many continuous test signals have a finite chance of having a very large amplitude, thus causing assumptions regarding system linearity to be violated. For these and other reasons specially shaped test signals are often used for identification purposes. The most important class of such signals is the *pseudo-random binary sequence*, which is a two-level signal with many of the properties of white noise. The generation and use of such signals and other discrete-level pseudo-random signals for identification for adaptive control are described, for example, in Davies (1970).

14.4 Steady-state accuracy in linear feedback control systems

Once some description of the basic system dynamics is available it is then possible for the engineer to consider questions of accuracy, stability and dynamic response in the closed loop situation. This section considers the static accuracy of continuous, lumped systems although similar consideration can be extended to sampled systems as well.

In many real situations the process to be controlled can be considered to have an open loop transfer function of the form $KG(s)/s^n$, where K is a constant (the *open loop gain*) and $G(s)$ is a rational function for which both the numerator and denominator polynomials have a constant term of unity [see I, §14.9]. When unity feedback without any form of compensation is used to control this process the system is said to be of *type n*, where n is the number of integrations in the loop. The steady-state errors of such systems are computed by finding the Laplace transforms of the errors $e(t)$ in response to a unit step of 'position', a unit ramp (a unit step of 'velocity'), a unit step of 'acceleration', etc., and using the *final value theorem* of Laplace transforms [see IV, §13.4], viz.

$$\lim_{t \to \infty} e(t) = \lim_{s \to 0} sE(s)$$

to find the value of the error as $t \to \infty$. ('Position', 'velocity', 'acceleration' are taken as the input signal and its derivatives in the general case.)

The following table gives the results of these calculations:

System type	Steady-state error to step changes of		
	Position	Velocity	Acceleration
0	$1/K$	∞	∞
1	0	$1/K$	∞
2	0	0	$1/K$
3	0	0	0
⋮			

It can be seen that the steady-state accuracy is improved if n is large and/or K is large. Unfortunately, in most cases as n and K are increased the closed loop system becomes less stable.

14.5 Stability of control systems

The amplifications, delays and other dynamic effects involved in closed loop control can cause the system to become unstable. The control engineer is concerned with obtaining satisfactory static and dynamic performance while avoiding instability.

If the open loop transfer function is $KH(s)$ then the closed loop transfer function is $KH(s)/[1 + KH(s)]$. A linear system can be said to be stable if its impulse response function tends to zero as $t \to \infty$. The closed loop impulse response function can be found by evaluating the inverse Laplace transform of the closed loop transfer function. If the transfer function is rational then the impulse response function is the sum of *exponential functions* [see IV, §2.11], with *real* or *complex argument* [see IV, §9.5], plus perhaps some *discontinuity functions*. If one or more of these exponential components grows with time as the result of having a positive real part to its exponent then the system is unstable. It can be seen, either from the use of *partial fractions* [see I, §14.10] or *contour integration* [see IV, §9.3] to invert the Laplace transform which is the closed loop transfer function, that the exponents of the exponential components of the impulse response are the roots of the equation

$$1 + KH(s) = 0.$$

This equation is called the *characteristic equation* and its roots are the *closed loop poles*. If there is any value of s which satisfies this characteristic equation and has a positive real part then the system is unstable.

If $H(s)$ is a rational function and is known explicitly then the stability can be examined using the *Routh stability criterion* (Barnett, 1975; Jacobs, 1974; Lambert, 1973) and the equation does not need to be solved. By writing the characteristic equation in the form [see IV, §9.5]

$$\sum_{m=0}^{n} a_{n-m} s^m = 0 \qquad (a_0 > 0), \tag{14.6}$$

then the Routh stability criterion for this system is that all the numbers r_{i1}, $i = 0, 1, 2, \ldots, n$ in the first column of the following *Routh array* be positive. The first two rows of the array are defined by

$$\{r_{01}, r_{02}, r_{03}, \ldots\} = \{a_0, a_2, a_4, \ldots\},$$

$$\{r_{11}, r_{12}, r_{13}, \ldots\} = \{a_1, a_3, a_5, \ldots\}$$

and subsequent rows by [see I, (6.8.7)]

$$r_{ij} = - \begin{vmatrix} r_{i-2,1} & r_{i-2,j+1} \\ r_{i-1,1} & r_{i-1,j+1} \end{vmatrix} / r_{i-1,1}, \qquad \text{for } i = 2, 3, \ldots, n.$$

For a sampled system governed by the characteristic equation $\sum_{m=0}^{n} a_{n-m} z^m = 0$, the requirement for stability is that all roots must have a modulus less than unity. This can be tested using the *Jury–Marden stability criterion* which requires for stability $d_{21} > 0$, $d_{i1} < 0$, $i = 3, \ldots, n + 1$, where the d_{i1} are the *alternate* first column elements in the following array. The first two rows are defined by

$$\{c_{11}, c_{12}, \ldots, c_{1,n+1}\} = \{a_0, a_1, a_2, \ldots, a_n\},$$

$$\{d_{11}, d_{12}, \ldots, d_{1,n+1}\} = \{a_n, a_{n-1}, \ldots, a_1, a_0\}$$

and subsequent *pairs* of rows by

$$c_{ij} = \begin{vmatrix} c_{i-1,1} & c_{i-1,j+1} \\ d_{i-1,1} & d_{i-1,j+1} \end{vmatrix}, \qquad \text{for } i = 2, 3, \ldots, n + 1,$$

$$d_{ij} = c_{i,n-j-i+3}, \qquad \text{for } i \geqslant 2.$$

In geometrical terms, the requirement for stability is that the roots of the characteristic equation lie in the left half of the complex plane for a continuous system and within the unit circle for a sampled system [see I, §2.7]. It is possible by graphical techniques to plot the change in the position of these roots as the gain K changes by means of the methods of the *root locus*. If the change of the position of the roots is to be plotted with the variations of a parameter other than gain it is called the method of *root contours*. Both these techniques rely on construction rules based on the properties of the roots of *polynomial equations* with real coefficients [see I, §14.5] (Saucedo and Schiring, 1968).

If the model of the open loop system only exists as data points describing the frequency response function, none of the above methods is suitable for assessing closed loop stability. In this case it is necessary to infer the position of the roots of the characteristic equation $KH(s) = -1$ when only $KH(j\omega)$ is known, either as an analytical function or as data points. This method of assessing stability is based on the *Nyquist criterion* (Barnett, 1975; Jacobs, 1974). This criterion states that if there are P open loop poles in the right half-plane ($P = 0$ for a system which is stable in the open loop) then the number of closed loop poles in the right half-plane is $P - N$, where N is the number of times a certain contour Γ encircles the point $(-1, j0)$ in the anticlockwise direction. This contour Γ is derived by *mapping* the contour in the s plane shown in Figure 14.1 into the plane of $KH(s)$. In the construction of this contour, Γ, it is assumed that the contour shown in Figure 14.1 avoids any singularities of $KH(s)$ on the imaginary axis by infinitesimally small semicircles in the right half-plane [see IV, §9.8].

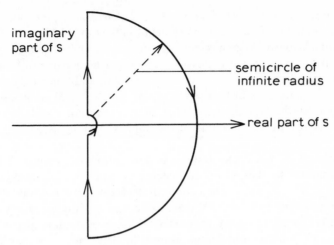

Figure 14.1: The *s* plane contour.

In practice the contour Γ consists of the open loop frequency response (the map of the positive half of the imaginary axis) plus its reflection in the real axis. The remainder of the contour can be established by consideration of the mapping process around singularities on the imaginary axis. An open loop hole at the origin (an integrator), for example, will map into a segment of a circle of infinite radius.

For systems which are stable in the open loop the engineer thus only needs to have some data describing the open loop frequency response function in the region of $(-1, j0)$ and a knowledge of the number of open loop poles at the origin to enable assessment of the closed loop stability.

Measures of the degree of stability are conventionally derived from the Nyquist stability criterion by consideration of how close the open loop frequency response function $KH(j\omega)$ comes to the point $(-1, j0)$. These measures are called the *gain margin* and the *phase margin*, the former being the amount by which the gain can be increased before instability sets in at the frequency where the phase shift is $-180°$ and the latter the amount by which the phase lag can be increased before instability sets in at that frequency where the gain is unity.

The characteristic equation can be written $H(s) = -1/K$ and thus questions of stability can equally well be resolved by comparing the contour $H(j\omega)$ with the point $(-1/K, j0)$ as by comparing $KH(j\omega)$ with $(-1, j0)$. This is particularly useful when a variation of open loop gain is being considered. By extension, if an element in the loop can be considered as a varying gain controlled by some signal within the system then it is often useful to examine the relationship between the locus of $H(j\omega)$ and the locus of $-1/K$. This is the basis of the design of certain non-linear systems using the method of *describing functions* mentioned in Section 14.2 and discussed in more detail in Section 14.13.

If a feedback control system is being considered which has a single- or double-valued non-linearity then it is possible to assess whether or not it is likely to oscillate or be otherwise unstable by using this method. If there is no value of K, as defined for sinusoids in Section 14.13, for which $-1/K$ is enclosed by the locus $H(j\omega)$ (in the Nyquist sense) then asymptotic (non-oscillatory) stability can be expected within the limits of the approximations associated with this method.

The stability of non-linear systems is conventionally defined in a slightly different manner to that of linear systems in that continuous oscillation is considered as a form of stable behaviour in the former case but not in the latter. Systems exhibiting this form of behaviour can usually be analysed using describing functions. However, more general forms of stability analysis in non-linear systems require a different approach. The only reasonably general methods for examining the stability of non-linear systems are due to Liapunov and to Popov. Most other methods are either restricted to a small class of non-linear systems or else a type of linear or quasi-linear behaviour is assumed within a restricted range [see IV, §7.11]. Liapunov's method can be most simply described for situations in which a vector of state variables, $\mathbf{x}$, of the system is related to their derivatives by the set of non-linear functions $\mathbf{f}$ and

$$\dot{\mathbf{x}} = \mathbf{f}(\mathbf{x}), \tag{14.7}$$

where the components $f_1, \ldots, f_n$ of the vector function $\mathbf{f}$ satisfy conditions for existence and uniqueness of solution. By a suitable translation of coordinates, if necessary, an *equilibrium* or *critical* point of the system can be transferred to the origin, i.e. $\mathbf{f}(0) = \mathbf{0}$. The state $\mathbf{x} = \mathbf{0}$ is (i) *stable in the sense of Liapunov* if for any $\varepsilon > 0$ there exists a $\delta > 0$ such that $\|\mathbf{x}(t_0)\| < \delta$ implies $\|\mathbf{x}(t)\| < \varepsilon$, $t \geqslant t_0$ ($\|\mathbf{x}\| = (x_1^2 + \cdots + x_n^2)^{1/2}$); (ii) *asymptotically stable* if it is stable and in addition $\lim_{t \to \infty} \mathbf{x}(t) = \mathbf{0}$; (iii) unstable if it is not stable.

A *Liapunov function* $V(x)$ is defined as being continuous and having continuous and partial derivatives $\partial V/\partial x_i$ [see IV, §5.3]; positive definite, i.e. $V(0) = 0$ and $V(\mathbf{x}) > 0$ for $\mathbf{x} \neq 0$, $\|\mathbf{x}\| \leqslant k$; and having its time derivative, namely

$$\frac{dV}{dt} = \sum_{i=1}^{n} \frac{\partial V}{\partial x_i} f_i$$

negative semidefinite [see I, Definition 9.2.1]. Then the origin is stable if there exists a Liapunov function and asymptotically stable if there exists a Liapunov function having a negative definite derivative. The origin is unstable if $\dot{V}$ is negative definite and in some neighbourhood containing the origin V takes negative values.

When applied to the constant linear system

$$\dot{\mathbf{x}} = \mathbf{A}\mathbf{x},$$

a *quadratic form* $V = \mathbf{x}^T\mathbf{P}\mathbf{x}$ [see I, Chapter 9] can be used as a Liapunov function and is asymptotically stable if and only if for any real symmetric positive definite matrix $\mathbf{Q}$ the solution for the symmetric matrix $\mathbf{P}$ of the *continuous Liapunov matrix equation*

$$\mathbf{A}^T\mathbf{P} + \mathbf{P}\mathbf{A} = -\mathbf{Q}$$

is also *positive definite*. For the discrete time system

$$\mathbf{x}_{k+1} = \mathbf{A}\mathbf{x}_k$$

the equation $\mathbf{A}^T\mathbf{P} + \mathbf{P}\mathbf{A} = -\mathbf{Q}$ is replaced by

$$\mathbf{A}^T\mathbf{P}\mathbf{A} - \mathbf{P} = -\mathbf{Q}.$$

The stability criteria in terms of the characteristic polynomial of $\mathbf{A}$ were given earlier. With non-linear systems it is usually only possible to make statements about regions of stability since the analysis does not usually hold true for all values of the state variables. Furthermore, an inability to find a Liapunov function does not mean the system is not stable. Liapunov's method is a generalization of the idea that for a system to be stable the energy stored within the system during a free response must decrease with time. The Liapunov function is analogous to a generalized 'energy' function. There are variants of Liapunov's theorem which apply in different situations. The reader is recommended to consult Willems (1970) for further information on these topics.

A non-linear control system described by the state equations

$$\dot{\mathbf{x}} = \mathbf{f}(\mathbf{x}, \mathbf{u}), \qquad \mathbf{y} = \mathbf{g}(\mathbf{x}, \mathbf{u}),$$

where the vectors $\mathbf{x}$ and $\mathbf{u}$ are the state variables and the control vector respectively, with $\mathbf{f}(0, 0) = 0$, is called *bounded input–bounded output stable* if a bounded input $\|\mathbf{u}\| < k_1$ produces a bounded output $\|\mathbf{y}\| < k_2$ for $t \geq t_0$, regardless of the initial state.

A result of *Popov* relates to the system

$$\dot{\mathbf{x}} = \mathbf{A}\mathbf{x} - \mathbf{b}u$$

subject to non-linear feedback $u = f(y)$, where $y = \mathbf{c}\mathbf{x}$, $\{\mathbf{A}, \mathbf{b}, \mathbf{c}\}$ is a minimal realization (i.e. having the lowest dimensionality) of a scalar transfer function $g(s)$, and $f(0) = 0$. The theorem states that the origin is asymptotically stable provided

$$0 < \frac{f(y)}{y} < k, \qquad \text{for all } y \neq 0,$$

where k is a positive constant and if, in addition, there exists a real number α

such that

$$(1 + \alpha s)g(s) + 1/k$$

is positive real. For further results on non-linear feedback see Willems (1970). Further consideration of the stability of closed loop non-linear systems is given in Section 14.13.

14.6 Frequency domain design of linear feedback control systems

If the dynamics of a process are known in terms of its frequency response function (usually graphically) then it is possible to establish the frequency response function when the process is controlled by negative feedback. In this way it is possible to predict and modify its behaviour. This section outlines the important main ideas in this area.

Prediction of closed loop frequency response and performance measures

If the open loop frequency response function is $KH(j\omega)$ then the corresponding closed loop function (unity feedback) is

$$\frac{KH(j\omega)}{1 + KH(j\omega)}.$$

By writing $KH(j\omega) = x + jy$ in $|KH(j\omega)/[1 + KH(j\omega)]| = M$, contours of constant closed loop gain in the plane of $H(j\omega)$ can be shown to be circles, known as M circles. Similarly, contours of constant closed loop phase N can also be shown to be circular (N circles).

The plane of $H(j\omega)$ with the M and N circles superimposed is known as a *Hall chart*. This chart can be used to plot the function $KH(j\omega)/[1 + KH(j\omega)]$ from the graph of $KH(j\omega)$. In this way measures of performance such as the closed loop bandwidth, resonant frequency, maximum gain and cut-off rate can be extracted and, together with other measures such as the gain and phase margins (Section 14.5) which are available directly from $H(j\omega)$ and the steady-state errors (Section 14.4), used to assess the design.

Design of cascade compensators

These performance measures can be adjusted by the use of *cascade compensators* which are dynamical devices in the forward path of the control system, often in the form of electronic or digital filters. A most convenient method of designing such compensators is by the use of the *Bodé approximations* to frequency response characteristics.

Bodé diagrams are graphs of gain in decibels against frequency and phase in degrees against frequency, the frequency axis in both cases being on a logarithmic scale. In the case of first-order poles and zeros it is possible to use two straight-line asymptotes to approximate the gain with no error greater than 3 decibels and three straight lines to approximate the phase with no error greater than 6 degrees. Second-order poles must be plotted in more detail to get equivalent accuracy. Details of the method of Bodé diagram construction are given in Saucedo and Schiring (1968). In this way it is possible to propose a compensator and, by transferring its frequency response data from the Bodé diagram to the Hall chart, check the effect on the closed loop system. This in turn will suggest a modification to the compensator, thus providing an iterative procedure.

An improvement on the use of Bodé and Hall charts is the use of Bodé and Nichols charts. A *Nichols chart* is a plot of $H(j\omega)$ with M and N contours superimposed as is the Hall chart. However, it is not a polar plot but a plot of gain in decibels against phase in degrees. This allows much easier transfer of data to and from the Bodé diagrams and simplifies construction because gains are now additive as a result of the logarithmic scale of gain. This method is also described in Saucedo and Schiring (1968).

Design of feedback compensators

When a compensator with frequency response $G(j\omega)$ is introduced into the feedback path, the closed loop frequency response becomes

$$F(j\omega) = \frac{KH(j\omega)}{1 + KG(j\omega)H(j\omega)},$$

from which it can be seen that the complex function $1/F(j\omega)$ is given by

$$\frac{1}{F} = \frac{1}{KH(j\omega)} + G(j\omega).$$

Thus if the open loop frequency response data are plotted on the complex plane in its inverse (known as an *inverse Nyquist plot*) the effect of feedback compensation can be seen as being a directly added complex function.

Here again it is possible to make use of M and N contours, these being circles centred on $(-1, j0)$ and straight lines through $(-1, j0)$ respectively. The design method is, as before, to adjust the plot of $1/F$ until satisfactory performance measures in the closed loop are obtained.

14.7 Root locus and root contour methods

As has already been stated (Section 14.4), the stability of a closed loop control system depends on the position of the roots of the characteristic equation $1 + KH(s) = 0$, where $KH(s)$ is the open loop transfer function. For stable systems the dynamic performance is also largely defined by the position of these roots. The choice of the gain K effects the steady state accuracy of the control system (Section 14.4) and will also effect the position of the roots. A diagram which shows the way these roots change their position as K varies is known as a *root locus* and can be used to analyse performance and to design compensators.

It is clear that the roots of the characteristic equation for $K = 0$ are the poles of $H(s)$, and for K very large they are the zeros of $H(s)$. Thus the root locus starts at the open loop poles and finishes at the open loop zeros. There are a set of rules arising from the nature of the complex equation $1 + KH(s) = 0$ which allows the locus to be plotted for all intermediate values of K without specifically solving the equation (Saucedo and Schiring, 1968). A similar set of rules applies when the system is digital and z-transfer functions are used. Unlike the frequency response techniques this method requires a knowledge of the transfer function. However, the graphical techniques are somewhat easier, particularly when construction aids such as the *Spirule* are used. (This aid is the copyright, 1951, of North American Aviation Inc.)

Both methods only give indirect evidence of the system performance in the time domain but gross features of transient responses, such as a tendency to oscillation and the frequency and damping factor associated with these oscillations, can often be predicted.

For a specific value of K the frequency response can be established from the position of the roots. The reverse is, however, not quite so easy.

In situations where parameters other than the gain can be varied, or else when there is more than one loop and the gain does not appear as a straight multiplier, it is still possible to plot contours of the roots of the characteristic equations. These contours are called *root contours* and give the same design information as root loci. Furthermore, similar construction rules can be made to apply to root contours in many cases (Saucedo and Schiring, 1968).

Design of compensators using root locus and root contour methods

There is no systematic method of obtaining an optimum compensator and, as in the case of frequency domain design, the engineer uses an iterative technique by proposing a compensator and observing its effect on the root locus (or contour). Normally a compensation transfer function model is assumed. Its parameters are chosen to have zeros which offset the effect of destabilizing poles, or poles causing an unacceptably slow response in the region of the locus for which the gain gives the required steady-state performance. A composite root locus (or contour) is then drawn to see if the performance is acceptable.

14.8 Static optimization

The design techniques considered in Sections 14.6 and 14.7 are not optimum. The designer has a set of specifications and tolerances which he may or may not be able to satisfy. In any case it is only possible to obtain a 'satisfactory' rather than a 'best' design in this way. This section is concerned with introducing methods of design which fix the system to optimize some mathematical performance criterion with respect to a particular type of signal. This group of methods is called static optimization and simply consists of choosing a suitable performance criterion and then adjusting all available parameters (usually by computer) to optimize it.

Optimum transient response

If it is required to optimize the response to a defined transient input, say a step, the performance criterion must be sensibly chosen. The criterion used generally in this situation is the ITAE (integral of time absolute error) criterion in which the performance criterion $E = \int_0^\infty t|e(t)|\,dt$ is minimized. More flexibility is available when more complex criteria are used, e.g. $E = \int_0^\infty t^\alpha |e(t)|^\beta\,dt$. The step response would become more oscillatory as the ratio β/α is increased since this gives more weight to the early part of the response. The choice of performance criterion is difficult and depends on the particular application.

Optimum response to continuous random signals

When a control system is subject to an ergodic random signal of known power–density spectrum $\Phi_{rr}(\omega)$, the mean square value of the error signal can be computed and then minimized with respect to the design parameters. This is done by first of all finding the input to error transfer function

$$\frac{E}{R} = \frac{1}{1 + KG(s)}$$

for a unit feedback system with an open loop transfer function $KG(s)$, for example, and thus finding the error power–density spectrum $\Phi_{ee}(\omega)$ by using the *power-gain theorem*

$$\Phi_{ee}(\omega) = \Phi_{rr}(\omega) \left| \frac{E(j\omega)}{R(j\omega)} \right|^2 .$$

where Φ_{rr} is the input power–density spectrum.

The mean square value of the error $e(t)$ can then be computed by using *Parseval's theorem* [see IV, Theorem 20.6.4], which, in this case, gives

$$E\{[e(t)]^2\} = \frac{1}{2\pi} \int_{-\infty}^{\infty} \Phi_{ee}(\omega)\,d\omega.$$

In many practical situations Φ_{ee} is either a rational function in ω^2 or can be approximated by such. In this case the error power (mean square value of the error) can be computed by substituting $\omega = s/j$ and using standard integrals, sometimes called *noise integrals* (Newton, Gould and Kaiser, 1957).

In some situations it is possible to obtain an optimum by differentiating with respect to the free parameters; otherwise hill climbing methods should be used. Complications arise in this method if constraints are put on the system. Many constrained design problems can, however, be solved by techniques related to the one described. These are discussed further in Newton, Gould and Kaiser (1957).

14.9 Adaptive control systems

Adaptive control systems contain computers which are programmed to alter certain of the parameters of the system (usually of feedforward or feedback controllers or input signal filters) in order to prevent the total system performance from deteriorating as the controlled process characteristics change.

Input signal adaption

If the input to a system is statistically stationary [see II, §19.6], it is possible to optimize the system in some static sense. (For example, use Parseval's theorem to find the mean square value of the error, knowing the input power spectrum, and minimize this with respect to the free parameters of the controller—e.g. gain, time constants, etc.—as discussed earlier in Section 14.8.) Should the statistics of the input now change (e.g. the power spectrum of the disturbance) the controller will no longer be at the optimum setting to keep the power of the error at a minimum. The free parameter settings can be changed to suit the new input signal by repeatedly solving the static design equations. Note that this is essentially an open loop system.

Input shaping adaption

In this system the ideal response to a given input change is computed rapidly and the input signal is shaped (filtered) to produce as near the desired overall response as possible. The response computer is effectively a model of the system working much faster than the real system. The model is updated by some learning process, e.g. by finite time correlation as described in 14.4.

Plant adaptive systems

Here one of two sets of data can be prescribed and used as a reference set of optimum values:

(i) System dependent variables, or 'outputs'. The controller parameters are then adjusted to maintain these values in the face of plant variations (the short-term fluctuations being accounted for by the normal feedback loops). Obviously the computer must be programmed so that the adaptive loop is stable.

(ii) System transfer functions (as a set of poles and zeros) or frequency responses or impulse responses (as a set of data points). This is often called 'model reference adaptive control'.

The corresponding actual values are measured, e.g. by correlation methods. The errors are computed and the controller variables (gains, etc.) are adjusted to keep the errors to a minimum. The correlations can be done using the normal operating signals or injected (low-level) signals such as a pseudo-random binary sequence discussed at the conclusion of Section 14.3. Again the multivariable minimization technique ('hill climbing', e.g. the method of steepest descent [see IV, §9.14]) must converge.

In the above cases the set of desired values or some combination of them are called *figures of merit* and the corresponding measured values are called *performance indices*. The figures of merit could be derived from the actual input signal via a model.

Further details on adaptive control are given in Davies (1970).

14.10 State-space methods

The basic mathematical description of a continuous-time linear control system is now

$$\dot{\mathbf{x}}(t) = \mathbf{A}\mathbf{x}(t) + \mathbf{B}\mathbf{u}(t), \tag{14.8}$$

where $\mathbf{x} = [x_1, x_2, \ldots, x_n]^T$ is the *state vector*, describing the state of the system, and the associated n-dimensional space is called the *state space* [see I, §5.4]. The *control vector* $\mathbf{u} = [u_1, u_2, \ldots, u_m]^T$ contains the variables which can be manipulated so as to make the system behave in some desired fashion. The *output vector* $\mathbf{y} = [y_1, y_2, \ldots, y_r]^T$ of the system is also assumed linear to the state, i.e.

$$\mathbf{y}(t) = \mathbf{C}\mathbf{x}(t) \tag{14.9}$$

and the matrices $\mathbf{A}, \mathbf{B}, \mathbf{C}$ have real, constant elements and dimensions $n \times n$, $n \times m$ and $r \times n$ respectively [see I, §6.2]. By direct integration [see IV, §4.1] the solution of (14.8) subject to $\mathbf{x}(t_0) = \mathbf{x}_0$ can be expressed as

$$\mathbf{x}(t) = \mathbf{\Phi}(t, t_0)\left[\mathbf{x}_0 + \int_{t_0}^{t} \mathbf{\Phi}(t_0, \tau)\mathbf{B}\mathbf{u}(\tau)\, d\tau \right], \tag{14.10}$$

where the *state transition matrix* is defined by

$$\mathbf{\Phi}(t, t_0) = \exp\left[\mathbf{A}(t - t_0) \right] \tag{14.11}$$

and has the properties

$$\frac{d}{dt}\,\mathbf{\Phi}(t, t_0) = \mathbf{A}\mathbf{\Phi}(t, t_0),$$

$$\mathbf{\Phi}(t, t) = \mathbf{I}, \tag{14.12}$$

$$\mathbf{\Phi}(t_0, t) = \mathbf{\Phi}^{-1}(t, t_0),$$

$$\mathbf{\Phi}(t, t_0) = \mathbf{\Phi}(t, t_1)\mathbf{\Phi}(t_1, t_0),$$

where $\mathbf{I}$ denotes the *unit matrix* [see I, (6.2.7)]. For the uncontrolled case when $\mathbf{u} = \mathbf{0}$, (14.10) can be represented as in Figure 14.2. If the elements of $\mathbf{A}$ and $\mathbf{B}$ are

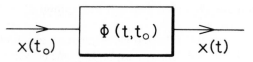

Figure 14.2: State transition matrix.

continuous functions of time then (14.11) no longer applies but (14.10) and (14.12) still hold. In the discrete-time case, (14.8) is replaced by

$$\mathbf{x}_{k+1} = \mathbf{A}\mathbf{x}_k + \mathbf{B}\mathbf{u}_k, \tag{14.13}$$

where $k = 0, 1, 2, \ldots$ is the time variable (i.e. $\mathbf{x}_k \equiv \mathbf{x}(kT)$) and the analogues of equations (14.10) to (14.12) are respectively

$$\mathbf{x}_k = \mathbf{\Phi}(k, k_0)\left[\mathbf{x}_0 + \sum_{i=k_0}^{k-1} \mathbf{\Phi}(k_0, i+1)\mathbf{B}\mathbf{u}(i) \right],$$

$$\mathbf{\Phi}(k, k_0) = \mathbf{A}^{k-k_0},$$

$$\mathbf{\Phi}(k+1, k_0) = \mathbf{A}\mathbf{\Phi}(k, k_0),$$

$$\mathbf{\Phi}(k, k) = \mathbf{I},$$

$$\mathbf{\Phi}(k_0, k) = \mathbf{\Phi}^{-1}(k, k_0), \text{ provided det } \mathbf{A} \neq 0,$$

$$\mathbf{\Phi}(k, k_0) = \mathbf{\Phi}(k, k_1)\mathbf{\Phi}(k_1, k_0), \qquad k \geqslant k_1 \geqslant k_0.$$

The system (14.8) is *completely controllable* (c.c.) if, for any $\mathbf{x}_0$ and any final state $\mathbf{x}_1$, there exists a finite time $t_1 > t_0$ and a control $\mathbf{u}(t)$ defined on $t_0 \leqslant t \leqslant t_1$ such that $\mathbf{x}(t_1) = \mathbf{x}_1$. A necessary and sufficient condition is that the *controllability matrix*

$$\mathscr{C}(\mathbf{A}, \mathbf{B}) = [\mathbf{B}, \mathbf{A}\mathbf{B}, \mathbf{A}^2\mathbf{B}, \ldots, \mathbf{A}^{n-1}\mathbf{B}] \tag{14.14}$$

has *rank n* [see I, §5.6]; the pair $[\mathbf{A}, \mathbf{B}]$ is also said to be c.c. A related notion is that of *complete observability*, which requires that, for any $\mathbf{x}_0$ and t_1, a knowledge of $\mathbf{u}(t)$ and $\mathbf{y}(t)$ on $t_0 \leqslant t \leqslant t_1$ suffices to determine $\mathbf{x}_0$. A necessary and sufficient condition for the system (14.8) and (14.9) or the pair $[\mathbf{A}, \mathbf{C}]$ to be completely observable (c.o.) is that the *observability matrix*

$$\mathcal{O}(\mathbf{A}, \mathbf{C}) = \begin{bmatrix} \mathbf{C} \\ \mathbf{CA} \\ \mathbf{CA}^2 \\ \vdots \\ \mathbf{CA}^{n-1} \end{bmatrix} \tag{14.15}$$

has rank n [see I, §5.6]. A link between the two concepts is provided by the fact that the system (14.8) and (14.9) is c.c. if and only if the *dual system*, defined by

$$\dot{\mathbf{x}} = -\mathbf{A}^T\mathbf{x} + \mathbf{C}^T\mathbf{u}, \qquad \mathbf{y} = \mathbf{B}^T\mathbf{x}$$

is c.o.—and conversely. Conditions for controllability and observability can be derived for the time-varying case, when the elements of $\mathbf{A}$, $\mathbf{B}$ and $\mathbf{C}$ are functions of t and involve integrals (see Barnett, 1975). The criteria (14.14) and (14.15) carry over directly to the discrete-time case provided $\mathbf{A}$ is non-singular [see I, Definition 6.4.2], so some modification is necessary if det $\mathbf{A} = 0$ since then $\mathbf{\Phi}(k, k_0)$ is singular [see I, §6.12]. Other types of controllability and observability can be defined (see Rosenbrock, 1970).

When $m = 1$, so that $\mathbf{B}$ is a column vector ($\mathbf{b}$, say) and (14.8) is c.c., the transformation

$$\mathbf{x} = \mathscr{C}(\mathbf{A}, \mathbf{b})[\mathscr{C}(\mathbf{C}, \mathbf{d})]^{-1}\mathbf{z}$$

puts (14.8) into the *controllable canonical form*

$$\dot{\mathbf{z}} = \mathbf{C}\mathbf{z} + \mathbf{d}u, \tag{14.16}$$

where $\mathbf{C}$ is a *companion matrix* in the form

$$\mathbf{C} = \begin{bmatrix} 0 & 1 & 0 & & & \\ 0 & 0 & 1 & & & \\ \cdot & & \cdot & \cdot & \cdot & 0 \\ \cdot & & \cdot & & \cdot & 0 \\ \cdot & & \cdot & & \cdot & 1 \\ -k_n & -k_{n-1} & & \cdots & & -k_1 \end{bmatrix}$$

having the coefficients of the characteristic polynomial (with $k_0 = 1$) of $\mathbf{A}$ along the last row and $\mathbf{d} = [0, 0, \ldots, 0, 1]^T$. In particular, a scalar equation similar to the differential equation of Section 14.2,

$$\frac{d^n y}{dt^n} + a_1 \frac{d^{n-1}y}{dt^{n-1}} + \cdots + a_{n-1}\frac{dy}{dt} + a_n y = u, \tag{14.17}$$

can be written in the form of (14.16) by taking

$$z_1 = y, \qquad z_2 = \frac{dy}{dt}, \qquad \cdots, \qquad z_n = \frac{d^{n-1}y}{dt^{n-1}}$$

and the z_i are in this case called *phase variables*. Thus (14.17) is always equivalent to a c.c. system, so that even for the single-input case the concept of controllability cannot be encountered until state-space representations are used. Many other canonical forms are possible, including extensions to the case $m > 1$ (see Chen, 1970). Corresponding expressions for c.o. systems are obtained by the duality principle.

Taking the Laplace transform of both sides of (14.8) and (14.9) and assuming $\mathbf{x}(0) = \mathbf{0}$, gives

$$\mathbf{Y}(s) = \mathbf{G}(s)\mathbf{U}(s)$$

and   (14.18)

$$\mathbf{G}(s) = \mathbf{C}(s\mathbf{I} - \mathbf{A})^{-1}\mathbf{B},$$

the *transfer function matrix*. The state-space *realization* problem is a converse one: given an $r \times m$ matrix $\mathbf{G}(s)$ whose elements are rational functions of s [see I, §14.9], find constant matrices $\mathbf{A}$, $\mathbf{B}$ and $\mathbf{C}$ satisfying (14.18). Amongst all such realizations those including matrices $\mathbf{A}$ having least dimensions are termed *minimal*. A necessary and sufficient condition for a realization to be minimal is that it be c.c. and c.o. Any two minimal realizations $\{\mathbf{A}, \mathbf{B}, \mathbf{C}\}$ and $\{\hat{\mathbf{A}}, \hat{\mathbf{B}}, \hat{\mathbf{C}}\}$ are related via similarity, i.e.

$$\hat{\mathbf{A}} = \mathbf{PAP}^{-1}, \qquad \hat{\mathbf{B}} = \mathbf{PB}, \qquad \hat{\mathbf{C}} = \mathbf{CP}^{-1}, \qquad (14.19)$$

where $\mathbf{P}$ is a constant non-singular matrix. The same result holds for the discrete-time system (14.13) with output $\mathbf{y}_k = \mathbf{Cx}_k$, since taking the Z-transform [see IV, §13.5] gives in this case

$$\mathbf{Y}(z) = \mathbf{G}(z)\mathbf{U}(z),$$

where $\mathbf{G}(z) = \mathbf{C}(z\mathbf{I} - \mathbf{A})^{-1}\mathbf{B}$, the same expression as (14.18).

For any constant linear system, there exists a transformation (14.19) such that the system can be decomposed into four mutually exclusive parts respectively: c.c. and c.o.; c.c. but unobservable; c.o. but uncontrollable; uncontrollable and unobservable. The transfer function matrix (14.18) depends *only* upon the c.c. and c.o. sub-system (Zadeh and Desoer, 1963).

14.11 Dynamic optimization

The general problem can be stated as follows: given a system description

$$\dot{\mathbf{x}}(t) = \mathbf{f}(\mathbf{x}, \mathbf{u}, t), \qquad \mathbf{x}(t_0) = \mathbf{x}_0,$$

where the components of the vector function $\mathbf{f}$ satisfy the same conditions as for

(14.7), choose $\mathbf{u}(t)$ so as to minimize the *performance index*

$$J(\mathbf{u}) = \phi[\mathbf{x}(t_1), t_1] + \int_{t_0}^{t_1} F(\mathbf{x}, \mathbf{u}, t)\, dt. \qquad (14.20)$$

The scalar functions ϕ and F are assumed to be continuous and to have continuous first partial derivatives [see IV, §5.3]. When there are no constraints on the control functions $u_i(t)$ the problem can be solved by the *calculus of variations* [see IV, Chapter 12]; necessary conditions for $\mathbf{u}^*$ to be an extremal are that

$$\dot{p}_i = -\frac{\partial H}{\partial x_i},$$

$$p_i(t_1) = \left(\frac{\partial \phi}{\partial x_i}\right)_{t = t_1}, \qquad (14.21)$$

$$\left(\frac{\partial H}{\partial u_i}\right)_{u = u^*} = 0, \qquad t_0 \leqslant t \leqslant t_1, i = 1, 2, \ldots, n$$

where the *Hamiltonian* is defined by

$$H(\mathbf{x}, \mathbf{u}, t) = F(\mathbf{x}, \mathbf{u}, t) + \sum_{i=1}^{n} p_i f_i.$$

When the $u_i(t)$ are allowed to be *piecewise continuous* and subject to contraints of the form $\|u_i(t)\| \leqslant k_i$, then *Pontryagin's principle* states that (14.21) is replaced by the condition that $\mathbf{u}^*$ minimizes H, i.e.

$$H(\mathbf{x}, \mathbf{u}^* + \delta\mathbf{u}, \mathbf{p}, t) \geqslant H(\mathbf{x}, \mathbf{u}^*, \mathbf{p}, t)$$

for all admissible $\delta\mathbf{u}$. The method of *dynamic programming* [see IV, Chapter 16] provides an alternative way for solving the optimal control problem (see Kirk, 1970).

For the linear system (14.8), when the performance index (14.20) is taken to be *quadratic*, namely

$$\mathbf{x}^T(t_1)\mathbf{M}\mathbf{x}(t_1) + \int_{0}^{t_1} (\mathbf{x}^T\mathbf{Q}\mathbf{x} + \mathbf{u}^T\mathbf{R}\mathbf{u})\, dt,$$

where $\mathbf{M}$ and $\mathbf{R}$ are *symmetric positive definite* and $\mathbf{Q}$ is *symmetric positive semidefinite* [see I, §9.2], then the optimal control is linear state feedback in the form

$$\mathbf{u} = -\mathbf{R}^{-1}\mathbf{B}^T\mathbf{P}\mathbf{x}, \qquad (14.22)$$

where the symmetric matrix $\mathbf{P}(t)$ is the solution of the *matrix Riccati differential equation*

$$\dot{\mathbf{P}} = \mathbf{P}\mathbf{B}\mathbf{R}^{-1}\mathbf{B}^T\mathbf{P} - \mathbf{A}^T\mathbf{P} - \mathbf{P}\mathbf{A} - \mathbf{Q} \qquad (14.23)$$

subject to $P(t_1) = M$. When $t_1 \to \infty$, the aim is to regulate the system so that $x(t_1) \to 0$. Setting $M = 0$, provided (14.8) is c.c. and the pair $[A, Q]$ is c.o., then P is the unique *constant* positive definite solution of the *algebraic* Riccati equation obtained by setting the left-hand side of (14.23) equal to the zero matrix and dropping the condition on $P(t_1)$. In this case the closed loop system obtained from (14.22) and (14.8), namely

$$\dot{x} = (A - BR^{-1}B^T P)x,$$

is asymptotically stable. For further details see Kwakernaak and Sivan (1972).

14.12 Multivariable control systems

Feedback systems and stability

Consider a multivariable linear system represented by the block diagram in Figure 14.3, where $\bar{v}(s)$, $\bar{e}(s)$, $\bar{y}(s)$, $\bar{z}(s)$ are vectors having appropriate dimensions representing transformed functions. The *open loop transfer function* matrix is

$$Q(s) = L(s)G(s)K(s). \tag{14.24}$$

The overall *closed loop* transfer function matrix is

$$H(s) = (I + QF)^{-1}Q \equiv Q(I + FQ)^{-1}. \tag{14.25}$$

If the closed loop in Figure 14.3 were broken just before F and a signal $\bar{w}$ injected into F, then the signal which would return to the other side of the broken loop is $-LGKF\bar{w}$. Therefore the difference between the two signals is $(I + QF)\bar{w}$, whence $I + QF$ is termed the *return difference* matrix. The ratio of the closed loop and open loop characteristic polynomials is equal to the determinant of the return difference matrix, which thus determines the difference in stability properties between the open and closed loop systems. In particular, the latter is asymptotically stable if and only if the Nyquist plot of $\det(I + QF)$,

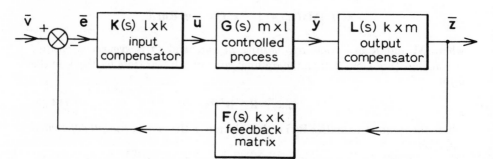

Figure 14.3: Multivariable system block diagram.

traced out as s traverses D in Figure 14.4, encircles the origin $-p$ times, where the open loop system has p poles in $Re(s) \geqslant 0$ (see Section 14.5).

In Figure 14.4, the contour D in the complex s plane is such that the semicircle is sufficiently large so as to enclose all right half-plane poles and zeros of $\det(\mathbf{I} + \mathbf{QF})$, and is indented so as to include any on the imaginary axis.

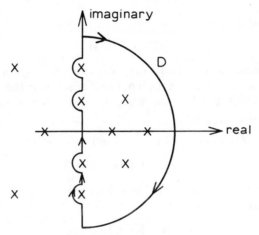

Figure 14.4: The s contour.

When $\mathbf{F}$ is a diagonal matrix [see I, §6.7(iv)] of constant non-zero feedback gains $f_1, f_2, \ldots, f_k$, then the following result is more convenient and provides a generalization of the criterion in Section 14.5. If the Nyquist plot of $q_{ii}(s)$ encircles the point $(-1/f_i + j0)$ n_i times, and if for each point s on D

$$\left| q_{ii}(s) + \left(\frac{1}{f_i} \right) \right| > \delta_i(s), \qquad i = 1, 2, \ldots, k,$$

where

$$\delta_i(s) = \sum_{\substack{j=1 \\ j \neq i}}^{k} |q_{ij}(s)|$$

$$\delta_i(s) = \sum_{\substack{j=1 \\ j \neq i}}^{k} |q_{ji}(s)| \qquad\qquad (14.26)$$

$$\delta_i(s) = \frac{1}{2} \sum_{\substack{j=1 \\ j \neq i}}^{k} |q_{ij}(s) + q_{ji}(s)|,$$

then the closed loop system is asymptotically stable if and only if $\sum_{i=1}^{k} n_i = -p$. An equivalent statement to the conditions (14.26) is that the matrix $\mathbf{Q}(s) + \mathbf{F}^{-1}$

is *diagonal dominant* [see III, §4.7] for every s on D. A sub-set of the complete stability region with respect to the gains f_i can thus be found by drawing the Nyquist plot of each $q_{ii}(s)$ surrounded by circles of radius $\delta_i(s)$. For further results and other graphical stability criteria, see Rosenbrock (1974).

Decoupling

A multivariable system

$$\mathbf{Y}(s) = \mathbf{G}(s)\mathbf{U}(s)$$

is called *decoupled* if $\mathbf{G}(s)$ is square, diagonal and non-singular [see I, §6.7(iv) and Definition 6.4.2], so that the ith input $u_i(t)$ affects *only* the ith output $y_i(t)$, $i = 1, 2, \ldots, m$. Let $\{\mathbf{A}, \mathbf{B}, \mathbf{C}\}$ be an nth order realization of $\mathbf{G}(s)$, and define positive integers $d_1, \ldots, d_m$ by

$$d_i = \min\{j \,|\, \mathbf{c}_i \mathbf{A}^j \mathbf{B} \neq 0, \quad j = 1, 2, \ldots, n - 1\}$$

$$= n - 1 \qquad \text{if } \mathbf{c}_i \mathbf{A}^j \mathbf{B} = 0, \text{ for all } j,$$

where $\mathbf{c}_i$ denotes the ith row of $\mathbf{C}$. Then there exists matrices $\mathbf{K}, \mathbf{L}$ such that linear state feedback

$$\mathbf{u} = \mathbf{Kx} + \mathbf{Lr}$$

produces a closed loop system which is decoupled if and only if the matrix having rows

$$\mathbf{c}_1 \mathbf{A}^{d_1} \mathbf{B}, \; \mathbf{c}_2 \mathbf{A}^{d_2} \mathbf{B}, \ldots, \mathbf{c}_m \mathbf{A}^{d_m} \mathbf{B}$$

is non-singular.

Pole assignment

For a c.c. system

$$\dot{\mathbf{x}} = \mathbf{Ax} + \mathbf{Bu}, \qquad \mathbf{y} = \mathbf{Cx}$$

it is always possible to determine a matrix $\mathbf{K}$ such that with linear state feedback $\mathbf{u} = \mathbf{Kx}$, the characteristic roots [see I, §7.1] of the closed loop matrix $\mathbf{A} + \mathbf{BK}$ can be assigned arbitrarily (subject only to the proviso that since $\mathbf{A} + \mathbf{BK}$ is real, any complex roots must occur in conjugate pairs). These roots are the poles of the closed loop transfer function matrix.

If only linear output feedback

$$\mathbf{u} = \mathbf{Ky} = \mathbf{KCx}$$

can be applied, then at least max(rank $\mathbf{B}$, rank $\mathbf{C}$) [see I, §5.6] roots can be arbitrarily assigned. An alternative approach in this case is to construct a *state observer*, which is a system whose inputs are **u** and **y** and whose output is an approximation to the state vector **x**(*t*), which can then be used for state feedback. For further details on observers, see Chen (1970), and for a discussion on the implications of pole assignment, see Munro (1979).

System matrices

An important framework for dealing with multivariable system analysis and design was introduced by Rosenbrock (1970). The linear system equations need not now be assumed to be first order, so after Laplace transformation [see IV, §13.4] these become

$$\mathbf{T}(s)\bar{\boldsymbol{\xi}} = \mathbf{U}(s)\bar{\mathbf{u}},$$

$$\bar{\mathbf{y}} = \mathbf{V}(s)\bar{\boldsymbol{\xi}} + \mathbf{W}(s)\bar{\mathbf{u}},$$

(14.27)

where **T**, **U**, **V**, **W** are polynomial matrices having dimensions $r \times r, r \times l, m \times r$ and $m \times l$ respectively, and $\bar{\boldsymbol{\xi}}(s)$, $\bar{\mathbf{u}}(s)$, $\bar{\mathbf{y}}(s)$ represent (Laplace-transformed) state, control and output vectors. The equations (14.27) can be written

$$\mathbf{P}(s)\begin{bmatrix} \boldsymbol{\xi} \\ -\bar{\mathbf{u}} \end{bmatrix} = \begin{bmatrix} \mathbf{0} \\ -\bar{\mathbf{y}} \end{bmatrix},$$

where

$$\mathbf{P}(s) = \begin{bmatrix} \mathbf{T} & \mathbf{U} \\ -\mathbf{V} & \mathbf{W} \end{bmatrix}$$

(14.28)

is the *polynomial system matrix*. The associated transfer function matrix is

$$\mathbf{G}(s) = \mathbf{V}\mathbf{T}^{-1}\mathbf{U} + \mathbf{W}.$$

(14.29)

In particular, if the system equations are

$$\dot{\mathbf{x}} = \mathbf{A}\mathbf{x} + \mathbf{B}\mathbf{u}, \qquad \mathbf{y} = \mathbf{C}\mathbf{x} + \mathbf{D}\mathbf{u}$$

then **P**(*s*) takes the *state-space* form

$$\begin{bmatrix} s\mathbf{I} - \mathbf{A} & \mathbf{B} \\ -\mathbf{C} & \mathbf{D} \end{bmatrix}.$$

(14.30)

Analogous results are obtained for discrete-time systems by using Z-transforms [see IV, §13.5].

When **G**(*s*) is constructed from (14.29) there may be cancellations of factors in numerators and denominators, i.e. of system zeros and poles. These are termed *decoupling zeros*, since they correspond to exponential terms ('modes') in the solution which are decoupled from the input and/or output. In particular, if **M**(*s*) is a *greatest common left divisor* [see I, §§4.1.3 and 6.2(vi)] of **T**(*s*) and **U**(*s*), and $N(s)$ is a *greatest common right divisor* of $T(s)$ and $V(s)$, then the zeros of det $M(s)$ and det $N(s)$ are respectively the *input decoupling* (i.d.) and *output decoupling* (o.d.) zeros of the system. The input–output decoupling (i.o.d.) zeros are those o.d. zeros which disappear when the i.d. zeros are removed. When the state space is decomposed as mentioned in Section 14.10, the dimensions of the uncontrollable, the unobservable, and the uncontrollable and unobservable subspaces are respectively equal to the numbers of i.d., o.d. and i.o.d. zeros. When the system is in state-space form (14.30) the numbers of i.d. and o.d. zeros are respectively equal to the rank defects of $\mathscr{C}(A, B)$ and $\mathcal{O}(A, C)$.

For further details on the mathematical background to the system matrix approach, see Barnett (1971).

Design techniques

Suppose that, in Figure 14.3, $l = k = m$ and $L = I_m$, the unit matrix of order m [see I, (6.2.7)]. The *commutative controller* method is to set

$$K(s) = E(s) \operatorname{diag} [k_1(s), \ldots, k_m(s)]E^{-1}(s),$$

$$F(s) = E(s) \operatorname{diag} [f_1(s), \ldots, f_m(s)]E^{-1}(s),$$

where $E(s)$ is a matrix of eigenvectors [see I, §7.1] of $G(s)$, which is assumed to be diagonalizable [see I, Definition 7.4.1], i.e.

$$G(s) = E(s) \operatorname{diag} [g_1(s), \ldots, g_m(s)]E^{-1}(s)$$

and also

$$H(s) = E \operatorname{diag} [h_1(s), \ldots, h_m(s)]E^{-1},$$

where

$$h_i = \frac{g_i k_i}{1 + g_i k_i f_i}$$

so the design procedure is to choose each pair $k_i(s), f_i(s)$, so that each $h_i(s)$ is a satisfactory closed loop transfer function; this is done using classical techniques (Sections 14.6 and 14.7).

For the *inverse Nyquist array* method, it is also assumed that $F(s)$ is diagonal and that det $K(s) \not\equiv 0$, det $G(s) \not\equiv 0$, with all poles of $K(s)$ and $G(s)$ lying in $Re(s) < 0$. Then

$$K(s) = K_1 K_2(s) K_3(s), \tag{14.31}$$

where K_1 is a matrix having just one non-zero element equal to ± 1, in every row and column, $K_2(s)$ represents a sequence of elementary row operations [see I, §5.3] and $K_3(s) = \operatorname{diag} [k_1(s), \ldots, k_m(s)]$ with each $k_i(s)$ non-zero and having all its poles and zeros in $Re(s) < 0$. The object is to choose K_1, K_2, K_3 so that (i) the *inverse* [see I, §6.4] of $\mathbf{Q}(s)$ in (14.24) is diagonal dominant [see III, §4.7] and (ii) the matrix $\mathbf{Q}^{-1}(s) + \mathbf{F}(s)$ is also diagonal dominant. This involves Nyquist plots of the diagonal elements of the inverse matrix $\mathbf{Q}^{-1}(s)$, further details being found in Rosenbrock (1974).

The *sequential design* method involves a sequential procedure for determining a suitable matrix $\mathbf{K}_3(s)$ in (14.31) when $\mathbf{F}(s) \equiv \mathbf{I}_m$. Munro (1979) gives a comparative account of these and other design techniques.

14.13 Non-linear control systems

Some consideration has already been given to mathematical models for non-linear systems (Section 14.2) and the stability of non-linear systems (Section 14.5). This section is concerned with the analysis of the specific situation of non-linear systems with feedback, such as is almost invariably the case in control engineering. It is clear, however, that when the closed loop characteristics are modelled as one unit, many of the previously mentioned ideas are also relevant to the feedback situation directly. Otherwise it should be taken that the non-linear systems previously referred to are to be regarded as elements in a feedback loop (which is often otherwise linear).

Systems incorporating models in which small signal linearization (Section 14.2) has been adopted will not be considered further, as linear analysis is appropriate in these cases. This section describes instead situations in which the signals are large compared with the range of linearity.

Many particular classes of non-linear systems have been extensively analysed in engineering physics and mathematics literature, e.g. systems controlled by equations of the *Van der Pol* type. This section, however, briefly summarizes techniques which have reasonably general application and apply to feedback control specifically.

Piecewise linearization

If a system is such that each non-linear element can be approximated by a sequence of linear segments, then linear analysis can be applied which can be used to predict performance within certain regions of the variable's total range. The change from one linear region to another can be made when the variables in question reach points at which a new linear approximation is required to avoid the accumulation of error. Apart from these step changes in the mathematical models being used the analysis is essentially linear.

Quasi-linearization

A quasi-linear approximation to a non-linear element is a gain or a vector of gains which is a function of some characterizing variable of the signal (say its amplitude or mean square value). Control system analysis using these methods can be considered in two groups classified in terms of the nature of the signals which appear at the input of the non-linear element. These two classes will be summarized in the next two sub-sections.

(a) *Periodic signals.* The method of *describing functions* can be used to predict the amplitude, frequency and d.c. bias of an oscillation in a non-linear autonomous feedback controller. If the open loop transfer function is $H(s)$ then the method can be summarized by stating that the components of the system response can be predicted from the vector of components of the signal at the input of the non-linear element $\mathbf{A}^T = (A_0, A_1, A_2, \ldots, A_N)$ by solving the equations

$$H(j\omega_n) = -\frac{1}{K_n(\mathbf{A}^T)},$$

where $\omega_n = n\omega_0$ (n an integer) $(0 < n < N)$ is the frequency of the component with amplitude A_n. This is called the method of *harmonic balance*. In many cases only the d.c. and fundamental components are considered since most control system components are sufficiently 'low-pass' in their filtering action to allow the higher harmonics to be ignored.

If $N \geqslant 2$ then the generation of sum and difference terms in the output can result in an effective phase shift occurring to one or more of the components, even if the non-linearity is single-valued and memoryless. Further consideration is given to this when the frequency response of closed loop systems is considered later in this section.

A further complication arises when the non-linearity is double-valued (e.g. when hysteresis is exhibited by a relay). In this situation the set of gains K_n must be treated as complex functions of the component amplitudes.

The frequency response of a closed loop non-linear system can be predicted in cases where single-valued non-linearities are contained within feedback loops. This method, developed by Singh (1965), assumes that just the fundamental components of the output of the non-linearity is of significance and that there is no autonomous oscillation. The system input $r = Re^{j\omega t}$, output $c = Ce^{j(\omega t + \phi)}$ and the error $e(t) = Ee^{j(\omega t + \phi)}$ are related by a *phasor diagram* which is a rotating vector diagram. Three relationships are built up: one via the non-linearity $Y = XK(X)$, where X is the input amplitude, Y the output amplitude and K the gain to the fundamental (the d.c. component has been ignored for the sake of simplicity here but can be included if required); and the others via the linear parts of the loop $E|G_1(j\omega)| = X$ and $Y|G_2(j\omega)| = C$, where $G_1(s)$ is the transfer function of linear elements preceding the non-linearity and $G_2(s)$ is that of those following the non-linearity. The equations are used in the form

$$\log \frac{E}{R} + \log |G_1| = \log X - \log R \tag{14.32}$$

and

$$\log \frac{C}{R} - \log |G_2| = \log Y - \log R. \qquad (14.33)$$

The graph of log Y against log X is available by means of the describing function model of the non-linearity. The graph of log (E/R) against log (C/R) is available from the phasor diagram. However, the key point is that the second graph is independent of the system elements and can be represented as a set of parametric curves (one for each open loop phase shift ϕ) with the closed loop phase shift θ as the parameter. This set of curves is known as a *universal chart*, the appropriate curve for a particular input frequency being computed from arg $G_1(j\omega)$ + arg $G_2(j\omega)$ [see I, §2.7.3]. The intersections of this curve and log Y versus log X gives the solution(s) to the problem once the origins have been shifted to satisfy equations (14.32) and (14.33). More than one solution is possible in many situations (always an odd number, say $2n - 1$). Of these n represents stable solutions. In this situation the system can jump from one mode to another, giving a type of instability known as *jump resonance*.

This method of analysis does not allow for multiple inputs into the non-linearity and this possibility must be ruled out first. If more than one sinusoid can exist within the loop at significant amplitude at any one time, one of them being the forcing signal and the other being a natural loop resonance, then the possibility of *sub-harmonic resonance* must be investigated. This is a form of instability due to the phase shift arising because of the cross-modulation terms previously mentioned, causing an otherwise stable feedback loop to become unstable.

(b) *Random signals.* The performance of non-linear feedback systems subject to random disturbances can be analysed under certain circumstances. Such analysis is possible if the loop contains only single-valued non-linearities $y = f(x)$ and if the inputs $x(t)$ to these non-linearities can be considered as ergodic random signals with a *Gaussian* (i.e. normal) amplitude–probability distribution $(p(x))$ [see II, §11.4].

It has been stated (Section 14.2) that it is possible to model a single-valued non-linear element by an equivalent linear gain $K_1(\sigma, d)$, where σ is the root mean square value (standard deviation [see II, Definition 9.2.2]) of the input signal and d is its mean value. This implies that the signal must be capable of description by first- and second-order statistics only [see II, §15.1]. In this situation, by a procedure in some ways analogous to that for predicting the frequency response of the control system, it is possible to find the mean square value of any signal associated with the system. This is achieved by obtaining equations from the linear and non-linear parts of the circuit separately and using the loop balance as before to relate them. Using the same notation we have the power–density spectrum of the input to the non-linearity given by

$$\Phi_{xx}(\omega) = \Phi_{ee}(\omega)|G_1(j\omega)|^2$$

and the power–density spectrum of the system output given by

$$\Phi_{ee}(\omega) = \Phi_{yy}(\omega)|G_2(j\omega)|^2$$

where Φ_{ee} is the power–density spectrum of the error signal and Φ_{yy} that of the output from the non-linearity, $y(t)$.

The relationships defined by the non-linearity are $\overline{y^2} = \sigma^2 K_1(\sigma, d)$, which gives the mean square value of y, the output from the non-linearity and

$$\bar{y} = dK_0(\sigma, d),$$

giving the mean value of y in terms of the gain K_0;

$$K_0 = \frac{1}{d} \int_{-\infty}^{\infty} yq(y)\, dy$$

where $q(y)$ is the distribution of $y(t)$, i.e.

$$q(y) = \frac{p(x)}{dy/dx} = \frac{p(x)}{f'(x)}$$

where $y = f(x)$ is defined by the non-linearity. When the mean square values of the signals are computed from the power–density spectra using *Parseval's theorem* [see IV, §20.6.3] the equations can be combined to solve for σ and d.

The requirements that the signals are to be linearly filtered (low pass) and that the form of the amplitude–probability–density distribution needs to be known at the input to the non-linearity must be added to the fact that the distribution of the error is the convolution of the input and output distribution. It follows that this analysis is restricted to cases for which the system input has a Gaussian amplitude distribution [see II, §11.4].

It is possible to extend this form of analysis of non-linear feedback control systems to cover cases of mixed periodic and random inputs. This situation can only be analysed in general when there is one single-valued non-linearity in the loop and when the random part of the input signal is stationary Gaussian noise for the same reasons as already stated. The procedure is as before in that a multiple input describing a function vector is defined which is a function of the amplitudes of all the significant periodic components as well as of the mean square value of the random component and the d.c. level. Here again it is not possible generally to apply a 'universal' method, as can be done for systems forced by pure sine waves. Again, care must also be taken to examine all possible significant modes arising from signal interaction through the non-linearity. The principle of the method is, however, the same, except that different procedures are used to estimate the signal characteristics for the deterministic and the random components.

N.B.J. and S.B.

References

Atherton, D. P. (1975). *Non-linear Control Engineering*, Van Nostrand Reinhold.

Barnett, S. (1971). *Matrices in Control Theory*, Van Nostrand Reinhold.

Barnett, S. (1975). *Introduction to Mathematical Control Theory*, Oxford University Press.

Bose, A. G. (1956). A theory of nonlinear systems. *Tech. Rep. 309. Res. Lab. of Electronics. M.I.T.* Cambridge, Mass.

Chen, C. T. (1970). *Introduction to Linear System Theory*, Holt, Rinehart and Winston.

Davies, W. D. T. (1970). *System Identification for Self Adaptive Control*, Wiley.

Eykhoff, P. (1974). *System Identification, Parameter and State Estimation*, Wiley.

Jacobs, O. L. R. (1974). *Introduction to Control Theory*, Clarendon Press.

Kirk, D. E. (1970). *Optimal Control Theory*, Prentice-Hall.

Kwakernaak, H. and Sivan, R. (1972). *Linear Optimal Control Systems*, Wiley Interscience.

Lambert, J. D. (1973). *Computational Methods in Ordinary Differential Equations*, Wiley.

McGillem, C. D., and Cooper, G. R. (1974). *Continuous and Discrete Signal and System Analysis*, Holt, Rinehart and Winston.

Munro, N. (Ed.) (1979). *Modern Approaches to Control System Design*, Peter Peregrinus/IEE.

Newton, G. C., Gould, L. A., and Kaiser, J. F. (1957). *Analytical Design of Linear Feedback Controls*, Wiley.

Rosenbrock, H. H. (1970). *State-space and Multivariable Theory*, Nelson.

Rosenbrock, H. H. (1974). *Computer-aided Control System Design*, Academic Press.

Saucedo, R., and Schiring, E. E. (1968). *Introduction to Continuous and Digital Control Systems*, Macmillan.

Schwartz, R. J., and Friedland, B. (1965). *Linear Systems*, McGraw-Hill.

Singh, Y. P. (1965). Graphical method for finding the closed loop frequency response of non-linear feedback control systems, *Proc. IEE*, **112**, 2167–2170.

Smith, H. W. (1966). *Approximate Analysis of Randomly Excited Non-linear Controls*, MIT Press.

Stewart, G. W. (1973). *Introduction to Matrix Computations*, Academic Press.

Willems, J. L. (1970). *Stability Theory of Dynamical Systems*, Nelson.

Wiener, N. (1942). Response of a nonlinear device to noise. *Tech. Rep. 129. Radiation Lab. M.I.T.* Cambridge, Mass.

Zadeh, L. A., and Desoer, C. A. (1963). *Linear System Theory*, McGraw-Hill.

Mathematical Methods in Engineering
Edited by G. A. O. Davies
© 1984, John Wiley & Sons, Ltd.

15

The Finite Element Method

15.1 Introduction

The finite element method is one of the most widely used techniques for computing approximations to the solutions of *initial-* and *boundary-value problems for partial differential equations* [see IV, §8.1]. Our purpose, in this chapter, is to introduce some of the basic ideas of the method without any undue attempt at generality or mathematical rigour. Much fuller accounts of both theoretical and practical aspects of the method can be found in the references given at the end of the chapter.

What characterizes the method is the type of approximant used. The region on which the unknown function is to be determined is sub-divided into non-overlapping sub-regions (elements). The approximants are then taken to be linear combinations of functions which vanish in all but a few (possibly one) contiguous elements. We shall describe the application of the method to a second-order *self-adjoint elliptic* boundary-value problem [see III, §9.5, and IV, (7.6.16)], with mixed boundary conditions, and will indicate briefly the application of the method to other classes of problems.

Notations and terminology

Notations and terms used in this chapter, which might possibly be unfamiliar to the reader, are explained briefly below.

Let S denote a *set* [see I, §1.1], e.g. $(a. b)$ the set of points x, $a < x < b$, or $C[a, b]$, the set of all functions which are continuous on the closed interval $[a, b]$ [see IV, Definition 2.10.1 and §2.1]. Then:

(i) $s \in S$ means that s belongs to the set S, and $s \notin S$ means that s is not a member of S. Thus, for example, $f \in C[a, b]$ means that f is a function, continuous on $[a, b]$.

(ii) $T \subset S$ means that every member of the set T is a member of S; $S \cap T$ is the set, the members of which belong to both S and T; and $S \cup T$ is the set, the members of which belong to at least one of the sets S and T.

(iii) $\{v: v \in S, B(v)\}$, where $B(v)$ is a statement which, for every $v \in S$, is either true

or false, denotes the set consisting of those members of S for which $B(v)$ is
true. Thus $\{v: v \in C[a, b], v(a) = 0\}$ denotes the set of functions which are
continuous on $[a, b]$ and vanish at $x = a$.

(iv) $\forall$ means 'for all'. Thus $\forall v \in S$ is the statement 'for all v belonging to S'.

A set of points (region) S in a plane is said to be *open* if every point belonging
to S is the centre of a circle, of positive radius, which is entirely contained in S
[see IV, §11.2]. For example, the set of points $S_1 \equiv \{(x, y): x^2 + y^2 < 1\}$ is open,
the set $S_2 \equiv \{(x, y): x^2 + y^2 \leqslant 1\}$ is not open (a point (x, y) with $x^2 + y^2 = 1$
belongs to S_2 but there is no circle of positive radius, centred at (x, y), which is
contained in S_2). An *accumulation point*, p, of S is a point with the property that
every circle of positive radius centred at p contains a point p', $p' \neq p$, which
belongs to S. The set S is said to be *closed* if it contains all its accumulation
points. The set consisting of S and all its accumulation points, called the *closure*
of S, is denoted by $\bar{S}$. The closure of S is a closed set. The boundary, $\partial\Omega$, of an
open set Ω is the set of points which belong to $\bar{\Omega}$ but not to Ω. Thus, for example,
the closure of the open region S_1, defined above, in the sets S_2, i.e. $\bar{S}_1 = S_2$ and
∂S_1 in the circle $\{(x, y): x^2 + y^2 = 1\}$. $C(S)$ denotes the set of functions which
are continuous on S and, if S is an open set, $C^r(\bar{S}), r = 1, 2, \ldots$, denotes the set of
functions which, together with all their derivatives up to the rth, are uniformly
continuous and bounded in S [see IV, Definition 2.10.1]. For example, if Ω is an
open region in the (x, y) plane, $v \in C^2(\bar{\Omega})$ means that v, $\partial v/\partial x$, $\partial v/\partial y$, $\partial^2 v/\partial x^2$ and
$\partial^2 v/\partial y^2$ are uniformly continuous and bounded in Ω. Let S and T be sets and let
there be associated with each member s of S a unique member t of T. The set of
ordered pairs $\{(s, t): s \in S$ and $t \in T$ is the member of T associated with $s\}$ is
called a *mapping* of S into T [see I, §1.4]. If the mapping is denoted by f, say, and
$(s, t) \in f$, then one writes $t = f(s)$. If $f(s_1) = f(s_2)$ implies $s_1 = s_2$, then the
mapping is said to be *one to one*, and if, corresponding to every member t of T,
there exists a member s of S such that $t = f(s)$, then the mapping is said to
be *onto*. A mapping which is one to one and onto is said to be *bijective* [see I,
§1.4.2]. The *range of* f, $R(f)$, is the sub-set $\{t: t = f(s), s \in S\}$ of T. For example,
let S_1 be the set $\{(x, y): x^2 + y^2 \leqslant 1\}$ and let S_2 be the set $\{(x, y): x^2 + y^2 \leqslant 1,$
$x > 0, y > 0\}$. Let $u(x, y) = x^2 - y^2$, $v(x, y) = 2xy$ and let the point $(u = u(x, y),$
$v = v(x, y))$ in the (u, v) plane be associated with the point (x, y) in the (x, y) plane.
The mapping from S_1 into the (u, v) plane is *not* one to one (e.g. the point $(1,0)$ in
the (u, v) plane is associated with the points $(1, 0)$ and $(-1, 0)$ in the (x, y) plane).
The mapping from S_2 to the (u, v) plane is one to one but not onto, and the
mapping from S_2 to the range of the mapping is both one to one and onto, i.e. is
bijective.

15.2 Second-order elliptic boundary-value problems

This class of problem includes, of course, *Laplace's equation*, *Poisson's
equation* and *Navier's equation*, which occur in one guise or another in Chapters

2, 3, 6, 7, 8, 10 and 12. We consider, first, a one-dimensional two-point boundary-value problem, then a two-dimensional elliptic boundary-value problem for which the boundary of the region is a polygon and finally we shall give a brief account of *parametrically defined* elements and show how such elements may be used in the solution of problems in which the region has curved boundaries.

A two-point boundary-value problem

Consider the two-point boundary-value problem P [see IV, §7.1]: find u such that

$$L(u)(x) \equiv -\frac{d^2u}{dx^2}(x) + f(x)u(x) = g(x), \qquad x \in (a, b), \qquad (15.1)$$

$$u(a) = \gamma_1, \qquad (15.2)$$

$$\frac{du}{dx}(b) + \sigma u(b) = \gamma_2, \qquad (15.3)$$

where (i) u is continuous on the interval $[a, b]$ and has a quadratically integrable first derivative, i.e. $\int_a^b (du/dx)^2 \, dx < \infty$ [see IV, §4.1] and
(ii) f and $g \in C[a, b]$, $f \geqslant 0$, γ_1, γ_2 and σ ($\sigma \geqslant 0$) are given.

Thus u is required to satisfy a Dirichlet boundary condition at $x = a$ [see IV, §8.2] and a third type of boundary condition at $x = b$. To derive a finite element approximation scheme for this problem we first transform it to a *weak* (or *variational*) form.

A weak form of the problem

Let v belong to the class, H, of functions which are continuous on $[a, b]$ and have quadratically integrable first derivatives; then, from (15.1),

$$\int_a^b (L(u) - g)v \, dx = 0.$$

Using integration by parts [see IV, §4.3], the boundary condition (15.3) and assuming $v(a) = 0$ (i.e. that v satisfies the *homogeneous* Dirichlet boundary condition), one obtains

$$\int_a^b (u'v' + fuv) \, dx + \sigma u(b)v(b) = \int_a^b gv \, dx + \gamma_2 v(b).$$

Thus, since by hypothesis (i) $u \in H$, u is a solution of the problem

$$P_1: \qquad \text{find } u \in H(\gamma_1)$$

such that

$$a(u, v) = l(v), \qquad \forall v \in H(0),$$

where

$$a(v, w) \equiv \int_a^b (v'w' + fvw) \, dx + \sigma v(b)w(b),$$

$$l(v) \equiv \int_a^b gv \, dx + \gamma_2 v(b)$$

and $H(\gamma)$ is the class of functions belonging to H which have the value γ at $x = a$, i.e.

$$H(\gamma) \equiv \{v: v \in H, v(a) = \gamma\}.$$

The problem, P_1, is said to be a weak (or variational) form of the problem P. We shall show below that P_1 can have at most one solution. Thus if P has a solution it must be the unique solution of P_1. Notice that whereas P requires that u has a second derivative, P_1 requires only that u belongs to the class of functions which are continuous and have quadratically integrable derivatives and that whereas in P u is required to satisfy the third type of boundary condition, at $x = b$, this requirement is not placed in the solution of P_1; the boundary condition is called a *natural* boundary condition for the weak form of the problem. However, in P_1, u is required to satisfy the Dirichlet boundary condition $(u(a) = \gamma_1)$; the Dirichlet boundary condition is said to be *essential*. These observations are of particular relevance in the construction of an approximation scheme to P_1. As we shall see below, in constructing a finite element approximation to P_1, we do not impose on $\tilde{u}$, the approximation to u, a requirement that it should have a second derivative or that $\tilde{u}$ should satisfy the third type of boundary condition.

Properties of $a(.,.)$ and $l(.)$

To show that P_1 can have at most one solution we shall make use of the following properties of $a(.,.)$ and $l(.)$, which we shall also use later in the derivation and error analysis of the finite element approximation scheme:

(i) For all $v, w, v_1, v_2 \in H$ and all (real) α_1, α_2,

$$a(v, w) = a(w, v),$$

$$a(\alpha_1 v_1 + \alpha_2 v_2, w) = \alpha_1 a(v_1, w) + \alpha_2 a(v_2, w),$$

i.e. $a(.,.)$ is a *symmetric, bilinear functional* on H,

$$l(\alpha_1 v_1 + \alpha_2 v_2) = \alpha_1 l(v_1) + \alpha_2 l(v_2),$$

i.e. l is a *linear functional* on H [see IV, §19.2.1].

(ii) There exist constants $\alpha, \beta > 0$ such that for all $v \in H(0)$,

$$\alpha\|v\|^2 \leqslant a(v, v) \leqslant \beta\|v\|^2,$$

where

$$\|v\|^2 \equiv (v, v)$$

with

$$(v, w) \equiv \int_a^b (vw + v'w')\, dx.$$

$((.,.)$ defines an *inner product* on H and $\|.\|$ is the corresponding norm [see I, §10.1]. The completion of this space with respect to the given norm is a *Sobolev space* [see IV, Example 19.2.4] (Ciarlet, 1978); we therefore refer to $\|.\|$ as a *Sobolev norm*.)

The results (i) follow at once from the definitions of $a(.,.)$ and $l(.)$. To establish the result (ii), let

$$f_{\min} \equiv \min_{x \in [a, b]} f(x), \qquad f_{\max} \equiv \max_{x \in [a, b]} f(x),$$

$$m \equiv \min(1, f_{\min}) \qquad \text{and} \qquad M \equiv \max(1, f_{\max}).$$

Then, since

$$m[v^2 + (v')^2] \leqslant fv^2 + (v')^2 \leqslant M[v^2 + (v')^2],$$

it follows at once that

$$m\|v\|^2 \leqslant \int_a^b \{(v')^2 + fv^2\}\, dx \leqslant M\|v\|^2.$$

(a) *Case* $\sigma = 0$. Since $f \geqslant 0$, $f_{\min} \geqslant 0$ and hence one can take $\beta = M$. If $f_{\min} > 0$ one can take $\alpha = m$ and if $f_{\min} = 0$ the existence of $\alpha > 0$ can be established using the *Poincaré–Friedrich inequality* (Ciarlet, 1978).

(b) *The case* $\sigma > 0$. The existence of α follows *a fortiori* and the existence of $\beta > 0$ can be established using an appropriate Sobolev space inequality (see, for example, Adams, 1975, p. 114).

Uniqueness of the solution of the weak problem

Let u_1 and u_2 be solutions of the weak problem, P_1. Then, by definition,

$$u_1, u_2 \in H(\gamma_1), \tag{15.4}$$

$$\left.\begin{array}{l} a(u_1, v) = l(v) \\ a(u_2, v) = l(v) \end{array}\right\} \forall v \in H(0)$$

and hence, by subtraction and using the linearity of $a(.,.)$,

$$a(u_1 - u_2, v) = 0, \qquad \forall v \in H(0). \tag{15.5}$$

However, from (15.4) it follows that $w \equiv u_1 - u_2 \in H(0)$ and taking $v = w$ in (15.5) gives $a(w, w) = 0$. Hence, from (i) above, $0 = a(w, w) \geqslant \alpha \|w\|^2$, from which it follows that $\|w\| = 0$. But, since w is continuous, $\|w\| = 0$ implies $w = 0$. Thus, $u_1 = u_2$.

An equivalent minimization problem

Using the properties (i) and (ii) above, one can easily show that the problem P_1 is equivalent to the minimization problem

$$P_2: \qquad \qquad \text{find } u \in H(\gamma_1)$$

such that

$$I(u) \leqslant I(v), \qquad \forall v \in H(\gamma_1),$$

where

$$I(v) \equiv \tfrac{1}{2}a(v, v) - l(v),$$

in the sense that u is a solution of P_1 if and only if u is a solution of P_2.

The problem P_2 can be taken as the starting point for the derivation of a finite element approximation scheme. However, since the method we use to derive such a scheme for P_1 is equally applicable to non-self-adjoint problems [see IV, Definition 19.2.23], and there is no minimization problem corresponding to P_2 for the non-self-adjoint case, we base our derivation of an approximation scheme on P_1.

15.3 Approximation schemes

A scheme for computing an approximation $\tilde{u}$ to the solution u of the weak problem (assuming such exists!) involves:

 (i) the choice of a suitable class of approximants,
(ii) a scheme for selecting, from the approximants, the approximation $\tilde{u}$.

Choice of a class of approximants

The scheme we shall describe is one in which the class of approximants, V, consists of all possible linear combinations of a set of n linearly independent functions, $\psi_1, \psi_2, \ldots, \psi_n$, with $\psi_i \in H$, $i = 1, \ldots, n$. Thus $V = \text{span}(\psi_1, \ldots, \psi_n)$ is a *linear subspace* of H [see I, §5.5]. (The approximation scheme is thus a *conforming* scheme; see, for example, Ciarlet, 1978). Naturally V should be chosen so that the elements of V are easily computed and be such that u can be sufficiently closely approximated by an element of V. For the two-point boundary-value problem possible choices for V are (a) polynomials and (b) continuous piecewise linear functions. It is the latter choice which typifies the finite element method.

(a) *Polynomials.* We could take, for example,

$$\psi_i = w_i \equiv (x - a)^{i-1}, \qquad i = 1, \dots, n,$$

Then $V = \text{span}(\psi_1, \dots, \psi_n) = \mathscr{P}_{n-1}$, the class of all polynomials of degree $n - 1$.

(b) *Continuous piecewise linear functions.* Let $a = x_1 < x_2 < \cdots < x_n = b$. The set of points $x_1, \dots, x_n$ defines a sub-division of $[a, b]$ into sub-intervals (elements) (x_i, x_{i+1}), $i = 1, \dots, n - 1$, which we denote by τ. Let $\mathscr{L}(\tau)$ denote the class of functions which are continuous on $[a, b]$ and piecewise linear with respect to τ, i.e.

$$\mathscr{L}(\tau) \equiv \{v: v \in C[a, b], v \text{ linear on } (x_i, x_{i+1}), i = 1, \dots, n - 1\}.$$

Let $v_i \in \mathscr{L}(\tau)$ be such that

$$v_i(x_j) = \delta_{ij} \quad \text{(the Kronecker delta)}.$$

The functions v_i are, for obvious reasons (see Figure 15.1), called *hat functions*. Clearly $v_i \in H$, $i = 1, \dots, n$; it is easily shown that the set of functions $v_1, \dots, v_n$ is linearly independent and that

$$\mathscr{L}(\tau) = \text{span}(v_1, \dots, v_n).$$

Thus a possible choice for V is $\mathscr{L}(\tau)$.

Selection of the approximant

Let us consider first the case $V = L(\tau)$, i.e. the case in which $\tilde{u}$, the approximation to u, is to be a continuous function, piecewise linear with respect to some sub-division of $[a, b]$.

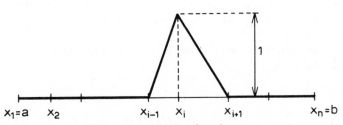

Figure 15.1: The hat function v_i.

Satisfaction of the essential boundary conditions

Since u satisfies the essential boundary condition, $u(a) = \gamma_1$, it seems natural to require that the approximation $\tilde{u} = \sum_i^n c_i v_i$ should also satisfy this condition. Since $v_i(x_j) = \delta_{ij}$ it follows that $u(a) = \gamma_1$ if and only if $c_1 = \gamma_1$, i.e. $\tilde{u}$ satisfies the essential boundary condition if and only if $c_1 = \gamma_1$.

Satisfaction of the weak form of the differential equation

There is, in general, no choice of the c_i such that $\tilde{u}$ will satisfy the weak form of the differential equation

$$a(\tilde{u}, v) = l(v), \qquad \forall v \in H(0).$$

A relaxation of this condition is that $\tilde{u}$ should satisfy

$$a(\tilde{u}, v) = l(v), \qquad \forall v \in V(0),$$

where $V(\gamma) \equiv \{v : v \in V, v(a) = \gamma\}$, and this suggests that we might define the approximation $\tilde{u}$ to u as the solution (if such exists!) of the problem

$$\tilde{P}_1 : \text{find } \tilde{u} \in V(\gamma_1) \tag{15.6}$$

such that

$$a(\tilde{u}, v) = l(v), \qquad \forall v \in V(0). \tag{15.7}$$

We shall show below that P_1 has a unique solution $\tilde{u}$; $\tilde{u}$ is called the *Galerkin* (V-) *approximation* to the solution of P_1. As has already been noted, $\tilde{u} = \sum_{i=1}^n c_i v_i$ satisfies (15.6), if and only if $c_1 = \gamma_1$, and a simple calculation (using the symmetry and bilinearity of $a(.,.)$ and the linearity of $l(.)$) shows that $\tilde{u}$ satisfies (15.7) if and only if

$$\sum_{j=1}^n a(v_i, v_j) c_j = l(v_i), \qquad i \neq 1.$$

Thus $\tilde{u} = \sum_{i=1}^n c_i v_i$ in the solution of P_1 if and only if

$$c_1 = \gamma_1$$

and

$$\sum_{j=1}^n a(v_i, v_j) c_j = l(v_i), \qquad i \neq 1.$$

This set of n simultaneous linear equations for the n unknowns $c_1, \ldots, c_m$ can be rewritten in the form

$$c_1 = \gamma_1,$$

$$\sum_{j \neq 1}^n a(v_i, v_j) c_j = l(v_i) - a(v_i, v_1) \gamma_1, \qquad i \neq 1.$$

If this latter system of equations is written in matrix notation as

$$\mathbf{Ac} = \mathbf{b}$$

then $\mathbf{A}$ is *symmetric, positive definite* and, for n large, *very sparse* [see I, §9.2]. In Section 15.5 we shall prove that the corresponding matrix $\mathbf{A}$ associated with a two-dimensional elliptic boundary-value problem also has these properties. Since the method of proof used there is immediately applicable to the one-dimensional problem we shall omit the proof of the results for the one-dimensional case.

Example. With *equally spaced* nodes $x_i = a + (i - 1)h$, $i = 1, \ldots, n$, with $h = (b - a)/(n - 1)$, $f =$ constant and $g =$ constant, a simple calculation gives $a(v_i, v_j) = 0$, $|i - j| > 1$,

$$a(v_i, v_i) = \begin{cases} 2\lambda, & i = 2, \ldots, n - 1, \\ \lambda, & i = 1, \\ \lambda + \sigma, & i = n, \end{cases}$$

where $\lambda \equiv 1/h + \frac{2}{6}hf$,

$$a(v_i, v_{i+1}) = \mu \equiv -\frac{1}{h} + \frac{1}{6}hf,$$

$$l(v_i) = \begin{cases} 2v, & i = 2, \ldots, n - 1, \\ v, & i = 1, \\ v + \gamma_2, & i = n, \end{cases}$$

where $v \equiv \frac{1}{2}hg$.
Hence,

$$\mathbf{A} = \begin{bmatrix} 1 & 0 & & & & & \\ 0 & 2\lambda & \mu & & & & \\ & \mu & 2\lambda & \mu & & & \\ & & \ddots & \ddots & \ddots & & \\ & & & \mu & 2\lambda & \mu \\ & & & & \mu & \lambda + \sigma \end{bmatrix},$$

$$\mathbf{b} = \begin{bmatrix} \gamma_1 \\ 2v - \mu\gamma_1 \\ 2v \\ \vdots \\ 2v \\ v + \gamma_2 \end{bmatrix}.$$

Thus, given f, g, σ, γ_1 and γ_2, one can easily compute $\mathbf{c}$, the solution of $\mathbf{Ac} = \mathbf{b}$, using Choleski's method for a tridiagonal matrix [see III, §4.2]. If f and g are not constants, $a(v_i, v_j)$ and $l(v_i)$ will, in general, have to be evaluated using *appropriate quadrature schemes* [see III, Chapter 7]. We shall return to this matter when we discuss the two-dimensional problem.

A Galerkin polynomial approximation

We define the Galerkin polynomial $(\mathscr{P}_{n-1})$ approximation, $\tilde{u}$, to u, as the solution of the problem,

$$\text{find} \quad \tilde{u} \in \mathscr{P}_{n-1}, \quad \tilde{u}(0) = \gamma_1$$

such that

$$a(\tilde{u}, v) = l(v), \qquad \forall v \in \mathscr{P}_{n-1}, \quad v(0) = 0.$$

Taking $\tilde{u} = \sum_{i=1}^{n} c_i w_i$ with $w_i = (x - a)^{i-1}$ we find, as before, that

$$c_1 = \gamma_1,$$

$$\sum_{j \neq 1}^{n} a(v_i, v_j)c_j = l(v_i) - a(v_i, v_1)\gamma_1, \qquad i \neq 1$$

and, if these equations are written in matrix notation as

$$\mathbf{Ac} = \mathbf{b}$$

then $\mathbf{A}$ is symmetric and positive definite, but *not* sparse.

15.4 A two-dimensional elliptic boundary-value problem

Let Ω be a bounded open region with boundary $\partial\Omega$, the boundary being made up of two parts $\partial\Omega_1$ and $\partial\Omega_2$ (see Figure 15.2). Consider the elliptic boundary-value problem

$$P: \qquad \text{find } u \text{ such that}$$

$$-\nabla^2 u + fu = g \qquad \text{in } \Omega, \tag{15.8}$$

$$u = \gamma_1 \qquad \text{on } \partial\Omega_1, \tag{15.9}$$

$$\frac{\partial u}{\partial n} + \sigma u = \gamma_2 \qquad \text{on } \partial\Omega_2, \tag{15.10}$$

where (i) u is continuous on $\bar{\Omega}$ and has quadratically integrable first derivatives on Ω and (ii) f, $g \in C(\bar{\Omega})$ $(f \geqslant 0)$, σ, $\gamma_2 \in C(\overline{\partial\Omega_2})$ $(\sigma \geqslant 0)$ and $\gamma_1 \in C(\overline{\partial\Omega_1})$. Note that (i) implies that u is quadratically integrable on $\partial\Omega_2$.

As with the one-dimensional problem, we first transform the problem to a weak form. In the derivation of this weak form we shall not adopt a particularly

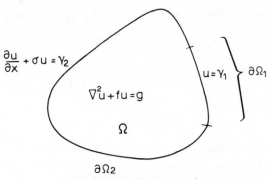

Figure 15.2: A two-dimensional elliptic boundary-
value problem.

rigorous approach: what is important is the weak form itself and the method by which it is derived.

For a discussion of questions relating to the existence and properties of solutions of elliptic boundary-value problems and a rigorous justification of the steps involved in the derivation of the weak form, the reader may consult, for example, Ciarlet (1978).

A weak form of the problem

Let $H = H(\Omega)$ denote the class of functions which are continuous on $\bar{\Omega}$ and have quadratically integrable first derivatives on Ω (and are, hence, quadratically integrable on $\partial\Omega_2$). Let us assume that the problem *does* have a solution u and let $v \in H$. Multiplying the differential equation (15.8) by v, using the identity $(-\nabla^2 u)v = -\nabla v \nabla u + \nabla u \nabla v$ and the *divergence theorem* [see V, Theorem 13.5.1] and assuming that $v = 0$ on $\partial\Omega_1$, one obtains

$$\int_\Omega (\nabla u \nabla v + fuv)\, d\Omega + \int_{\partial\Omega_2} \sigma uv\, ds = \int_\Omega gv\, d\Omega + \int_{\partial\Omega_2} \gamma_2 v\, ds,$$

which we can write more compactly as

$$a(u, v) = l(v),$$

where

$$a(v, w) \equiv \int_\Omega (\nabla v \nabla w + fvw)\, d\Omega + \int_{\partial\Omega_2} \sigma vw\, ds$$

and

$$l(v) \equiv \int_\Omega gv\, d\Omega + \int_{\partial\Omega_2} \gamma_2 v\, ds.$$

Hence, since, by hypothesis (i), $u \in H$, u is a solution of the weak problem

$$P_1: \qquad\qquad\qquad \text{find } u \in H(\gamma_1)$$

such that

$$a(u, v) = l(v), \qquad \forall v \in H(0),$$

where

$$H(\gamma) \equiv \{v: v \in H, v = \gamma \text{ on } \partial\Omega_1\}.$$

The third type of boundary condition (15.10) is not imposed on u in P_1—it is a *natural boundary condition*. The Dirichlet boundary condition (15.9) is imposed on u—it is an *essential boundary condition*. It follows at once from the definition of $a(.,.)$ and $l(.)$ that, as in the one-dimensional case,

(i) $a(.,.)$ is symmetric and bilinear and $l(.)$ is linear on H and, using an obvious generalization of the proof sketched in Section 15.2 for the one-dimensional case, that

(ii) there exist constants $\alpha, \beta > 0$ such that

$$\alpha\|v\|^2 \leqslant a(v, v) \leqslant \beta\|v\|^2, \qquad \forall v \in H(0), \tag{15.11}$$

where

$$\|v\|^2 \equiv (v, v)$$

and

$$(v, w) \equiv \int_\Omega (vw + \nabla v \nabla w) \, d\Omega.$$

The remark made following the corresponding definitions for the one-dimensional case also apply here. We shall therefore refer to $\|.\|$ as a Sobolev norm.

Thus, using the same argument as for the one-dimensional case, it follows that P_1 can have at most one solution, and hence that *if P has a solution it will be the unique solution of P_1*. By replacing H by a suitably enlarged space (a Sobolev space) and interpreting the derivatives in $a(.,.)$ as generalized derivatives, one can ensure that the corresponding weak problem *does* have a (unique) solution (see, for example, Ciarlet, 1978). The finite element method described below for the problem P_1 is also applicable to this (Sobolev space) weak problem, and the finite element method may be thought of as a method for computing an approximation to the solution of the Sobolev space weak problem.

15.5 A finite element approximation scheme

We consider, first, the case in which Ω has a polygonal boundary and τ is a triangulation of Ω (see Figure 15.3). Let $\mathscr{L}(\tau)$ denote the set of functions which

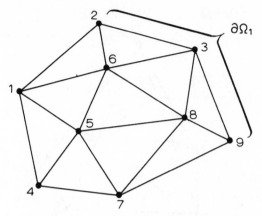

Figure 15.3: A triangulation of a polynomial
region Ω.

are continuous on $\bar{\Omega}$ and piecewise linear with respect to τ. Let the vertices
(nodes) be numbered $1, 2, \ldots, n$ (as, for example, in Figure 15.3), the coordinates
of the ith node be denoted by $\mathbf{x}^{(i)}$ and $v_i \in \mathscr{L}(\tau)$ denote the function which has the
value 1 at the ith node and vanishes at all other nodes, i.e.

$$v_i(\mathbf{x}^{(j)}) = \delta_{ij}.$$

The function v_i which vanishes on all triangles (elements) except those
containing the nodes i is, for obvious reasons, called the *pyramid function*
associated with the ith node (see Figure 15.4). It is not difficult to show that
$v_i \in H$, $i = 1, \ldots, n$, the set of functions $v_1, \ldots, v_n$, is linearly independent and
$\mathscr{L}(\tau) = \mathrm{span}\,(v_1, \ldots, v_n)$. We can now construct a finite element approximation
to the problem P_1, following closely the approach used for the one-dimensional
problem.

   Let the class of approximants be $\mathscr{L}(\tau)$; then, since $u = \gamma_1$ on $\partial\Omega_1$, it seems
reasonable to require that the approximation $\tilde{u}$ to u should satisfy $\tilde{u} = \gamma_1$ at the
set of nodes $\mathscr{B}$ which belong to $\partial\Omega_1$, i.e. that $\tilde{u}$ should *collocate* u at $\mathbf{x}^{(i)}$, $i \in \mathscr{B}$.
(For the case illustrated in Figure 15.3, $\mathscr{B} = \{2, 3, 9\}$). As in the one-dimensional
case we cannot, in general, find $\tilde{u} \in \mathscr{L}(\tau)$ such that

$$a(\tilde{u}, v) = l(v), \qquad \forall v \in H(0). \tag{15.12}$$

Since, for $v \in \mathscr{L}(\tau)$, $v = 0$ on $\partial\Omega_1$ if and only if $v(\mathbf{x}^{(i)}) = 0$, $\forall i \in \mathscr{B}$, it follows that
$V_{\mathrm{coll}}(0) \subset H(0)$, where

$$V_{\mathrm{coll}}(\gamma) \equiv \{v : v \in \mathscr{L}(\tau), v(\mathbf{x}^{(i)}) = \gamma(\mathbf{x}^{(i)}), \forall i \in \mathscr{B}\},$$

where γ is any function defined on $\partial\Omega_1$. Thus it seems reasonable to relax the

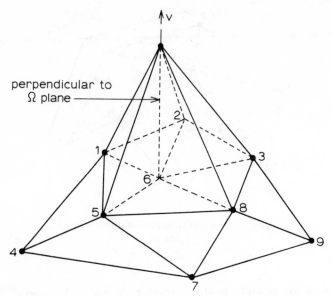

Figure 15.4: A pyramid function v_6 for the triangulation of
Figure 15.3.

condition (15.12) above to

$$a(\tilde{u}, v) = l(v), \qquad \forall v \in V_{\text{coll}}(0).$$

We demand, then, that $\tilde{u}$ should be a solution of the problem

$$\tilde{P}_1: \qquad\qquad \text{find } \tilde{u} \in V_{\text{coll}}(\gamma_1)$$

such that $a(\tilde{u}, v) = l(v), \forall v \in V_{\text{coll}}(0)$.

Let $u = \sum_{i=1}^{n} c_i v_i$; then it follows at once, using the properties of $a(.,.)$ and $l(.)$ and the fact that $\mathscr{L}(\tau) = \text{span}(v_1, \ldots, v_n)$, that u satisfies $\tilde{P}_1$ if and only if $c_1, \ldots, c_n$ satisfy

$$c_i = \gamma_1(\mathbf{x}^{(i)}), \quad \forall i \in \mathscr{B}$$

and   (15.13)

$$\sum_{j=1}^{n} a(v_i, v_j)c_j = l(v_i), \qquad \forall i \notin \mathscr{B}.$$

This is a system of n simultaneous linear equations for the n unknowns $c_1, \ldots, c_n$.

Whereas in the one-dimensional case the finite element approximation $\tilde{u}$ satisfies the Dirichlet boundary conditions exactly, in the scheme just described u is not, in general, equal to γ_1 on $\partial\Omega_1$, though it does of course collocate γ_1 at $\mathbf{x}^{(i)}$, $i \in \mathscr{B}$. Thus the scheme for selecting u incorporates Galerkin and collocation-like features and might be described as a finite element *Galerkin–collocation* scheme.

The linear equations (15.13) can be rewritten as

$$c_i = \gamma_1(\mathbf{x}^{(i)}), \qquad\qquad \forall i \in \mathscr{B}$$

and

$$\sum_{j \notin \mathscr{B}} a(v_i, v_j)c_j = l(v_i) - \sum_{j \in \mathscr{B}} a(v_i, v_j)\gamma_1(\mathbf{x}^{(j)}), \qquad \forall i \notin \mathscr{B},$$

and if these equations are written in matrix notation as

$$\mathbf{Ac} = \mathbf{b}$$

then the $n \times n$ matrix A is symmetric, positive definite and, for n large, very sparse.

Symmetry, positive definiteness and sparseness of the matrix **A**

The symmetry of **A** [see I, §6.2] follows at once from $a(v_i, v_j) = a(v_j, v_i)$. To prove that **A** is positive definite [see I, §9.2] let us assume, without loss of generality, that the nodes are numbered so that $\mathscr{B} = \{m + 1, m + 2, \ldots, m + (n - m) = n\}$. Then

$$\mathbf{A} = \begin{bmatrix} \mathscr{A} & \mathbf{O} \\ \mathbf{O} & \mathbf{I} \end{bmatrix} \begin{matrix} \updownarrow m \\ \updownarrow n - m \end{matrix}$$

where $\mathbf{A} = [a(v_i, v_j)]$, $1 \leq i, j \leq m$ and **I** is the $(n - m) \times (n - m)$ unit matrix. Let

$$\mathbf{x} = \begin{bmatrix} \mathbf{u} \\ \mathbf{v} \end{bmatrix}$$

when **u** is an $m \times 1$ vector and **v** an $(n - m) \times 1$ vector. Then

$$\mathbf{x}^T \mathbf{Ax} = \mathbf{u}^T \mathscr{A} u + \mathbf{v}^T \mathbf{v}$$

and it follows immediately that **A** is positive definite if and only if $\mathscr{A}$ is positive definite [see I, §9.2]. By definition,

$$\mathbf{u}^T \mathscr{A} \mathbf{u} = \sum_{i=1}^{m} \sum_{j=1}^{m} u_i a(v_i, v_j) u_j$$

and hence, using the bilinearity of $a(.,.)$,

$$\mathbf{u}^T \mathscr{A} \mathbf{u} = a(v, v),$$

where $v = \sum_{i=1}^{m} u_i v_i$.

Since v_i vanishes on $\partial\Omega_1$ for $i \notin \mathcal{B}$, $v \in H(0)$ and hence, using (15.11),

$$\mathbf{u}^T \mathscr{A}\mathbf{u} = a(v, v) \geqslant \alpha\|v\|^2.$$

Now, for $v \in H$, $\|v\| \geqslant 0$ and $\|v\| = 0$ if and only if $v = 0$, and (since the set of functions $v_1, \ldots, v_n$ is linearly independent) $v = 0$ if and only if $u_i = 0$, $i = 1, \ldots, m$; it then follows that $\mathbf{u}^T \mathscr{A}\mathbf{u} \geqslant 0$ and $\mathbf{u}^T \mathscr{A}\mathbf{u} = 0$ if and only if $\mathbf{u} = 0$. Thus $\mathscr{A}$ and hence $\mathbf{A}$ is positive definite. The *sparseness* of $\mathbf{A}$ follows from the fact that the integrands (and hence the integrals) occurring in $a(v_i, v_j)$ are zero unless nodes i and j both belong to the same element. Thus for a triangulation in which each node belongs to, at most, a few elements, the matrix $\mathbf{A}$ will be very sparse for large values of n.

15.6 Computation of the finite element approximation

The simplest computational scheme involves four main stages:

1. triangulation of the region (this stage will normally be carried out using a 'grid generation' program),
2. computation of $\mathbf{A}^* = [a(v_i, v_j)]_{n \times n}$ and $\mathbf{b}^* = [l(v_1), \ldots, l(v_n)]^T$,
3. modification of $\mathbf{A}^*$ and $\mathbf{b}^*$ using the boundary condition data to give $\mathbf{A}$ and $\mathbf{b}$, and
4. solution of the linear equations $\mathbf{Ac} = \mathbf{b}$.

(In the *frontal solution* method, stages 2, 3 and the forward elimination phase of 4 proceed in parallel.)

Because of the symmetry, positive definiteness and sparsity of $\mathbf{A}$ the matrix $\mathbf{A}$ can be stored in a very compact form in which, essentially, only non-zero upper (or lower) triangular elements of the matrix are stored, and the equations can be solved by some form of *Choleski decomposition* [see III, §4.2] which exploits this storage scheme.

Whatever method is used to compute the finite element approximation, one of the main tasks is the evaluation of $\mathbf{A}^*$ and $\mathbf{b}^*$.

Computation of $\mathbf{A}^*$ and $\mathbf{b}^*$

Let the elements be numbered $1, 2, \ldots, N$, the interior of the element e be denote by Ω_e ($\Omega = \cup_e \bar{\Omega}_e$) and the boundary of Ω_e denoted by $\partial\Omega_e$. Then

$$a(v_i, v_j) = \sum_e a^e(v_i, v_j)$$

and

$$l(v_i) = \sum_e l^e(v_i),$$

where

$$a^e(v_i, v_j) \equiv \int_{\Omega_e} (\nabla v_i \, \nabla v_j + f v_i v_j) \, d\Omega + \int_{\partial\Omega_e \cap \partial\Omega_2} \sigma v_i v_j \, ds$$

and

$$l^e(v_i) \equiv \int_{\Omega_e} g v_i \, d\Omega + \int_{\partial\Omega_e \cap \partial\Omega_2} \gamma_2 v_i \, ds.$$

Thus the element e makes a contribution $a^e(v_i, v_j)$ to $a(v_i, v_j)$ and a contribution $l^e(v_i)$ to $l(v_i)$. Since the integrands in $a^e(v_i, v_j)$ vanish unless both nodes i and j belong to e, so also does $a^e(v_i, v_j)$ vanish unless nodes i and j belong to e. Similarly, $l^e(v_i)$ vanishes unless node i belongs to e.

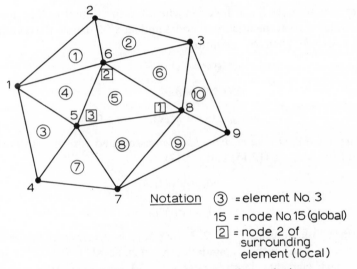

Figure 15.5: Element and element-node numbering.

Let the nodes of element e be given a 'local' numbering, 1, 2, 3, and let the node number of the rth node of e be denoted by elnod(e, r) (referring to Figure 15.5, with the local numbering of the nodes of element 5 as shown, elnod$(5, 1)$ = 8, etc.).

It follows then that $a^e(v_i, v_j)$ and $l^e(v_i)$ are zero unless $i, j \in \{\text{elnod}(e, 1), \text{elnod}(e, 2), \text{elnod}(e, 3)\}$ and the contribution that e makes to $\mathbf{A}^*$ and $\mathbf{b}^*$ is completely determined by the (ordered) set $\{\text{elnod}(e, 1), \text{elnod}(e, 2), \text{elnod}(e, 3)\}$ and the *element stiffness matrix*, $\mathbf{k}^e = [k^e_{rs}]$, where

$$k^e_{rs} = a(v_{\text{elnod}(e,r)}, v_{\text{elnod}(e,s)}) \tag{15.14}$$

and the *element load vector*, $l^e = (l_1^e, l_2^e, l_3^e)$, where

$$l_r^e = l^e(v_{\text{elnod}(e,r)}). \tag{15.15}$$

Thus A^* and b^* can be computed by the following (schematic) algorithm.

Set to zero the arrays in which the (compact form of) A^* and b^* are to be computed.
Then, for $e = 1, \ldots, N$, compute k^e and l^e and add, to the appropriate locations in the arrays, the contribution which k^e and l^e make to A^* and b^*.

In this algorithm A^* and b^* are 'assembled' from the element contributions. The process is known as the *assembly process*.

Calculation of the element stiffness matrix and the element load vector

Let ϕ_r^e denote the linear function which has the value 1 at the rth node of e and the value zero at the other two nodes and let $x^{(e,r)}$ denote the coordinates of the rth node of e; then, by definition,

$$x^{(e,r)} = x^{\text{elnod}(e,r)},$$

$$\phi_r^e(x^{(e,s)}) = \delta_{rs} \tag{15.16}$$

and

$$v_{\text{elnod}(e,r)}(x) = \phi_r^e(x), \qquad x \in \Omega_e \tag{15.17}$$

(ϕ_r^e is the linear *Lagrangian shape function* associated with the rth node of e).

Thus, from (15.14), (15.15) and (15.17),

$$k_{rs}^e = a^e(\phi_r^e, \phi_s^e)$$

and

$$l_r^e = l^e(\phi_r^e).$$

The linear functions ϕ_r^e, $r = 1, 2, 3$, are uniquely determined by the equation (15.16) above. Thus, if the triangulation is given, i.e. $x^{(i)}$ ($i = 1, \ldots, n$) are given, and, for each element, elnode(e, r) ($r = 1, 2, 3$) are given, then, for all e, ϕ_r^e, $r = 1, 2, 3$ and hence k^e and l^e can be computed. If f, g, σ and γ_2 are constants, the required integrals can be evaluated analytically. In general some form of *numerical quadrature* must be used [see III, Chapter 7]. The use of numerical quadrature will be explained, briefly, in Section 15.9.

15.7 Error analysis

We derive, below, a bound on the error in the finite element approximation $\tilde{u}$ to u. We treat the case where there are no errors attributable to collocation and we shall assume that $\tilde{u}$ is the exact solution of $\tilde{P}_1$. Thus we ignore the effect of arithmetical rounding errors and the effect of replacing $a(v_i, v_j)$ and $l(v_i)$ by their quadrature approximations.

In the following analysis we assume that $\tilde{u} = \gamma_1$ on $\partial\Omega_1$, i.e. that γ_1 is continuous and linear between consecutive nodes of $\partial\Omega_1$.

Let u be the solution of P_1 and $\tilde{u}$ the solution of $\tilde{P}_1$. Since $V_{\mathrm{coll}}(0) \subset H(0)$ it follows that

$$a(u - \tilde{u}, v) = 0, \qquad \forall v \in V_{\mathrm{coll}}(0). \tag{15.18}$$

Using the symmetry and bilinearity of $a(.,.)$ one has, with $p \equiv u - \tilde{u}$ and $q \equiv \tilde{u} - v$,

$$a(u - v, u - v) = a(p, p) + 2a(p, q) + a(q, q). \tag{15.19}$$

Let $v \in V_{\mathrm{coll}}(\gamma_1)$; then $q \in V_{\mathrm{coll}}(0)$ and hence, using (15.18), the second term on the right-hand side of (15.19) vanishes. Further, since $V_{\mathrm{coll}}(0) \subset H(0)$,

$$a(q, q) \geqslant \alpha\|q\|^2,$$

and it follows, then, that

$$a(u - \tilde{u}, u - \tilde{u}) \leqslant a(u - v, u - v), \qquad \forall v \in V_{\mathrm{coll}}(\gamma_1). \tag{15.20}$$

Now it is a result of approximation theory that if $w \in C^2(\bar{\Omega})$ and w_{I} is the $\mathscr{L}(\tau)$ interpolant of w ($w_{\mathrm{I}} \in \mathscr{L}(\tau)$, $w_{\mathrm{I}}(\mathbf{x}^{(i)}) = w(\mathbf{x}^{(i)})$, $i = 1, \ldots, n$), then

$$\|w - w_{\mathrm{I}}\| \leqslant \frac{C(w)h}{\sin\theta}, \tag{15.21}$$

where the constant $C(w)$ depends on w and Ω but not on τ, h is the longest triangle side and θ the smallest angle of triangles in the triangulation τ. Since $u_{\mathrm{I}} \in V_{\mathrm{coll}}(\gamma_1)$ we can take $v = u_{\mathrm{I}}$ in (15.20), and hence we obtain, using equations (15.11), (15.20) and (15.21),

$$\alpha\|u - \tilde{u}\|^2 \leqslant a(u - \tilde{u}, u - \tilde{u}) \leqslant a(u - u_{\mathrm{I}}, u - u_{\mathrm{I}})$$

$$\leqslant \beta\|u - u_{\mathrm{I}}\|^2 \leqslant \beta[C(u)h/\sin\theta]^2,$$

i.e.
$$\|u - \tilde{u}\| = \sqrt{\frac{\beta}{\alpha}}\,\frac{C(u)h}{\sin\theta}.$$

Thus, if every triangulation, τ, of Ω is restricted by the condition $\theta \geqslant \theta_0$ where $\theta_0 > 0$ is fixed, one has

$$\|u - \tilde{u}\| = O(h)$$

and it can be shown, using the 'Aubin–Nitsche trick' (see, for example, Strang and Fix, 1973, or Ciarlet, 1978) that

$$\|u - \tilde{u}\|_{L_2} = O(h^2),$$

where $\|v\|_{L_2}^2 \equiv \int_\Omega v^2 \, d\Omega$ is the square of the L_2 norm of v [see IV, §11.6].

Thus the difference between u and $\tilde{u}$, measured by the Sobolev (respectively

L_2) norm of $u - \tilde{u}$, decreases at least as fast as a multiple of h (respectively h^2) as h tends to zero. Ciarlet (1978, pp. 105, 106) makes some pertinent remarks concerning error estimates of this kind.

15.8 Parametrically defined elements and functions

In Sections 15.4 to 15.7 we have described the finite element method for the case in which the approximants are defined on a polygonal region. If one wishes to use this scheme to compute an approximation to the solution of a problem with a curved boundary, the region must be approximated by a polygon and a large number of elements will be required to obtain a good approximation to the region.

We have taken the approximants from sets of functions which are continuous and piecewise linear. Could one, perhaps, do better by taking, say, continuous piecewise quadratic functions? We shall show below how, using parametrically defined elements, one can construct a more general set of approximants (the functions are continuous and piecewise quadratic when the mesh is appropri-ately restricted) which are well suited to dealing with problems with curved boundaries.

Parametrically defined elements

We consider a special case of schemes in which the elements are defined to be the images of a standard reference element under mappings which are defined using *polynomial Lagrangian shape functions*. Specifically, we consider the case in which the reference element is a triangle and the shape functions are quadratic.

Let the region defined by the interior and boundary of the triangle (element) with vertices $(1, 0)$ $(0, 1)$ and $(0, 0)$ in the (ξ_1, ξ_2) plane be denoted by T^0. Let these vertices be numbered 1, 2 and 3 respectively, let the mid-side points $(\frac{1}{2}, \frac{1}{2})$, $(0, \frac{1}{2})$ and $(\frac{1}{2}, 0)$ be numbered 4, 5 and 6 respectively (see Figure 15.6) and let the coordinates of the point (node) i be denoted by $\xi^{(i)}$. Let $N_i(\xi)$ denote the quadratic function associated with node i, with the property

$$N_i(\xi^{(j)}) = \delta_{ij}.$$

$N_i(\xi)$ is the quadratic Lagrangian shape function associated with node i:

$$N_i(\xi) = \xi_i(2\xi_i - 1), \qquad i = 1, 2, 3,$$

$$N_4(\xi) = 4\xi_1\xi_2, \qquad N_5(\xi) = 4\xi_2\xi_3, \qquad N_6(\xi) = 4\xi_3\xi_1,$$

where $\xi_3 = (1 - \xi_1 - \xi_2)$.

Let Ω be an open region with boundary $\partial\Omega$, where $\partial\Omega$ may be curved and may have corners (see Figure 15.7a). Let $\tilde{\tau}$ be a triangulation of Ω in which the triangles (elements) may have curved sides (an element, one side of which is

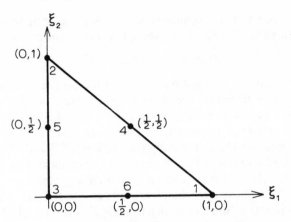

Figure 15.6: The reference element T^0.

formed from a segment of a curved part of $\partial\Omega$, will of course have a curved side)
and let an (intervertex) node be chosen on each triangle side. Let the set of all
vertex and intervertex nodes be numbered $1, 2, \ldots, n$, the coordinates of node i
be $\mathbf{x}^{(i)}$ and, for each element, let the nodes be numbered from one of the vertex
nodes—in anticlockwise order, 1, 4, 2, 5, 3, 6, (cf. the ordering of the nodes in the
element T^0). Let the elements be numbered $1, \ldots, N$ and let the rth node in the
element e be elnod(e, r), $r = 1, \ldots, 6$. For the triangulation $\tilde{\tau}$, shown in Figure
15.7b, $N = 8$ and a choice of element node numbers for element 6 is indicated.

Consider the mapping x^e from the ξ plane to the $\mathbf{x}$ plane, defined by

$$\mathbf{x}^e(\boldsymbol{\xi}) = \sum_{r=1}^{6} N_r(\boldsymbol{\xi})\mathbf{x}^{(e,r)},$$

where $\mathbf{x}^{(e,r)} = \mathbf{x}^{(\text{elnod}(e,r))}$.

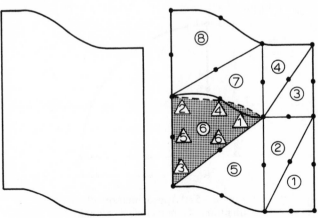

Figure 15.7: (a) The region Ω. (b) A triangulation, $\tilde{\tau}$ of Ω.

It follows at once, from the definition of the $N_r(\xi)$, that $\mathbf{x}^e$ maps the *sth* node of T^0 into the *sth* node of e. Let T^e be the image of T^0 under this mapping, i.e.

$$T^e \equiv \{\mathbf{x}: \mathbf{x} = \mathbf{x}^e(\xi)\ \xi \in T^0\}.$$

Provided that the nodes 4, 5 and 6 of e are not too far removed from half-way between the vertex nodes of the sides to which they belong, the mapping $\mathbf{x}^e: T^0 \to T^e$ will be bijective (one to one and onto) and T^e will be a triangular region with, in general, curved sides—the vertices being $\mathbf{x}^{(e,r)}$, $r = 1, 2, 3$, and $\mathbf{x}^{(e, 4)}$ being a point on the side joining the vertices $\mathbf{x}^{(e, 1)}$ and $\mathbf{x}^{(e, 2)}$, etc. (T^6 is shown as the shaded region in Figure 15.7b). In what follows we assume that the nodes $\mathbf{x}^{(j)}, j = 1, \ldots, n$, and the element node numbering, elnod$(e, r), e = 1, \ldots, N$ and $r = 1, \ldots, 6$, is chosen so that for every e, $e = 1, \ldots, N$, the associated mapping, $\mathbf{x}^e$, is bijective. Since a side of a triangle T^e is completely determined by the nodes belonging to the side, triangles sharing two vertices and an intervertex node will not overlap except at their common boundary and there will be no gaps between such triangles. Any two triangles T^e, $T^{e'}$, $e \neq e'$, are therefore either (i) disjoint or (ii) have a single vertex node in common or (iii) have a side in common. The nodes and elnod array thus determine a region, $\tilde{\Omega}$, the interior of $\cup_e T^e$, which approximates Ω and a triangulation of this region (which we denoted by τ) into parametrically defined elements $T^e, e = 1, \ldots, N$. Henceforth, we shall refer to the parametrically defined element T^e as the element e. Note that the triangulation $\tilde{\tau}$ was introduced simply to suggest one way that the data defining the triangulation τ (the $\mathbf{x}^{(j)}$ and elnod array) might be obtained.

That one can obtain a good approximation to a region with a curved boundary using surprisingly few elements is illustrated in the following simple example.

Let Ω be a quadrant of a unit circle. Let the number of elements be 1 and let the element nodes be as shown in Figure 15.8. The sides 2–5–3 and 3–6–1 of the

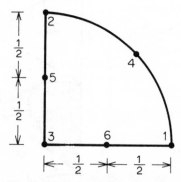

Figure 15.8: Approximation of a quadrant of a unit circle by a single parametrically defined element.

parametrized element coincide with the corresponding sides of the quadrant, and the distance of points on the side 1–4–2 of the parametrized element from the node 3 differs from unity by not more than 0.012.

Parametrically defined finite element functions

One can define element functions v_i, $i = 1, \ldots, n$, on the triangulation τ, in the following way. Since, by hypothesis, $\mathbf{x}^e$ is bijective, it follows that there exists an inverse mapping ξ^e,

$$\xi^e \colon T^e \to T^0,$$

such that

$$\mathbf{x}^e(\xi^e(\mathbf{x})) = \mathbf{x}, \qquad \forall \mathbf{x} \in T^e,$$

$$\xi^e(\mathbf{x}^e(\xi)) = \xi, \qquad \forall \xi \in T^0.$$

We define ϕ_r^e, the *shape function* associated with the rth node of e (i.e. the node elnod(e, r)), by

$$\phi_r^e(\mathbf{x}) = N_r(\xi^e(\mathbf{x})), \qquad \mathbf{x} \in T^e$$

and we define v_j, the finite element function associated with node j, by

$$v_j(\mathbf{x}) = \begin{cases} 0, & \text{if } \mathbf{x} \text{ does not belong to an element containing the node } j, \\ \phi_r^e(\mathbf{x}), & \text{if } \mathbf{x} \text{ does belong to an element } e \text{ containing } j \text{ and } j \text{ is the } r\text{th} \\ & \text{node of } e, \text{ i.e. } j = \text{elnod}(e, r). \end{cases}$$

Notice that if $\mathbf{x}$ lies on the boundary of two elements, e and e' say, containing the node j (so that $j = \text{elnod}(e, r) = \text{elnod}(e', r')$), there is an apparent ambiguity in the definition. However, in this situation, as the reader may easily verify, the two values $\phi_r^e(\mathbf{x})$ and $\phi_{r'}^{e'}(\mathbf{x})$ are equal and the ambiguity does not arise.

It follows from the definition that

(i) v_j is continuous, and piecewise differentiable with respect to τ [see IV, §2.9],
(ii) $v_i(\mathbf{x}^{(j)}) = \delta_{ij}$ [see I, (6.2.8)],
(iii) $v_j(\mathbf{x}) = 0$ unless $\mathbf{x}$ belongs to an element containing the node j,
(iv) the set of functions $v_1, \ldots, v_n$ is linearly independent [see I, §5.3],
(v) every function of the form $v = \alpha_1 x_1 + \alpha_2 x_2 + \beta$, $\mathbf{x} = (x_1, x_2) \in \tilde{\Omega}$, can be expressed as a linear combination of the v_i's.

The result (v) ensures that the constant strain condition (see, for example, Zienkiewicz, 1977)—a crucial ingredient in establishing convergence results for finite element approximations to second-order elliptic problems—is satisfied.

If in the triangulation τ the intervertex nodes are mid-way between the vertex nodes, the elements will, of course, be straight-sided triangles; the ϕ_r^e will be quadratic functions; and span $(v_1, \ldots, v_n)$ will be the set of functions which are continuous and piecewise quadratic with respect to τ—a set of functions which clearly contains $\mathscr{L}(\tau)$.

If the intervertex nodes are not too far removed from mid-way between the vertex nodes then $v_j \in H(\tilde{\Omega}), j = 1, \ldots, n$. A triangulation which has the property that $v_j \in H(\tilde{\Omega}), j = 1, \ldots, n$, we shall call 'acceptable'.

15.9 A finite element scheme for an elliptic boundary-value problem, using parametrically defined finite element functions

In this and the following section we describe a finite element approximation scheme for problem P_1 of Section 15.4, in which the boundary of Ω may be curved. To motivate the derivation of the scheme we first consider a natural extension of the Galerkin–collocation scheme derived in Section 15.5, in which the finite element functions are the parametrically defined functions of Section 15.9. This extension suffers, however, from serious defects; in attempting to overcome these defects we consider a modification of the scheme in which the element stiffness matrices and element load vectors are evaluated by numerical quadrature. The resulting finite element approximation scheme is practical (the computations can be carried out using the assembly process described in Section 15.6) and has good convergence properties.

Derivation of the scheme

Given an acceptable triangulation τ, we might, following the argument of Section 15.5, tentatively try to define an approximation $\tilde{u}$ to the solution u of P_1 by

$$\tilde{u} = \sum_{i=1}^{n} c_i v_i,$$

where

$$c_i = \gamma_1(\mathbf{x}^{(i)}), \qquad i \in \mathscr{B},$$

$$\sum_{j=1}^{n} a^*(v_i, v_j)c_j = l^*(v_i), \qquad i \notin \mathscr{B},$$

where $a^*(.,.)$ and $l^*(.)$ are defined in exactly the same way as $a(.,.)$ and $l(.)$ respectively, but with Ω replaced by $\tilde{\Omega}$.

For the case in which Ω has a curved boundary, this first attempt at an approximation scheme suffers from two obvious defects (the first of which is fatal!). If Ω has a curved boundary, Ω and $\tilde{\Omega}$ will not, in general, coincide—there will, in general, be points of Ω which are not in $\tilde{\Omega}$, and vice versa. Hence:

(i) The scheme is not properly defined; f and g are not defined at points of $\tilde{\Omega}$ not belonging to Ω, and σ and γ_2 are not defined at points of $\partial\tilde{\Omega}_2$ not belonging to $\partial\Omega_2$.

(ii) Even when f and g are defined in $\tilde{\Omega}$ and σ and γ_2 on $\partial\tilde{\Omega}_2$, $\tilde{u}$ can only be taken as an approximation to u on $\tilde{\Omega} \cap \Omega$.

The second defect is, in a certain sense, not too serious: if $\mathbf{x} \in \Omega$ then by taking a sufficiently fine triangulation, $\mathbf{x} \in \tilde{\Omega}$. Thus it appears that if the tentative scheme is to be the basis of a practical approximation scheme, practical schemes for the computation of σ and γ_2 on $\partial\tilde{\Omega}_2$, f and g at points in $\tilde{\Omega}$ but not in Ω, and of $a^*(v_i, v_j)$ and $l^*(v_i)$, must be specified. One way of dealing with the problem concerning σ and γ_2 is to use interpolants of these functions on $\partial\tilde{\Omega}_2$ (the problem does not, of course, arise if $\sigma = \gamma_2 = 0$ on $\partial\tilde{\Omega}_2$). Henceforth we shall assume that such an interpolation scheme is (if needed) used.

Even if we had a recipe for extending the definition of f and g to $\tilde{\Omega}$, the tentative scheme would not be a practical proposition unless the matrix $\mathbf{A} \equiv [a^*(v_i, v_j)]$ and the vector $\hat{\mathbf{b}} \equiv [l^*(v_1), \ldots, l^*(v_n)]^T$ could be computed.

It follows from the properties of the v_i that, as in the case of the $\mathscr{L}(\tau)$ approximation scheme, $\mathbf{A}$ and $\hat{\mathbf{b}}$ could be computed by the assembly process from the element stiffness matrix $\mathbf{k}^e$ and element load vector $\mathbf{l}^e$ for the element e, if these were known.

Let us then consider the problem of computing $\mathbf{k}^e$ and $\mathbf{l}^e$ (we restrict our discussion to the case $\sigma = \gamma_2 = 0$ on $\partial\Omega_2$). By definition,

$$k_{rs}^e = \int_{T^e} \Phi_{rs}^{(1)} \, d\Omega$$

and

$$l_r^e = \int_{T^e} \Phi_r^{(2)} \, d\Omega,$$

where

$$\Phi_{rs}^{(1)} = \nabla\phi_r^e \, \nabla\phi_s^e + \phi_r^e \phi_s^e$$

and

$$\Phi_r^{(2)} = g\phi_r^e.$$

we have

$$\int_{T^e} F(\mathbf{x}) \, d\Omega = \int_{T^0} F(\mathbf{x}^e(\boldsymbol{\xi}))| \, J^e(\boldsymbol{\xi})| \, d\Omega,$$

where $J^e(\boldsymbol{\xi})$ is the jacobian of the mapping $\mathbf{x}^e$ [see IV, §5.12]. Thus

$$k_{rs}^e = \int_{T^0} \Psi_{rs}^{(1)} \, d\Omega$$

and

$$l_r^e = \int_{T^0} \Psi_r^{(2)} \, d\Omega,$$

where

$$\Psi_{rs}^{(1)}(\xi) \equiv \Phi_{rs}^{(1)}(\mathbf{x}^e(\xi)) | J^e(\xi) |$$

and

$$\Psi_r^{(2)}(\xi) \equiv \Phi_r^{(2)}(\mathbf{x}^e(\xi)) | J^e(\xi) |.$$

To evaluate the integrals $\int_{T^0} \Psi_{rs}^{(1)} \, d\Omega$ and $\int_{T^0} \Psi_r^{(2)} \, d\Omega$ one must, in general, resort to a *numerical quadrature* scheme [see III, Chapter 7]:

$$\int_{T^0} F(\xi) \, d\Omega \simeq \sum_{j=1}^{q} w_j^{(q)} F(\xi^{(q,j)}),$$

where $w_j^{(q)}$ and $\xi^{(q,j)}$, $j = 1, \ldots, q$, are the *weights* and *quadrature points*. Using such a scheme, one can calculate approximations to $\mathbf{k}^e$ and $\mathbf{l}^e$, provided that the integrands $\Psi_{rs}^{(1)}$ and $\Psi_r^{(2)}$ can be calculated at the quadrature points.

Now, for ξ given, one can compute $N_r(\xi)$ and $\partial N_r / \partial \xi_k(\xi)$ for $r = 1, \ldots, 6$ and $k = 1, 2$, immediately, and hence

$$\phi_r^e(\mathbf{x}^e(\xi)) \equiv N_r(\xi),$$

$$\mathbf{x}^e(\xi) \equiv \sum_{r=1}^{6} N_r(\xi) \mathbf{x}^{(e,r)},$$

$$\frac{\partial x_i}{\partial \xi_j}(\xi) \equiv \sum_{r=1}^{6} \frac{\partial N_r}{\partial \xi_j}(\xi) x_i^{(e,r)}$$

and

$$J^e(\xi) = \det J, \qquad \text{where } J \equiv \left[\frac{\partial x_i^e}{\partial \xi_j}(\xi) \right],$$

can be computed without difficulty.

Since, by definition, $N_r(\xi) = \phi_r^e(\mathbf{x}^e(\xi))$ one has

$$\frac{\partial N_r}{\partial \xi_k}(\xi) = \sum_{j=1}^{2} \frac{\partial \phi_r^e}{\partial x_j}(\mathbf{x}^e(\xi)) \frac{\partial x_j^e}{\partial \xi_k}(\xi)$$

and hence, putting

$$B \equiv \left[\frac{\partial \phi_r^e}{\partial x_i}(\xi) \right] \equiv [\nabla \phi_1^e(\xi), \ldots, \nabla \phi_6^e(\xi)]$$

and

$$B^* \equiv \left[\frac{\partial N_j}{\partial \xi_i} \right],$$

it follows that

$$B = (J^T)^{-1} B^*$$

from which B, and hence the $\nabla \phi_r^e(\xi)$, can be computed. Thus there remains the problem of computing $f(\mathbf{x}^e(\xi))$ and $g(\mathbf{x}^e(\xi))$ for e given and $\xi = \xi^{(q,j)}, j = 1, \ldots, q$.

When $\mathbf{x}^e(\xi^{(q,j)}) \in \bar{\Omega}$, these functions can, of course, be computed from their definitions. Hence, provided the quadrature points $\mathbf{x}^e(\xi^{(q,j)})$ lie in $\bar{\Omega}$ the numerical quadrature approximations to $\mathbf{k}^e$ and $\mathbf{l}^e$ can be computed. If every node $\xi^{(q,j)}$ of the quadrature scheme either coincides with a node of T^0 or lies in the interior of T^0, and if the nodes of τ are appropriately chosen, then for sufficiently fine triangulations *all* quadrature points $\mathbf{x}^e(\xi^{(q,j)})$ belong to $\bar{\Omega}$ (see, for example, Ciarlet, 1978, pp. 253, 254).

15.10   The finite element scheme

Consider, then, the approximation scheme,

$$\tilde{u} = \sum_{i=1}^{n} c_i v_i,$$

the approximation to the solution u, of P_1, where

$$c_i = \gamma_1(x^{(i)}), \qquad i \in \mathcal{B},$$

$$\sum_{j=1}^{n} \tilde{a}(v_i, v_j)c_j = \tilde{l}(v_i), \qquad i \notin \mathcal{B},$$

(15.22)

where $\tilde{a}(v_i, v_j)$ and $\tilde{l}(v_i)$ are the quadrature approximations to $a^*(v_i, v_j)$ and $l^*(v_i)$ obtained by replacing $\mathbf{k}^e$ and $\mathbf{l}^e$ by their quadrature approximations.

If τ is an acceptable triangulation, the nodes of which are appropriately chosen, the quadrature scheme satisfies the conditions described above and τ is sufficiently fine, then the scheme is properly defined and $\tilde{a}(.\,,.)$ is a symmetric bilinear form. If, in addition, there exists $\tilde{\alpha} > 0$ such that

$$\tilde{a}(v, v) \geqslant \tilde{\alpha} \|v\|^2,$$

(15.23)

$\forall v \in \text{span } (v_i, i \notin \mathcal{B})$, where $\| . \|$ is the Sobolev norm over $\tilde{\Omega}$, then the equations (15.22) have a unique solution and the computation of $\tilde{u}$ can be carried out using the assembly process described in Section 15.6. Even when the quadrature scheme is chosen to ensure that the condition (15.23) holds, conditions must be imposed on τ (in addition to that of acceptability) to ensure that $\tilde{u} \to u$ as $h \equiv \max_e \text{diam } (T^e) \to 0$ (cf. the condition $\theta > \theta_0 > 0$ for the $\mathcal{L}(\tau)$ approximation scheme).

If (i) the quadrature scheme on T^0 is exact for quadratics,
  (ii) the triangulation is suitably restricted, and
  (iii) u has a sufficiently smooth extension, u^*, to a region $\hat{\Omega}$ with $\tilde{\Omega} \subset \hat{\Omega}$ and $\Omega \subset \hat{\Omega}$,

then (see Ciarlet, 1978, p. 266) condition (15.23) holds (and hence $\tilde{u}$ is defined) and $\|u^* - u\| < Ch^2$ where C is a constant independent of the triangulation, i.e.

$\|u^* - u\| = O(h^2)$ (compare the corresponding result for the $\mathscr{L}(\tau)$ approximation given in Section 15.7).

In practical applications of the scheme the triangulation might well, for computational simplicity, be such that all interelement sides are straight lines.

15.11 Extensions and generalizations

The method used to obtain the finite element approximation scheme for the boundary-value problem, P, of Section 15.4 can easily be applied to much more general self-adjoint elliptic boundary-value problems (e.g. boundary-value problems in linear elastomechanics) and to non-self-adjoint problems. For simplicity, we consider a generalization of the two-dimensional problem of Section 15.4. Let Ω, $\partial\Omega$, $\partial\Omega_1$, $\partial\Omega_2$, f $(f \geqslant 0)$, g, σ $(\sigma \geqslant 0)$, γ_1 and γ_2 have the meanings given in that section and let u be a solution of the differential equation

$$L(u) \equiv -\partial_j(a_{ij}\,\partial_i u) + fu = g \text{ in } \Omega,$$

where ∂_j denotes $\partial/\partial x_j$, $a_{ij} = a_{ij}(\mathbf{x})$ are sufficiently smooth functions in Ω [see IV, §2.10] and the *summation convention* is used for repeated suffices [see V, §7.1].

Using the identity

$$\partial_j[(a_{ij}\,\partial_i u)v] = \partial_j a_{ij}\,\partial_i u + a_{ij}\,\partial_i u\,\partial_j v$$

and the divergence theorem [see V, Theorem 13.5.1]

$$\int_\Omega \operatorname{div} \mathbf{F}\, d\Omega = \int_{\partial\Omega} \mathbf{F}\mathbf{n}\, ds,$$

i.e.

$$\int_\Omega \partial_j F_j\, d\Omega = \int_{\partial\Omega} F_i n_i\, ds,$$

where $\mathbf{F}$ is a vector-valued function on Ω with components F_i, and n_i is the ith component of the outward unit normal $\mathbf{n}$ to $\partial\Omega$, one obtains

$$\int_\Omega (L(u) - g)\, d\Omega = \int_\Omega (a_{ij}\,\partial_i u\,\partial_j v + fuv - gv)\, d\Omega - \int_{\partial\Omega} (a_{ij}\,\partial_i u)n_j v\, ds$$

and hence, if u is a solution of the boundary-value problem

$$P: \qquad\qquad\qquad L(u) = g \qquad \text{in } \Omega,$$

$$u = \gamma_1 \qquad \text{on } \partial\Omega_1,$$

$$(a_{ij}\,\partial_i u)n_j + \sigma u = \gamma_2 \qquad \text{on } \partial\Omega_2,$$

then it is also a solution of the (weak) problem,

$$P_1\text{: find } u \in H(\gamma_1)$$

such that
$$a(u, v) = l(v), \qquad \forall v \in H(0),$$
where $H(\gamma)$ is as defined in Section 15.4,
$$a(v, w) \equiv \int_\Omega (a_{ij}\, \partial_i v\, \partial_j w + fvw)\, d\Omega + \int_{\partial\Omega_2} \sigma vw\, ds$$
and
$$l(v) \equiv \int_\Omega gv\, d\Omega + \int_{\partial\Omega_2} \gamma_2 v\, ds.$$

The functional $a(.,.)$ is certainly *bilinear* and, if $a_{ij}(\mathbf{x}) = a_{ji}(\mathbf{x})$, $\mathbf{x} \in \Omega$, then it is also symmetric and the boundary-value problem is *self-adjoint* [see IV, §19.3].

If there exists a constant $\hat\alpha > 0$ such that
$$\sum_{i,j=1}^2 a_{ij}(\mathbf{x})\xi_i\xi_j \geqslant \hat\alpha \sum_{i=1}^2 \xi_i^2, \qquad \forall \mathbf{x} \in \Omega,$$
then the differential equation is elliptic, and if the differential equation is elliptic and $\partial\Omega_1 \neq \phi$ then it can be shown that there exists a constant $\alpha > 0$ such that
$$\alpha\|v\|^2 \leqslant a(v, v), \qquad \forall v \in H(0).$$

(Thus the use of the terms 'elliptic' and 'self-adjoint' to describe the boundary-value problem of Section 15.2 is justified.) It follows, then, by the argument used in Section 15.2 that if $\partial\Omega_1 \neq \phi$, then the (weak) problem P_1, and hence P, can have at most one solution [see IV, §8.2].

For the case when Ω is a polygonal region we can define, exactly as in Section 15.5, the $\mathscr{L}(\tau)$ Galerkin–collocation approximation scheme for P_1 by

$$\tilde P: \text{find } \tilde u \in V_{\text{coll}}(\gamma_1)$$

such that
$$a(\tilde u, v) = l(v), \qquad \forall v \in V_{\text{coll}}(0),$$
where $V_{\text{coll}}(\gamma)$ is defined as in Section 15.5.

It follows at once that $\tilde u = \sum_{i=1}^n c_i v_i$ (where $v_1, \ldots, v_n$ are the pyramid functions of Section 15.5) is a solution of $\tilde P$ if and only if
$$c_i = \gamma_1(x^{(i)}), \qquad \forall i \in \mathscr{B},$$
$$\sum_{j=1}^n a(v_i, v_j)c_j = l(v_i), \qquad \forall i \notin \mathscr{B}.$$

For a region Ω with a boundary which may be curved we can define, as in Section 15.10, a numberical quadrature Galerkin–collocation approximation scheme in which the v_i are parametrically defined finite element functions and the integrals are evaluated using an appropriate quadrature scheme. In this scheme, and the $\mathscr{L}(\tau)$ approximation scheme above, the matrix of the system of equations defining $c_1, \ldots, c_n$ (which will, of course, be *sparse* but, in general, non-symmetric) may be computed by a variant of the assembly process described in Section 15.6. If the boundary-value problem is elliptic and self-adjoint and the essential boundary conditions are incorporated as in Section 15.5, then the matrix of the system is additionally *symmetric* and *positive definite*, and hence the remarks made in Section 15.6 concerning the solution of the resulting linear system apply to this more general case.

An initial-value problem for a parabolic partial differential equation

Finally, let us consider briefly a finite element scheme for the solution of an initial-value problem for a parabolic partial differential equation.

Let Ω, $\partial\Omega_1$ and $\partial\Omega_2$ have the meanings given in Section 15.4 and let $u = u(\mathbf{x}, t)$ be a solution of the initial-value problem,

$$u(\mathbf{x}, 0) = u_0(\mathbf{x}), \quad \text{given, and, for } t > 0,$$

$$\frac{\partial u}{\partial t} + L(u) = g, \qquad \mathbf{x} \in \Omega,$$

$$u = 0, \qquad \mathbf{x} \in \partial\Omega_1,$$

$$\frac{\partial u}{\partial n} + \sigma u = 0, \qquad \mathbf{x} \in \partial\Omega_2,$$

where

$$L(u) = -\nabla^2 u + fu,$$

$$\nabla^2 \equiv \left(\frac{\partial^2}{\partial x_1^2} + \frac{\partial^2}{\partial x_2^2} \right),$$

$f = f(\mathbf{x}, t) \geqslant 0$, $g = g(\mathbf{x}, t)$ and $\sigma = \sigma(\mathbf{x}, t) \geqslant 0$. Note that if f, g and σ are independent of t, the 'steady-state' problem, obtained by setting $\partial u/\partial t$ to zero, is the elliptic boundary-value problem considered in Section 15.4.

Let $v \in H(0)$, where $H(\gamma)$ is the set of functions defined in Section 15.4. Then

$$\int_\Omega \left[\frac{\partial u}{\partial t} + L(u) - g \right] v \, d\Omega = 0$$

and hence

$$\left(\frac{\partial u}{\partial t}, v \right)_{L^2} + a(u, v) = l(v),$$

where $(v, w)_{L^2} \equiv \int_\Omega vw \, d\Omega$ and $a(v, w)$ and $l(v)$ are as defined in Section 15.4. Thus, if u is a solution of the initial-value problem it is also a solution of the (weak) problem

$$u(\mathbf{x}, 0) = u_0(\mathbf{x}), \qquad \mathbf{x} \in \Omega,$$

and, for $t > 0$,

$$u(x, t) \in H(0),$$

$$(u_t, v)_{L^2} + a(u, v) = l(v), \qquad \forall v \in H(0),$$

where $u_t \equiv \partial u/\partial t$. This suggests that we might define an approximation $\tilde{u}$ to u by the approximation scheme

$$\tilde{u}(\mathbf{x}, 0) = (u_0)_I(\mathbf{x}), \qquad \mathbf{x} \in \Omega \tag{15.24}$$

and, for $t > 0$,

$$\tilde{u}(x, t) \in V_{\text{coll}}(0), \tag{15.25}$$

$$(\tilde{u}_t, v) + a(\tilde{u}, v) = l(v), \qquad \forall v \in V_{\text{coll}}(0), \tag{15.26}$$

where $(u_0)_I$ is the $\mathscr{L}(\tau)$ interpolant of u_0. Then

$$\tilde{u}(\mathbf{x}, t) = \sum_{j=1}^n Q_j(t) v_j(\mathbf{x})$$

satisfies (15.25) and (15.26) if and only if, for $t > 0$,

$$Q_i = 0, \qquad \forall i \in \mathscr{B},$$

$$\sum_{j \notin \mathscr{B}} [(v_j, v_i)_{L^2} \dot{Q}_j + a(v_j, v_i) Q_j] = l(v_i), \qquad \forall i \notin \mathscr{B},$$

and, since $(u_0)_I = \sum_{j=1}^n u_0(\mathbf{x}^{(j)}) v_j$, $u(\mathbf{x}, 0)$ satisfies (15.24) if and only if

$$Q_i(0) = u_0(\mathbf{x}^{(i)}), \qquad i = 1, \ldots, n.$$

Thus the calculation of the Q_i's, $i \notin \mathscr{B}$, for $t > 0$, requires the solution of the initial-value problem

$$Q_i = (u_0)(\mathbf{x}^{(i)}), \qquad \forall i \notin \mathscr{B},$$

$$\sum_{j \notin \mathscr{B}} [(v_j, v_i)_{L^2} \dot{Q}_j + a(v_j, v_i) Q_j] = l(v_i), \qquad \forall i \notin \mathscr{B},$$

For a useful discussion of this approximation scheme, see, for example, Strang and Fix (1973).

In this brief account of the finite element method we have restricted our discussion to elliptic boundary-value problems (and even here we have considered only one type of parametrically defined finite element functions) and an initial-value problem for a parabolic partial differential equation. For a further discussion of these topics and many other aspects of the method (e.g. non-conforming methods, sub-structuring, solution of hyperbolic partial differential equations, treatment of singularities, applications to non-linear problems, etc.) the reader should consult the references given below.

M.J.M.B.

References

Adams, R. A. (1975). *Sobolev Spaces*, Academic Press, New York.

Akin, J. E. (1982). *Application and Implementation of Finite Element Methods*, Academic Press, London.

Becker, E. B., Carey, G. F., and Oden, J. T. (1981). *Finite Elements*, Vol. I, Prentice-Hall, Englewood Cliffs.

Ciarlet, P. G. (1978). *The Finite Element Method for Elliptic Problems*, North-Holland.

Hinton, E., and Owen, D. R. (1977). *Finite Element Programming*, Academic Press, London.

Oden, J. T., and Reddy, J. N. (1976). *An Introduction to the Mathematical Theory of Finite Elements*, Wiley Interscience, New York.

Strang, G., and Fix, G. J. (1973). *An Analysis of the Finite Element Method*, Prentice-Hall, Englewood Cliffs.

Zienkiewicz, O. C. (1977). *The Finite Element Method*, 3rd ed., McGraw-Hill.

Index